T0206152

Introduction to Linear Algebra

Textbooks in Mathematics

Series editors:
Al Boggess, Kenneth H. Rosen

The Geometry of Special Relativity
Tevian Dray

Mathematical Modeling in the Age of the Pandemic
William P. Fox

Games, Gambling, and Probability
An Introduction to Mathematics
David G. Taylor

Linear Algebra and Its Applications with R
Ruriko Yoshida

Maple™ Projects of Differential Equations
Robert P. Gilbert, George C. Hsiao, Robert J. Ronkese

Practical Linear Algebra
A Geometry Toolbox, Fourth Edition
Gerald Farin, Dianne Hansford

An Introduction to Analysis, Third Edition
James R. Kirkwood

Student Solutions Manual for Gallian's Contemporary Abstract Algebra, Tenth Edition
Joseph A. Gallian

Elementary Number Theory
Gove Effinger, Gary L. Mullen

Philosophy of Mathematics
Classic and Contemporary Studies
Ahmet Cevik

An Introduction to Complex Analysis and the Laplace Transform
Vladimir Eiderman

An Invitation to Abstract Algebra
Steven J. Rosenberg

Numerical Analysis and Scientific Computation
Jeffery J. Leader

Introduction to Linear Algebra
Computation, Application and Theory
Mark J. DeBonis

https://www.routledge.com/Textbooks-in-Mathematics/book-series/CANDHTEXBOOMTH

Introduction to Linear Algebra

Computation, Application, and Theory

Mark J. DeBonis
Manhattan College, USA

CRC Press
Taylor & Francis Group
Boca Raton London New York

CRC Press is an imprint of the
Taylor & Francis Group, an **informa** business

A CHAPMAN & HALL BOOK

First edition published 2022
by CRC Press
6000 Broken Sound Parkway NW, Suite 300, Boca Raton, FL 33487-2742

and by CRC Press
2 Park Square, Milton Park, Abingdon, Oxon, OX14 4RN

Library of Congress Cataloging-in-Publication Data
Names: DeBonis, Mark J., author.
Title: Introduction to linear algebra : computation, application and theory / authored by Mark J. DeBonis, Manhattan College, USA.
Description: First edition.	Boca Raton : C&H/CRC Press, 2022.	Series: Textbooks in mathematics	Includes bibliographical references and index.
Identifiers: LCCN 2021042588 (print)	LCCN 2021042589 (ebook)	ISBN 9781032108988 (hardback)	ISBN 9781032109381 (paperback)	ISBN 9781003217794 (ebook)
Subjects: LCSH: Algebras, Linear.
Classification: LCC QA184 .D43 2022 (print)	LCC QA184 (ebook)	DDC 512/.5--dc23
LC record available at https://lccn.loc.gov/2021042588
LC ebook record available at https://lccn.loc.gov/2021042589

ISBN: 978-1-032-10898-8 (hbk)
ISBN: 978-1-032-10938-1 (pbk)
ISBN: 978-1-003-21779-4 (ebk)

DOI: 10.1201/9781003217794

Publisher's note: This book has been prepared from camera-ready copy provided by the authors.

To my dear daughter, Anna.

Contents

Preface

This text is meant as an introduction for students who have never been exposed to the topics in a linear algebra course. Although the text is filled with interesting and diverse application sections, it is ultimately a theoretical text with the goal of training students to do succinct computation, but in a knowledgeable way. In other words, after completing the course with this text the student will know the best and shortest ways to do linear algebraic computations, but will also know why such computations are both effective and successful. This text is best used in an undergraduate course in linear algebra. It is flexible enough so that an instructor can emphasize either theory, computation, and/or application and incorporate the use of technology. One hope of this author is that this text will eventually become a reliable reference/manual on the topic with a worn-out spine and cover due to its continual usage by students in their later career or studies.

The author's desire is to present the basic concepts of linear algebra in a palatable yet concise way. This is done primarily by illustrating each of the fundamental concepts of linear algebra with plenty of examples (something which students of the subject always demand in such a text). Another way the author makes these concepts reachable is to consistently present shortcuts for any computation. Therefore, although this text is often conceptual, it presents the material in a practical way so that the student who eventually will apply it in later classes will be versed in what the concepts mean and how to do quick and easy computations.

One of the primary reasons for writing this text was to reorder the topics presented in a standard linear algebra text in such a way that all the computational skills are presented first. For this reason, Chapters 1 and encapsulate for the reader all the necessary computational skills which will be used in later chapters (matrix and vector operations, Gaussian elimination, determinant and inverse of a matrix, etc.). Since these are the easiest notions for the student to understand, it is best to present them first so as to make the reader feel at ease with the subject matter. Besides, these computational skills will be used over and over again in the remaining chapters.

In Chapter 1, we also familiarize the reader with the notion of vector space by presenting two classical examples: Tuples and Matrices. These are concrete examples which will prepare the student for the general definition in Chapter 3. The author chose to treat scalars as real numbers and forgo the notion of a field (although it appears in some early exercises), even though later in the text scalars at times will be complex numbers as well.

Chapters 3 and 4 present the general notions of vector space and linear transformation. To make each linear algebraic concept understandable, after each definition we present the reader with examples from four classic vector spaces: Tuples, Matrices,

Polynomials, and Functions. These help the reader fully grasp each definition in its generality by observing how the definition *plays out* in each of these four settings. Other texts would first present all notions of vector space (span, linear independence, basis, etc.) and linear transformation (one-to-one, kernel, isomorphism, etc.) for Tuples in initial chapters before presenting them all once again in later chapters in a more general setting. There are even texts which first present these notions for $\mathbb{R}^2$, then $\mathbb{R}^3$, and then finally for $\mathbb{R}^n$! The goal of this text is to be concise, to the point, and practical.

In Chapter 3, we opted for **finite** dimensional vector spaces, since this is the arena where most of the applications reside, although we do at times suggest to the reader how the notion of vector space can be extended to the infinitely dimensional case.

Chapter 5 covers real inner product spaces (although a brief discussion of complex inner product spaces is found in Section 5.5) with such notions as orthogonality, the Gram-Schmidt Process, and best approximation. There are many important applications in this chapter. Chapter 7 covers quadratic forms with the goal of applying them in the context of finding extrema for multivariate real-valued functions.

Typically, linear algebra is the first subject where students are exposed to formulating proofs after having taken a course on mathematical proof. For this reason we include proofs, some of which are quite straightforward. But there are also proofs which require more thought as well. The instructor should carefully consider which and how many proof exercises to include in homework assignments. In addition, at the end of Chapter 4, there are some more theoretical ideas for the more advanced reader. These include the axioms of a determinant, quotient vector spaces, and dual vector spaces. These can be easily skipped without disrupting the continuity of the material, yet are there for instructors who feel the class can benefit from them. Indeed, if the majority of the class consists of mathematics majors, the author would strongly encourage the instructor to include these sections in their syllabus, for these topics will enrich the theoretical backbones of the students.

Sections beginning with the word *Application* have been put into the text for those who wish to cover relevant applications in linear algebra. The author tried to select applications which were easy enough to explain, diverse in subject matter, and ultimately would convince the reader that linear algebra has many deep and meaningful uses. Indeed, the author has included an entire chapter on the timely subject of machine learning and data analytics (Chapter 6). There are several sections on the topic of optimization, both linear and nonlinear. There are also several sections covering linear algebraic computational methods, such as the LU and QR factorizations. However, no such application section covers new theoretical information which might be used later in the text. For this reason, any application section can be easily skipped (at the instructor's discretion) without destroying any continuity. Furthermore, these application sections are inserted at the moment when a student can grasp their contents based on the appropriate background knowledge of prior sections. In addition, at the end of certain application sections, there is a *Project* that is intended as a challenging problem that encapsulates and synthesizes the ideas presented in the application section. These problems are generally more involved and realistically require some use of technology to solve them. Finally, the author wishes to point out that a number of the application sections require a certain amount of background in multi-variate calculus which students typically will have when beginning a course in linear algebra.

Although the intention of this text is for students to get their hands dirty and do all the computations by hand (something which the author feels helps the students actually understand the computations they are doing), these problems can also be done with the aid of technology. Indeed, once the student is comfortable with doing computations by hand in Chapters 1 and 2, in later chapters they could use technology to speed up their computation. By technology, the author had in mind some computer software package such as MATLAB®, Maple, Sage Math, or Mathematica (to name a few). If the goal is to prepare students for careers in applied mathematics, a programming language may be more appropriate, such as Python. Indeed, many of the examples and illustrations in this text were produced in Python. Of course, the use of hand-held technology can also be employed for these problems with certain limitations. However, this text is not a *How To* manual for technology usage. For this reason, the text does not include command lines for the various technology packages, for in the end, some packages would have to be favored over others. Furthermore, if we try to include all software packages, then this text will become needlessly thicker and thus destroy its intended conciseness. In any case, technology changes so quickly that anything written about it becomes almost instantly obsolete (perhaps even what was just stated in this paragraph!).

Therefore, in summary, whether the goal is to have the necessary background to pursue a career in applied mathematics or continue on to study mathematics in graduate school, this text can be used in either case. If the instructor carefully selects which chapters and sections to cover, either goal can be achieved.

CHAPTER 1

Examples of Vector Spaces

I N THIS CHAPTER, we introduce the reader to the notion of vector space by presenting two classical examples: Tuples and Matrices. These are concrete examples which will prepare the student for the general definition in Chapter 3. We also begin the process of providing the student with the computational skills necessary for the subject of Linear Algebra. In Section 1.1, the first example of a vector space, Tuples, is introduced. In Section 1.2, the operation dot product for Tuples is presented as a precursor for Chapter 5 which discusses general inner product spaces. Section 1.3 is our first application section and shows how one can use vectors in $\mathbb{R}^2$ or $\mathbb{R}^3$ in order to prove in a new way facts about Euclidean geometry. In Section 1.4, the second example of a vector space, Matrices, is introduced. Finally, in Section 1.5, we define multiplication of matrices.

1.1 FIRST VECTOR SPACE: TUPLES

Here now is our first example of what later will be called a **vector space**. A notion in linear algebra of some importance is the **scalar**. For most of our discussion, a scalar will just be a real number and, at times, a complex number. A more comprehensive and perhaps advanced treatise on linear algebra would assume a scalar to be a element of what is called a **field**. Roughly speaking, a field gathers together some of the essential properties (or axioms) of the real numbers. We list these properties below:

Definition 1.1 *A* **field** *is a set of objects F together with two operations $+$ and $\cdot$ (called* **addition** *and* **multiplication***) having the following properties:*

Closure*: For all $a, b \in F$, we have $a + b \in F$ and $a \cdot b \in F$.*

Commutativity*: For all $a, b \in F$, we have $a + b = b + a$ and $a \cdot b = b \cdot a$.*

Associativity*: For all $a, b, c \in F$, we have $a + (b + c) = (a + b) + c$ and $a \cdot (b \cdot c) = (a \cdot b) \cdot c$.*

Identity*: There exist $0, 1 \in F$ such that for all $a \in F$, we have $a + 0 = a$ and $a \cdot 1 = a$.*

DOI: 10.1201/9781003217794-1

Inverse: *For every $a \in F$ there exists $b \in F$ such that $a + b = 0$ (b is called the* **additive** *inverse of a) and for every $0 \neq a \in F$ there exists $b \in F$ such that $a \cdot b = 1$ (b is called the* **multiplicative** *inverse of a).*

Distribution: *For all $a, b, c \in F$, we have $a \cdot (b + c) = a \cdot b + a \cdot c$.*

The main examples of fields addressed in this text are the real numbers and the complex numbers (one can easily check that the properties above are satisfied in each example). At times we may want to prove results in more generality without assuming what field we have, but as stated, a **scalar** for the time being is simply another name for a real number. The standard notation for real numbers is $\mathbb{R}$.

Definition 1.2 *The vector space called n-**tuples** is defined as follows:*

- **Vectors** *are the elements of the set*

$$\mathbb{R}^n = \{[a_1, \ldots, a_n] \ : \ a_1, \ldots, a_n \in \mathbb{R}\}.$$

The scalars $a_1, \ldots, a_n$ in each vector are called its **components**.

- **Vector Addition** *is defined componentwise as follows:*

$$[a_1, \ldots, a_n] + [b_1, \ldots, b_n] = [a_1 + b_1, \ldots, a_n + b_n].$$

- **Scalar Multiplication** *is defined componentwise as follows: If c is a scalar, then*

$$c[a_1, \ldots, a_n] = [ca_1, \ldots, ca_n].$$

We will denote this vector space by the notation $\mathbb{R}^n$ and we often denote a vector by single letters such as u, v, w, etc., due to its brevity. Scalars will be typically denoted by letters a, b, c, etc.

Example 1.1 *Let $u = [3, -4]$ and $v = [5, -4]$. Then*

$$(-1)u + 3v = (-1)[3, -4] + (3)[5, -4] = [-3, 4] + [15, -12] = [12, -8].$$

Definition 1.3 *Two n-tuple vectors $[a_1, \ldots, a_n]$, $[b_1, \ldots, b_n] \in \mathbb{R}^n$ are* **equal** *if $a_1 = b_1, \ldots, a_n = b_n$. In other words they have equal components.*

We point out that if $n = 2$ or $n = 3$, then $\mathbb{R}^n$ has a geometric interpretation. For instance, when $n = 2$, $u = [a_1, a_2]$ is represented as an arrow originating from the origin, called the **initial point**, and ending at the ordered pair (a_1, a_2), called the

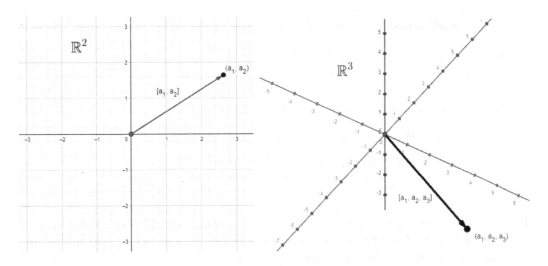

Figure 1.1 Geometric interpretation of 2-tuples and 3-tuples.

terminal point (Figure 1.1). Indeed, many books use the notation $\vec{v}$ to represent a vector to remind us of this interpretation.

There is a nice geometric interpretation of vector addition and subtraction in this same setting. If u and v are two adjacent sides of a parallelogram, then the sum and difference form the diagonals of the parallelogram formed by the vectors u and v (Figure 1.2). Scalar multiplication changes the length of a vector; i.e. it *scales* either up or down the length of the vector.

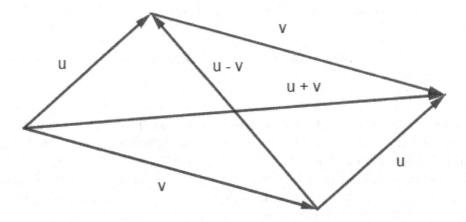

Figure 1.2 Geometric interpretation of the sum and difference of two vectors.

The following theorem proves that $\mathbb{R}^n$ is (what we will later call) a **vector space**:

Theorem 1.1 $\mathbb{R}^n$ *as defined above has the following properties:*

0. *For all* $u, v \in \mathbb{R}^n$ *and* $a \in \mathbb{R}$, *we have* $u + v$, $au \in \mathbb{R}^n$.

1. *For all* $u, v \in \mathbb{R}^n$, $u + v = v + u$.

2. *For all $u, v, w \in \mathbb{R}^n$, $(u + v) + w = u + (v + w)$.*

3. *There exists $0 \in \mathbb{R}^n$, such that for all $u \in \mathbb{R}^n$, $u + 0 = u$.*

4. *For each $u \in \mathbb{R}^n$, there is a $v \in \mathbb{R}^n$ such that $u + v = 0$.*

5. *For all $u, v \in \mathbb{R}^n$, and $a \in \mathbb{R}$, $a(u + v) = au + av$.*

6. *For all $u \in \mathbb{R}^n$, and $a, b \in \mathbb{R}$, $(a + b)u = au + bu$.*

7. *For all $u \in \mathbb{R}^n$, and $a, b \in \mathbb{R}$, $(ab)u = a(bu)$.*

8. *For all $u \in \mathbb{R}^n$, $1u = u$.*

Proof 1.1 *We shall prove some of the properties and leave the rest to the reader. Property 0 is clear from the properties of real numbers, since adding or multiplying real numbers yields another real number. For Property 1, let $u = [a_1, \ldots, a_n]$ and $v = [b_1, \ldots, b_n]$. Then by properties of real numbers,*

$$u + v = [a_1, \ldots, a_n] + [b_1, \ldots, b_n] = [a_1 + b_1, \ldots, a_n + b_n]$$

$$= [b_1 + a_1, \ldots, b_n + a_n] = [b_1, \ldots, b_n] + [a_1, \ldots, a_n] = v + u.$$

To verify Property 3, the 0 we are looking for is $[0, \ldots, 0]$. To verify Property 7, notice that

$$(ab)u = (ab)[a_1, \ldots, a_n] = [(ab)a_1, \ldots, (ab)a_n] = [a(ba_1), \ldots, a(ba_n)] =$$

$$a[ba_1, \ldots, ba_n] = a(b[a_1, \ldots, a_n]) = a(bu).$$

$\square$

Some remarks are in order here before we continue. The reader may not realize or perhaps find it confusing that we have named two different objects with the name 0 and two different operations by the name + (in both $\mathbb{R}$ and $\mathbb{R}^n$). But there is really no possibility of misunderstanding. For when one sees, for instance, the statement $a + 0$, we know from the context that both 0 and + refer to scalar arithmetic in $\mathbb{R}$ versus $u + 0$ which must by the context refer to vector arithmetic in $\mathbb{R}^n$.

The set of vectors $e_1 = [1, 0, \ldots, 0]$, $e_2 = [0, 1, \ldots, 0]$, $\ldots$, $e_n = [0, 0, \ldots, 1]$ are called the **standard basis** vectors and play a special role in linear algebra.

For $n = 2$ and $n = 3$, the notation is conventionally different: For $n = 2$ and $n = 3$,

$$\hat{\imath} = [1, 0] \quad \text{and} \quad \hat{\jmath} = [0, 1].$$

$$\hat{\imath} = [1, 0, 0], \quad \hat{\jmath} = [0, 1, 0] \quad \text{and} \quad \hat{k} = [0, 0, 1].$$

One can easily verify that in $\mathbb{R}^n$, $[a_1, a_2, \ldots, a_n] = a_1 e_1 + a_2 e_2 + \cdots + a_n e_n$. Furthermore, the standard basis vectors are (what we will later call) unit vectors (exercise).

Example 1.2 *In $\mathbb{R}^3$,*

$$[\pi, -1, \sqrt{2}] = [\pi, 0, 0] + [0, -1, 0] + [0, 0, \sqrt{2}] = \pi \hat{\imath} + (-1)\hat{\jmath} + (\sqrt{2})\hat{k}.$$

EXERCISES

1. Let $u = [2, -1]$ and $v = [-1, 4]$. Compute $u + v$, $-3u$ and $u - 2v$.

2. Let $u = [1, -2, 0]$ and $v = [3, -1, 1]$. Compute $u + v$, $2v$, $v - u$, $\frac{1}{2}u$ and $2u - 3v$.

3. List the standard vectors e_1, e_2, e_3, e_4 for $\mathbb{R}^4$.

4. Express $[\pi, -1, \sqrt{3}, 2]$ in terms of e_1, e_2, e_3, e_4.

5. Prove properties 2,4–6 and 8 of Theorem 1.1.

6. Prove that in $\mathbb{R}^n$, $[a_1, a_2, \ldots, a_n] = a_1 e_1 + a_2 e_2 + \cdots + a_n e_n$.

7. In the field of real numbers, what is the additive inverse of the scalar 2?

8. In the field of complex numbers, what is the additive inverse of the scalar $2 - 3i$?

9. In the field of real numbers, what is the multiplicative inverse of the scalar 2?

10. In the field of complex numbers, what is the multiplicative inverse of the scalar $2 - 3i$? (your answer should be represented as a complex number)

11. Show that the real numbers satisfy the axioms of a field.

12. Show that the complex numbers satisfy the axioms of a field.

13. Define $F = \{0, 1, 2, 3, 4\}$. Define addition, $n +_5 m$, to be the remainder when dividing $n + m$ by 5 (for instance, $3 +_5 4 = 2$, since the remainder when dividing 7 by 5 is 2). Define multiplication, $n \cdot_5 m$, to be the remainder when dividing $n \cdot m$ by 5.

    a. Complete the addition and multiplication table for the field F (we will not ask you to prove F is a field, although it is the case).

    b. What element of F is the additive inverse of 2?

    c. What element of F is the multiplicative inverse of 4?

1.2 DOT PRODUCT

Here we present another operation applicable in $\mathbb{R}^n$ in which the inputs are two vectors and the output is a scalar. The various names of this operation are dot, scalar or inner product. Although this is not an operation indicative of a vector space, it is an essential ingredient of what we will later call an **inner product space**.

Definition 1.4 *Let* $u = [a_1, \ldots, a_n]$, $v = [b_1, \ldots, b_n] \in \mathbb{R}^n$. *The* **dot product** *of* u *and* v, *written*

$$u \cdot v = a_1 b_1 + \cdots + a_n b_n.$$

Example 1.3 *In* $\mathbb{R}^4$,

$$[2, 25, -1, -1.3] \cdot [-3, 1/5, 3, 10] = (2)(-3) + (25)(1/5) + (-1)(3) + (-1.3)(10)$$

$$= -6 + 5 - 3 - 13 = -17.$$

The following result summarizes some elementary properties of the dot product:

Theorem 1.2 *If* $u, v, w \in \mathbb{R}^n$ *and* $a \in \mathbb{R}$, *then*

i. $u \cdot v = v \cdot u$.

ii. $u \cdot (v + w) = u \cdot v + u \cdot w$.

iii. $a(u \cdot v) = (au) \cdot v = u \cdot (av)$.

Proof 1.2 *The proofs of all three statements are straightforward. We will prove ii. First, set* $u = [a_1, \ldots, a_n]$, $v = [b_1, \ldots, b_n]$ *and* $w = [c_1, \ldots, c_n]$. *Then, using the distributive rule for real numbers, we have*

$$u \cdot (v + w) = [a_1, \ldots, a_n] \cdot [b_1 + c_1, \ldots, b_n + c_n]$$

$$= a_1(b_1 + c_1) + \cdots + a_n(b_n + c_n) = a_1 b_1 + a_1 c_1 + \cdots + a_n b_n + a_n c_n$$

$$= a_1 b_1 + \cdots + a_n b_n + a_1 c_1 + \cdots + a_n c_n$$

$$= [a_1, \ldots, a_n] \cdot [b_1, \ldots, b_n] + [a_1, \ldots, a_n] \cdot [c_1, \ldots, c_n] = u \cdot v + u \cdot w.$$

□

A definition that goes hand in hand with dot product is the following:

Definition 1.5 *The* **length** *(***magnitude** *or* **norm***) of a vector* $u = [a_1, \ldots, a_n] \in \mathbb{R}^n$, *written*

$$|u| = \sqrt{u \cdot u} = \sqrt{a_1^2 + \cdots + a_n^2}.$$

A vector u is called a **unit** *vector if* $|u| = 1$.

A couple of remarks are in order. First, observe that the context of the vertical bars $| * |$ will determine its significance, whether it is absolute value or length. Indeed, $|u|$ is the length of a tuple while $|a|$ is the absolute value of a scalar. Second, $|u|$ is called length for good reason, for if we think back to our geometric interpretation of vectors as arrows, then the length of those arrows are precisely $|u|$ (by the Pythagorean Theorem).

Next we give a list of basic properties involving magnitude.

Theorem 1.3 *Let* $u, v \in \mathbb{R}^n$ *and* $a \in \mathbb{R}$. *Then*

i. $|u| \geq 0$ *and* $|u| = 0$ *iff (i.e. if and only if)* $u = 0$.

ii. $|au| = |a||u|$.

iii. $\frac{1}{|u|}u$ *(or just* $u/|u|$*) is a unit vector.*

iv. $u \cdot v = |u||v| \cos \theta$, *where* θ *is the smaller of the two angles between the vectors u and v.*

v. (Cauchy-Schwartz Inequality) $|u \cdot v| \leq |u||v|$.

vi. (Triangle Inequality) $|u + v| \leq |u| + |v|$.

Proof 1.3 *Statements i ii and iii are left as exercises. To prove iv, consider the triangle formed by the vectors u, v and* $u - v$.

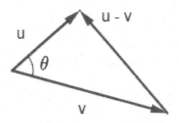

By the Law of Cosines,

$$|u - v|^2 = |u|^2 + |v|^2 - 2|u||v| \cos \theta.$$

Notice that

$$|u - v|^2 = (u - v) \cdot (u - v) = u \cdot u - (u \cdot v) - (v \cdot u) + (v \cdot v)$$

$$= |u|^2 - 2(u \cdot v) + |v|^2.$$

Equating these two observations and simplifying yields the desired result. To prove v, notice that

$$|u \cdot v| = |\,||u||v|\cos\theta\,| = |u||v||\cos\theta| \leq |u||v|.$$

To prove vi, notice that using v we have

$$|u + v|^2 = (u + v) \cdot (u + v) = u \cdot u + 2(u \cdot v) + v \cdot v = |u|^2 + 2(u \cdot v) + |v|^2$$

$$\leq |u|^2 + 2|u||v| + |v|^2 = (|u| + |v|)^2.$$

Now take the square root of both sides to get the desired result. □

In Theorem 1.3.iii, $u/|u|$ is called the **normalization** of u. Geometrically speaking, the normalization of u is a unit vector pointing in the same direction as u.

Example 1.4 *Set $u = [1, -2, 3]$ in $\mathbb{R}^3$. The magnitude of u,*

$$|u| = \sqrt{(1)^2 + (-2)^2 + (3)^2} = \sqrt{14},$$

so the normalization of u would be

$$\left[\frac{1}{\sqrt{14}}, \frac{-2}{\sqrt{14}}, \frac{3}{\sqrt{14}} \right].$$

We now introduce some new notation and terminology.

Definition 1.6 *Let u and v be vectors in $\mathbb{R}^n$. The **component** of v along u, written*

$$\text{comp}_u v = |v| \cos\theta,$$

where θ, as usual, is the smaller angle between u and v.

Component has a geometric interpretation. There are two cases to consider: When θ is acute and obtuse. Using basic trigonometry, one can verify that $|\text{comp}_u v|$ corresponds to the length of the thicker line indicated in each case in Figure 1.3.

In this second case $\text{comp}_u v$ is a negative quantity, although $-\text{comp}_u v$ is the length of the thicker line (again, simple trigonometry can verify this). Now there is a simpler way to compute components without having to find the angle θ. Notice that

$$\text{comp}_u v = |v| \cos\theta = |v| \left(\frac{u \cdot v}{|u||v|} \right) = \frac{u \cdot v}{|u|}.$$

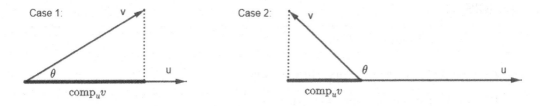

Figure 1.3 Geometric interpretation of component.

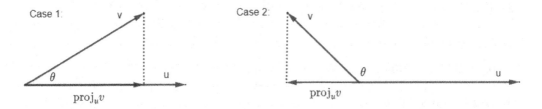

Figure 1.4 Geometric interpretation of projection.

Example 1.5 *We compute the component of* $v = [1, -2, 3]$ *along* $u = [2, -3, 1]$.

$$\text{comp}_u v = \frac{[2, -3, 1] \cdot [1, -2, 3]}{\|[2, -3, 1]\|} = \frac{11}{\sqrt{14}}.$$

Keep in mind that $\text{comp}_u v$ is a scalar (real number). A second definition is now in order.

Definition 1.7 *Let* u *and* v *be two vectors in* $\mathbb{R}^n$. *The* **projection** *of* v *along* u, *written* $\text{proj}_u v$, *is a vector of length* $|\text{comp}_u v|$ *parallel to* u. *More precisely,*

$$\text{proj}_u v = (\text{comp}_u v)\frac{u}{|u|}.$$

Projections are depicted in Figure 1.4.

Again, we simplify our computation as follows:

$$\text{proj}_u v = (\text{comp}_u v)\frac{u}{|u|} = \frac{u \cdot v}{|u|}\frac{u}{|u|} = \frac{u \cdot v}{|u|^2}u = \frac{u \cdot v}{u \cdot u}u.$$

Keep in mind that $\text{proj}_u v$ is a vector in $\mathbb{R}^n$.

Example 1.6 *In our previous example,*

$$\text{proj}_u v = \frac{[2, -3, 1] \cdot [1, -2, 3]}{[2, -3, 1] \cdot [2, -3, 1]}[2, -3, 1] = \frac{11}{14}[2, -3, 1] = [11/7, -33/14, 1/14].$$

EXERCISES

1. If $u = [1, -2]$ and $v = [3, 2]$,

   a. Compute $|u|$, $u \cdot v$, $-(u \cdot u)u$ and $|2u - v|$.

   b. Normalize u and v into unit vectors.

   c. Find the smaller angle between u and v.

2. Let $u = [2\sqrt{3}, 4, \sqrt{2}]$ and $v = [1, 2, \sqrt{2}]$.

   a. Normalize u into a unit vector.

   b. Find the smaller angle between u and v.

3. If $u = [\sqrt{3}, 1, 2]$ and $v = [1, -1, 2]$,

   a. Compute $|u|$, $u \cdot v$, $2(v \cdot v)v$ and $(u \cdot v)|3u - 2v|$.

   b. Normalize u and v into unit vectors.

   c. Find the smaller angle between u and v.

4. For each given set of vectors compute $\text{comp}_u v$ and $\text{proj}_u v$.

   a. $u = [1, -2]$ and $v = [2, 3]$.

   b. $u = [1, 0, -2]$ and $v = [1, 1, -1]$.

5. Let $u = [-\sqrt{5}, 2, \sqrt{3}]$ and $v = [0, 2, -\sqrt{3}]$.

   a. Find a unit vector pointing in the same direction as u.

   b. Find the smaller angle between u and v.

   c. Compute $\text{comp}_u v$ and $\text{proj}_u v$.

   d. Compute $3(u \cdot v)u - |u|v$ and $|3u - 4v|$.

6. Prove the following geometric facts:

   a. Two vectors u and v are perpendicular iff $u \cdot v = 0$.

   b. Two vectors u and v are parallel iff $u = av$ or $v = au$ for some scalar $a \in \mathbb{R}$.

   c. The vector $-u$ points in the opposite direction of u.

7. Prove that $|u + av| \geq |u|$ for all $a \in \mathbb{R}$ iff $u \cdot v = 0$

8. Prove that $|v_1 + v_2 + \cdots + v_n| \leq |v_1| + |v_2| + \cdots + |v_k|$, for any vectors $v_1, v_2, \ldots, v_k \in \mathbb{R}^n$

   (hint: induction)

9. Prove that $u \cdot v = \frac{1}{4}(|u+v|^2 - |u-v|^2)$ for any $u, v \in \mathbb{R}^n$.

10. Prove that $|u+v| = |u| + |v|$ iff u and v point in the same direction (i.e. there is an $a > 0$ with $v = au$).

11. Prove that $|u+v|^2 \leq |u|^2 + |v|^2$ with equality iff $u \cdot v = 0$.

12. Prove properties i and iii of Theorem 1.2.

13. Prove properties i–iii of Theorem 1.3.

1.3 APPLICATION: GEOMETRY

As we have already stated tuples in $\mathbb{R}^n$ along with their operations take on a geometric meaning. This section is devoted to further exploration of this observation. Recall briefly the following geometric facts about tuples:

1. A vector, u, can be viewed physically as an arrow.

2. The sum and difference of two vectors, $u+v$ and $u-v$, comprise the diagonals of a parallelogram whose adjacent sides are these two vectors.

3. The magnitude of a vector, $|u|$, corresponds to the length of the arrow representing u.

4. For vectors u and v, we have the equation $u \cdot v = |u||v| \cos \theta$, where θ is the smaller angle between u and v.

5. Two vectors u and v are parallel iff $u = av$ or $v = au$ for some real number a.

6. Two vectors u and v are perpendicular iff $u \cdot v = 0$.

7. The vector $-u$ points in the opposite direction of u.

Only in this section will we allow vectors which do not have their initial point at the origin so that we can derive some nice geometric results. In this case, we will say that two vectors are equal if they have the same length and are point in the same direction.

For instance, in Figure 1.5 we have depicted a collection of vectors which are all equal to each other.

We need to introduce some notation. If A and B are points in space, then $\overrightarrow{AB}$ denotes the vector with initial point A and terminal point B as depicted in Figure 1.6.

From our discussion of the parallelogram earlier, it is clear that if $u = [a_1, a_2, \ldots, a_n]$ is a vector with terminal point at A and $v = [b_1, b_2, \ldots, b_n]$ is a vector with terminal point at B, then

$$\overrightarrow{AB} = v - u = [b_1 - a_1, b_2 - a_2, \ldots, b_n - a_n].$$

With just these few facts we are capable of proving many standard geometric results.

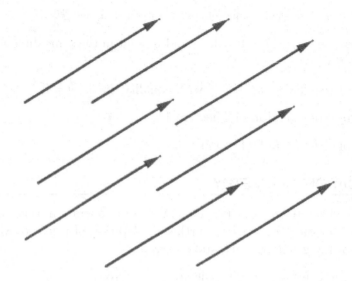

Figure 1.5 A collection of equal vectors.

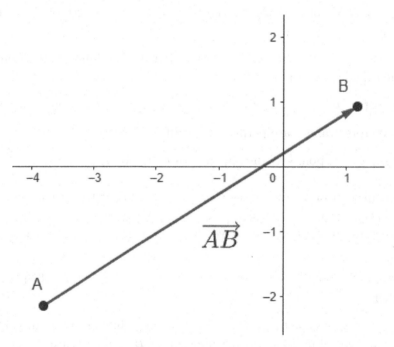

Figure 1.6 A vector with initial point at A and terminal point at B.

Example 1.7 *We prove that the midpoints of all the sides of any quadrilateral form a parallelogram.*

Proof 1.4 *Consider a quadrilateral with points P, Q, R and S. Set T to be the midpoint of $\overline{PQ}$, U to be the midpoint of $\overline{QR}$, V to be the midpoint of $\overline{RS}$ and W to be the midpoint of $\overline{SP}$ (see Figure 1.7).*

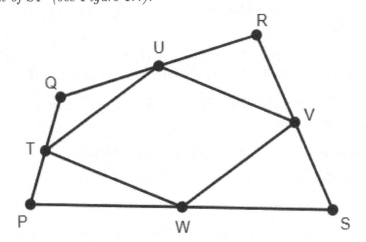

Figure 1.7 A quadrilateral $PQRS$ with parallelogram $TUVW$.

We leave as an exercise the fact that

$$\overrightarrow{PQ} + \overrightarrow{QR} + \overrightarrow{RS} + \overrightarrow{SP} = 0.$$

Notice also that

$$\overrightarrow{TU} = \frac{1}{2}\overrightarrow{PQ} + \frac{1}{2}\overrightarrow{QR} \quad and \quad \overrightarrow{VW} = \frac{1}{2}\overrightarrow{RS} + \frac{1}{2}\overrightarrow{SP}.$$

But then

$$\overrightarrow{TU} + \overrightarrow{VW} = \frac{1}{2}(\overrightarrow{PQ} + \overrightarrow{QR} + \overrightarrow{RS} + \overrightarrow{SP}) = \frac{1}{2}0 = 0,$$

which implies that $\overrightarrow{TU} = -\overrightarrow{VW}$. This last equation proves half the result, since

a. The length of $\overline{TU} = |\overrightarrow{TU}| = |-\overrightarrow{VW}| = |\overrightarrow{VW}| =$ the length of $\overline{VW}$ and

b. $\overrightarrow{VW}$ is a scalar multiple of $\overrightarrow{TU}$, implies $\overrightarrow{VW}$ and $\overrightarrow{TU}$ are parallel, and so $\overline{VW}$ and $\overline{TU}$ are parallel.

In a similar manner one derives the vector equation $\overrightarrow{UV} = -\overrightarrow{WT}$ to get that the remaining two opposite sides are equal in length and parallel. □

EXERCISES

1. Prove if $u = [a_1, a_2, \ldots, a_n]$ is a vector with terminal point at A and $v = [b_1, b_2, \ldots, b_n]$ is a vector with terminal point at B, then

$$\overrightarrow{AB} = [b_1 - a_1, b_2 - a_2, \ldots, b_n - a_n].$$

2. Use vectors to prove the following statement: The diagonals of a rhombus are perpendicular.

3. Use vectors to prove the following for any triangle ABC: If D is the midpoint of $\overline{AB}$ and E is the midpoint of $\overline{BC}$, then $\overline{DE}$ is parallel to $\overline{AC}$.

4. Use vectors to prove that the altitude of an isosceles triangle bisects the base (hint: use components).

5. Use vectors to prove the following (refer to the diagram): If $\triangle ABC$ is a right triangle, then $d = \frac{a^2}{c}$.

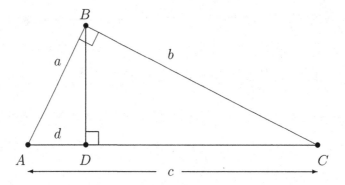

6. Referring to the diagram below, show by induction that the vector $v = v_1 + v_2 + \cdots + v_k$.

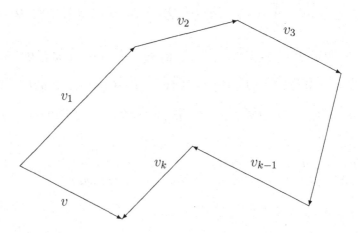

7. Verify that the standard basis vectors e_i ($i = 1, 2, \ldots, n$) for $\mathbb{R}^n$ are unit vectors.

8. Referring to the diagram below depicting an arbitrary closed polygonal set of vectors, show that $v_1 + v_2 + \cdots + v_k = 0$ (use the previous exercise).

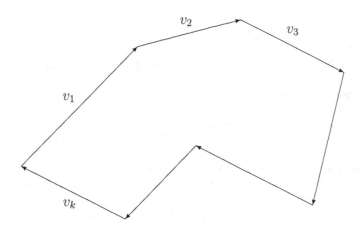

1.4   SECOND VECTOR SPACE: MATRICES

Here now is our second example of what later will be called a vector space. First we define a matrix.

Definition 1.8 *An $m \times n$ **matrix** is a rectangular array of scalars consisting of m rows and n columns. We say the **dimensions** of the matrix are "m-by-n or $m \times n$."*

Example 1.8 $\begin{bmatrix} -1 & \pi & 6 \\ \sqrt{3} & -1.2 & 3/4 \end{bmatrix}$ *is an example of a 2×3 matrix.*

There are several useful ways of representing a matrix. The most descriptive (and most cumbersome) is the following:

$$\begin{bmatrix} a_{11} & a_{12} & \cdots & a_{1n} \\ a_{21} & a_{22} & \cdots & a_{2n} \\ \vdots & \vdots & \ddots & \vdots \\ a_{m1} & a_{m2} & \cdots & a_{mn} \end{bmatrix}.$$

Each scalar a_{ij} is called the ijth **entry** of the matrix where $1 \le i \le m$ and $1 \le j \le n$. A simpler notation for a matrix is $[a_{ij}]$. We often represent a matrix simply by A. Another useful way to represent a matrix is by its rows or by its columns:

$$A = \begin{bmatrix} r_1 \\ r_2 \\ \vdots \\ r_m \end{bmatrix}, \text{ where } r_i = [a_{i1} \; a_{i2} \; \cdots \; a_{in}] \qquad (i = 1, 2, \ldots, m), \text{ or}$$

$$A = [c_1 \ c_2 \ \cdots \ c_n], \text{ where } c_j = \begin{bmatrix} a_{1j} \\ a_{2j} \\ \vdots \\ a_{mj} \end{bmatrix} \qquad (j = 1, 2, \ldots, n).$$

We are now ready to define our second vector space.

Definition 1.9 *Consider $m \times n$ matrices for fixed positive integers m and n. Then*

- **Matrices** *are the elements of the set*

$$M_{mn} = \{[a_{ij}] \ : \ a_{ij} \in \mathbb{R}\}.$$

- **Matrix Addition** *is defined entry-wise as follows:*

$$[a_{ij}] + [b_{ij}] = [a_{ij} + b_{ij}].$$

- **Scalar Multiplication** *is defined entry-wise as follows: If c is a scalar, then*

$$c[a_{ij}] = [ca_{ij}].$$

Example 1.9 *If $A = \begin{bmatrix} 1 & 2 & 3 \\ 4 & 5 & 6 \end{bmatrix}$ and $B = \begin{bmatrix} -1 & 3 & 2 \\ 2 & -5 & 1 \end{bmatrix}$, then $-A + 3B$ equals*

$$(-1) \begin{bmatrix} 1 & 2 & 3 \\ 4 & 5 & 6 \end{bmatrix} + (3) \begin{bmatrix} -1 & 3 & 2 \\ 2 & -5 & 1 \end{bmatrix}$$

$$= \begin{bmatrix} -1 & -2 & -3 \\ -4 & -5 & -6 \end{bmatrix} + \begin{bmatrix} -3 & 9 & 6 \\ 6 & -15 & 3 \end{bmatrix} = \begin{bmatrix} -4 & 7 & 3 \\ 2 & -20 & -3 \end{bmatrix}.$$

Definition 1.10 *Two matrices $[a_{ij}]$, $[b_{ij}] \in M_{mn}$ are **equal** if they have the same dimensions and $a_{ij} = b_{ij}$ for all i and j.*

The following theorem proves that M_{mn} is (what we will later call) a **vector space**:

Theorem 1.4 *For M_{mn} as defined above the following properties hold:*

0. *For all $A, B \in M_{mn}$ and scalar a, we have $A + B$, $aA \in M_{mn}$.*

1. *For all $A, B \in M_{mn}$, $A + B = B + A$.*

2. *For all $A, B, C \in M_{mn}$, $(A + B) + C = A + (B + C)$.*

3. *There exists $0_{mn} \in M_{mn}$, such that for all $A \in M_{mn}$, $A + 0_{mn} = A$.*

4. *For each $A \in M_{mn}$ there is a $B \in M_{mn}$ such that $A + B = 0_{mn}$.*

5. For all $A, B \in M_{mn}$ and scalar a, $a(A + B) = aA + aB$.

6. For all $A \in M_{mn}$ and scalars a, b, $(a + b)A = aA + bA$.

7. For all $A \in M_{mn}$ and scalars a, b, $(ab)A = a(bA)$.

8. For all $A \in M_{mn}$, $1A = A$.

Proof 1.5 *We prove a few items to give the reader insight into the proper approach. The rest are left as exercises. First, set $A = [a_{ij}]$ and $B = [b_{ij}]$. To prove 1, notice that*

$$A + B = [a_{ij}] + [b_{ij}] = [a_{ij} + b_{ij}] = [b_{ij} + a_{ij}] = [b_{ij}] + [a_{ij}] = B + A.$$

To prove 3, take 0_{mn} to be the $m \times n$ matrix filled with 0's. To prove 7, notice that

$$(ab)A = (ab)[a_{ij}] = [(ab)a_{ij}] = [a(ba_{ij})] = a[ba_{ij}] = a(b[a_{ij}]) = a(bA).$$

□

We remark that $\mathbb{R}^n$ is just a particular example of M_{mn}. Namely, $\mathbb{R}^n = M_{1n}$ or M_{n1}. This is illustrated below:

$$[a_1, a_2, \ldots, a_n] = [a_1 \ a_2 \ \cdots \ a_n] = \begin{bmatrix} a_1 \\ a_2 \\ \vdots \\ a_n \end{bmatrix}.$$

1.4.1 Special Matrix Families

We now introduce some terminology and designate names to some special matrices.

Definitions and Examples:

1. The $m \times n$ **zero** matrix, written 0_{mn}, is a matrix filled with 0's. For example, the 2×3 zero matrix is

$$0_{23} = \begin{bmatrix} 0 & 0 & 0 \\ 0 & 0 & 0 \end{bmatrix}.$$

2. The **principal diagonal** of a matrix is comprised of the entries of the form a_{ii}. In the following example, the asterisks form the principal diagonal:

$$\begin{bmatrix} * & 1 & 2 & 3 \\ 1 & * & 2 & 3 \\ 1 & 2 & * & 3 \end{bmatrix}.$$

3. A matrix is **square** if it has the same number of rows as columns. An example of a 3×3 square matrix is

$$\begin{bmatrix} 1 & 2 & 3 \\ 2 & 1 & 3 \\ 3 & 1 & 2 \end{bmatrix}.$$

4. A **diagonal** matrix is a square matrix with the property that every entry off the diagonal is zero. Formally, $a_{ij} = 0$ for $i \neq j$. An example of a 3×3 diagonal matrix is

$$\begin{bmatrix} 1 & 0 & 0 \\ 0 & -1 & 0 \\ 0 & 0 & 2 \end{bmatrix}.$$

5. The **identity** matrix, written I_n is an $n \times n$ diagonal matrix with 1's on the diagonal. Formally,

$$a_{ij} = \begin{cases} 0, & \text{if } i \neq j \\ 1, & \text{if } i = j \end{cases}$$

The 3×3 identity matrix is

$$I_3 = \begin{bmatrix} 1 & 0 & 0 \\ 0 & 1 & 0 \\ 0 & 0 & 1 \end{bmatrix}.$$

6. A **scalar** matrix is any matrix of the form aI_n where a is a scalar. An example of a 3×3 scalar matrix is

$$\begin{bmatrix} -2 & 0 & 0 \\ 0 & -2 & 0 \\ 0 & 0 & -2 \end{bmatrix} = -2I_3.$$

7. An **upper triangular** matrix is a square matrix with 0's below the diagonal. Formally, $a_{ij} = 0$ if $i > j$. An example of a 3×3 upper triangular matrix is

$$\begin{bmatrix} 1 & -2 & 4 \\ 0 & -1 & 0 \\ 0 & 0 & 2 \end{bmatrix}.$$

8. A **lower triangular** matrix is a square matrix with 0's above the diagonal. Formally, $a_{ij} = 0$ if $i < j$. An example of a 3×3 lower triangular matrix is

$$\begin{bmatrix} 1 & 0 & 0 \\ 4 & -1 & 0 \\ -1 & 5 & 2 \end{bmatrix}.$$

9. The **transpose** of an $m \times n$ matrix A, written A^T, is an $n \times m$ matrix with the property that its ijth entry is the jith entry of A. Formally, if $A = [a_{ij}]$ and $A^T = [b_{ij}]$, then $b_{ij} = a_{ji}$ for all i and j. Some examples of taking the transpose of a matrix are

$$\begin{bmatrix} 1 & 2 & 3 \\ 4 & 5 & 6 \\ 7 & 8 & 9 \end{bmatrix}^T = \begin{bmatrix} 1 & 4 & 7 \\ 2 & 5 & 8 \\ 3 & 6 & 9 \end{bmatrix}, \qquad \begin{bmatrix} 1 & 2 & 3 & 1 \\ 4 & 5 & 6 & 3 \\ 7 & 8 & 9 & 2 \end{bmatrix}^T = \begin{bmatrix} 1 & 4 & 7 \\ 2 & 5 & 8 \\ 3 & 6 & 9 \\ 1 & 3 & 2 \end{bmatrix}.$$

10. A **symmetric** matrix A has the property that $A^T = A$. Formally, $a_{ij} = a_{ji}$ for all i and j. Note that this definition implicitly requires that A be a square matrix. An example of a 3×3 symmetric matrix is

$$\begin{bmatrix} 1 & -2 & 4 \\ -2 & -1 & 0 \\ 4 & 0 & 2 \end{bmatrix}.$$

11. A **skew symmetric** matrix A has the property that $A^T = -A$. Formally, $a_{ij} = -a_{ji}$ for all i and j. Note that this definition also requires A to be a square matrix. Furthermore, one can easily show that such a matrix must have 0's on the diagonal. An example of a 3×3 skew symmetric matrix is

$$\begin{bmatrix} 0 & 1 & -2 \\ -1 & 0 & 3 \\ 2 & -3 & 0 \end{bmatrix}.$$

We list below some basic properties of transposition:

Theorem 1.5 *If $A, B \in M_{mn}$ and a is a scalar, then*

i. $(A^T)^T = A$.

ii. $(A + B)^T = A^T + B^T$.

iii. $(aA)^T = aA^T$.

Proof 1.6 *We prove iii. and leave the rest as exercises. Set $A = [a_{ij}]$. Then*

$$(aA)^T = [aa_{ij}]^T = [aa_{ji}] = a[a_{ji}] = aA^T.$$

□

EXERCISES

1. Consider the following matrices:

$$A = \begin{bmatrix} 1 & -2 & 3 \\ 0 & 1 & 0 \\ 1 & -1 & 3 \end{bmatrix}, \quad B = \begin{bmatrix} 2 & -1 & 1 \\ 1 & -1 & 3 \\ -1 & 1 & 0 \end{bmatrix}, \quad C = \begin{bmatrix} 1 & -1 \\ 2 & 0 \\ 0 & -3 \end{bmatrix},$$

$$D = \begin{bmatrix} 2 & -2 & 0 \\ -1 & 2 & 3 \end{bmatrix}, \quad E = \begin{bmatrix} 1 \\ 4 \\ -3 \end{bmatrix}, \quad F = \begin{bmatrix} 1 & -3 & 2 \end{bmatrix}.$$

   Compute (if possible) $A + B$, $C + D$, $-3C$, $2A - 3B$, $B + 2C$, $(A - B)^T$, and $E^T + 2F$.

2. Consider the following matrices:

$$A = \begin{bmatrix} 1 & 2 & 3 \\ 0 & 1 & 0 \end{bmatrix}, \quad B = \begin{bmatrix} 0 & 1 & -2 & 0 \\ 1 & 1 & 0 & 3 \\ -2 & 0 & 0 & 1 \\ 0 & 3 & 1 & 0 \end{bmatrix}, \quad C = \begin{bmatrix} 0 & 1 & -2 \\ -1 & 0 & -3 \\ 2 & 3 & 0 \end{bmatrix},$$

$$D = \begin{bmatrix} 1 & 1 \\ 0 & 1 \end{bmatrix}, \quad E = \begin{bmatrix} 1 & 0 & 0 \\ 2 & 0 & 0 \\ -3 & 0 & 0 \end{bmatrix}, \quad F = \begin{bmatrix} 2 & 3 & -1 \\ 0 & 1 & -4 \\ 0 & 0 & 1 \end{bmatrix},$$

$$G = \begin{bmatrix} -2 & 0 & 0 \\ 0 & 3 & 0 \\ 0 & 0 & 0 \end{bmatrix}, \quad H = \begin{bmatrix} 0 & 0 \\ 0 & 0 \end{bmatrix}, \quad K = \begin{bmatrix} -1 & 0 \\ 0 & 3 \end{bmatrix}.$$

   List the matrices which meet each of the following criterion: square, diagonal, upper triangular, lower triangular, symmetric and skew-symmetric.

3. Let A be a square matrix.

   a. Show that $\frac{1}{2}(A + A^T)$ is symmetric.

   b. Show that $\frac{1}{2}(A - A^T)$ is skew-symmetric.

   c. Use parts a. and b. to explain why any square matrix can be expressed as a sum of a symmetric matrix and a skew-symmetric matrix.

4. Using Exercise 3, express $B = \begin{bmatrix} 1 & 3 \\ 0 & -2 \end{bmatrix}$ as a sum of a symmetric and skew-symmetric matrix.

5. Prove that a skew-symmetric matrix has 0's on its diagonal.

6. Show that if $a \in \mathbb{R}$ and A, B are symmetric matrices, then so is $aA + B$.

7. Show that if A and B are skew-symmetric matrices, then so is $aA + bB$, for any scalars a, b.

8. Show that if A and B are skew-symmetric and $AB = BA$, then AB is symmetric.

9. Prove properties 0, 2, 4–6 and 8 of Theorem 1.4.

10. Prove properties i and ii of Theorem 1.5.

1.5 MATRIX MULTIPLICATION

Here, we present another operation applicable in M_{mn} in which the inputs are two matrices and the output is another matrix. Although this is not an operation indicative of a vector space, it is an essential ingredient in what will follow.

Definition 1.11 *Let $A = [a_{ij}] \in M_{mn}$ and $B = [b_{ij}] \in M_{nr}$. Then the product $C = [c_{ij}] = AB \in M_{mr}$ is defined as follows:*

$$c_{ij} = \sum_{k=1}^{n} a_{ik} b_{kj}.$$

Notice that to perform matrix multiplication on matrices, it is necessary that the number of columns in A be equal to the number of rows in B and the resulting matrix has the same number of rows as A and the same number of columns as B. Perhaps a simpler way to remember the entries of C is that the ijth entry of C is obtained by taking the dot product of the ith row of A with the jth column of B. Conversely, one can define dot product in terms of matrix multiplication. Indeed, if $v, w \in \mathbb{R}^n$, then $v \cdot w = v^T w$, where v and w are viewed as $n \times 1$ column matrices. This is sometimes a useful representation of dot product when demonstrating certain proofs.

Example 1.10

$$\begin{bmatrix} 1 & 2 & 3 \\ 4 & 5 & 6 \end{bmatrix} \begin{bmatrix} 1 & -1 & 1 \\ -1 & 0 & 1 \\ 0 & 1 & 1 \end{bmatrix}$$

$$= \begin{bmatrix} (1)(1) + (2)(-1) + (3)(0) & (1)(-1) + (2)(0) + (3)(1) & (1)(1) + (2)(1) + (3)(1) \\ (4)(1) + (5)(-1) + (6)(0) & (4)(-1) + (5)(0) + (6)(1) & (4)(1) + (5)(1) + (6)(1) \end{bmatrix}$$

$$= \begin{bmatrix} -1 & 2 & 6 \\ -1 & 2 & 15 \end{bmatrix}.$$

We now list some basic properties of matrix multiplication:

Theorem 1.6 *For matrices A, B, C and I (the identity matrix) of the appropriate dimensions the following statements are true:*

   i. $A(BC) = (AB)C$.

   ii. $A(B + C) = AB + AC$ and $(B + C)A = BA + CA$.

   iii. $a(AB) = (aA)B = A(aB)$.

   iv. $AI = A$ and $IA = A$.

   v. $(AB)^T = B^T A^T$.

Proof 1.7 *We prove some of the statements and leave the rest as exercises. First, we establish the notation $A = [a_{ij}]$, $B = [b_{ij}]$ and $C = [c_{ij}]$. To prove ii, we first set our dimensions: $A \in M_{mn}$ and $B, C \in M_{nr}$. Then*

$$A(B + C) = [a_{ij}]([b_{ij}] + [c_{ij}]) = [a_{ij}][b_{ij} + c_{ij}]$$

$$= \left[\sum_{k=1}^{n} a_{ik}(b_{kj} + c_{kj})\right] = \left[\sum_{k=1}^{n}(a_{ik}b_{kj} + a_{ik}c_{kj})\right] = \left[\sum_{k=1}^{n} a_{ik}b_{kj} + \sum_{k=1}^{n} a_{ik}c_{kj}\right]$$

$$= \left[\sum_{k=1}^{n} a_{ik}b_{kj}\right] + \left[\sum_{k=1}^{n} a_{ik}c_{kj}\right] = AB + AC.$$

Using the same notation and dimensions, we prove v. To ensure lucidity, we also establish the notation: $B^T = [d_{ij}]$ and $A^T = [e_{ij}]$ where $d_{ij} = b_{ji}$ and $e_{ij} = a_{ji}$ for all i and j. Then

$$(AB)^T = \left[\sum_{k=1}^{n} a_{ik}b_{kj}\right]^T = \left[\sum_{k=1}^{n} a_{jk}b_{ki}\right] = \left[\sum_{k=1}^{n} e_{kj}d_{ik}\right]$$

$$= \left[\sum_{k=1}^{n} d_{ik}e_{kj}\right] = [d_{ij}][e_{ij}] = B^T A^T.$$

□

We warn the reader that there are certain very basic properties which matrix multiplication does **not** enjoy. For one, we do not have the **commutative** property, i.e. that $AB = BA$ is not true in general, even for square matrices.

Example 1.11 *Set $A = \begin{bmatrix} 1 & 0 \\ 0 & 0 \end{bmatrix}$ and $B = \begin{bmatrix} 0 & 1 \\ 0 & 0 \end{bmatrix}$. Then one can easily check that $AB \neq BA$.*

A second property that we do not have for matrix multiplication is the **cancellation** property, i.e. $AB = AC$ does not necessarily imply that $B = C$ (exercise).

We state two facts whose proofs will be left as an exercise.

Lemma 1.1 *Let $A \in M_{mn}$ and $B \in M_{nr}$. If $B = [c_1\ c_2\ \ldots\ c_r]$ (in columns), then $AB = [Ac_1\ Ac_2\ \ldots\ Ac_r]$ (in columns).*

Lemma 1.2 *Let $A = [c_1 \cdots c_n] \in M_{mn}$ (in columns) and $u = [a_1, \ldots, a_n] = \begin{bmatrix} a_1 \\ \vdots \\ a_n \end{bmatrix} \in \mathbb{R}^n$. Then $Au = a_1 c_1 + \cdots + a_n c_n$.*

We are now in a position to define exponentiation for square matrices.

Definition 1.12 *Let $A \in M_{nn}$ and k be a positive integer. We define matrix **exponentiation** as follows:*

- $A^0 = I$

- $A^1 = A$

- $A^2 = AA$

- *In general, for positive integer k,* $A^k = \underbrace{AA \cdots A}_{k\ times}.$

We state without proof (although they can be easily verified by induction) the usual properties of exponentiation:

Theorem 1.7 *For a square matrix A the following are true:*

i. $A^k A^l = A^{k+l}$.

ii. $(A^k)^l = A^{kl}$.

EXERCISES

1. Consider the following matrices:

$$A = \begin{bmatrix} 1 & 0 & -1 \\ 2 & 3 & 1 \end{bmatrix} \qquad B = \begin{bmatrix} 1 & 3 \\ 0 & -2 \end{bmatrix}.$$

   a. If possible, compute AB and BA.

   b. Compute B^3.

2. Consider the matrices listed below:

$$A = \begin{bmatrix} 1 & 2 \\ -3 & 0 \\ 0 & 1 \end{bmatrix} \qquad B = \begin{bmatrix} 1 & 0 & -1 \\ -2 & 1 & 1 \\ 1 & -2 & 2 \end{bmatrix} \qquad C = \begin{bmatrix} 1 & 2 & -1 \\ -2 & 1 & 1 \end{bmatrix}.$$

Compute the following: $2A - C^T$ and $(AC)^T + B$.

3. Consider the following matrices:

$$A = \begin{bmatrix} 1 & 0 & -1 \\ 2 & 3 & 1 \\ 3 & -1 & 2 \end{bmatrix}, \ B = \begin{bmatrix} 0 & -1 \\ 2 & 1 \\ 3 & -1 \end{bmatrix}, \ C = \begin{bmatrix} 0 & -2 & 3 \\ -1 & 0 & 2 \end{bmatrix},$$

$$D = \begin{bmatrix} -1 & 1 \\ 0 & 2 \end{bmatrix}, \ E = \begin{bmatrix} 1 \\ 4 \\ -3 \end{bmatrix}, \ F = \begin{bmatrix} 1 & -3 \end{bmatrix}.$$

Compute (if possible) AB, AC, AD, AE, AF, BA, BC, BD, BE, BF, CA, CB, CD, CE, CF, DA, DB, DC, DE, DF, EA, EB, EC, ED, EF, FA, FB, FC, FD, FE, ABC, BDC, ACD, $B^T A$, AC^T, DB^T, DC^T, $2A(B + C^T)$, D^3, $2D^2$, and $(2D)^2$.

4. Give an example which illustrates that the Cancellation Property fails for matrix multiplication.

5. Prove that if A and B are symmetric and $AB = BA$, then AB is symmetric.

6. Prove that the product of two upper-triangular matrices is upper-triangular.

7. Let $A, B \in M_{mn}$. Prove that if for all $i = 1, \ldots, n$ we have $Ae_i = Be_i$, then $A = B$.

   (Hint: Use Lemma 1.1.)

8. Prove the following statement: Let A and B be square matrices. If $AB = A$ and $BA = B$ then $A^2 = A$.

9. Use induction to prove that for any integer $n \geq 0$ and square matrix A, we have $(-A)^n = (-1)^n A^n$.

10. Prove by induction that $(A^n)^T = (A^T)^n$, for any square matrix A and positive integer n.

11. Show that if $A \in M_{nn}$ and $AB = BA$ for all $B \in M_{nn}$, then $A = aI$ for some scalar a.

12. Prove properties i, iii and iv of Theorem 1.6.

13. Prove Lemma 1.1.

14. Prove Lemma 1.2.

15. Prove Theorem 1.7.

CHAPTER 2

Matrices and Linear Systems

I N THIS CHAPTER, we continue introducing computational skills relevant to linear algebra. In Section 2.1, system of equations are introduced. In Section 2.2, the main algorithm for solving systems of equations is presented called Gaussian Elimination. Section 2.3 is an application section on Markov Chains. A method is presented for solving in the case that the transition matrix is *regular*. Section 2.4 is another application section introducing the Simplex Method on a sub-collection of linear programming problems. In Section 2.5, the discussion is less computational and more theoretical where matrix equivalence is discussed, a notion necessary for further theoretical developments. In Section 2.6, the inverse of a matrix is presented, when it exists and how to find it. In Section 2.7, the Simplex Method is revisited and redone using matrix multiplication in place of elementary row operations. In Section 2.8, linear systems of equations are divided into homogeneous and non-homogeneous. The rank of a matrix is also presented including its theoretical significance. In Section 2.9, the determinant of a matrix is computed in several ways and a connection to the existence of an inverse for a matrix is made. In Section 2.10, certain linear systems are solved entirely in terms of determinant and the inverse of a matrix is computed entirely in terms of determinant, when it exists. One final numerical methods application is presented in Section 2.11 called the LU factorization.

2.1 SYSTEMS OF LINEAR EQUATIONS

Our main goals in this chapter are to give a systematic way of solving linear systems and show their intimate connection to matrices, but before we can do this we need to introduce relevant terminology and notation.

DOI: 10.1201/9781003217794-2

Definition 2.1 *An $m \times n$ system of linear equations has the form*

$$
\begin{array}{rcl}
a_{11}x_1 + a_{12}x_2 + \cdots + a_{1n}x_n &=& b_1 \\
a_{21}x_1 + a_{22}x_2 + \cdots + a_{2n}x_n &=& b_2 \\
&\vdots& \\
a_{m1}x_1 + a_{m2}x_2 + \cdots + a_{mn}x_n &=& b_m
\end{array},
$$

where the $x_1, x_2, \ldots, x_n$ represent the unknowns and the a_{ij}'s and b_i's are scalars. The number of equations is m and the number of unknowns is n.

There are several other ways of representing a system of linear equations which we present here.

1. **Vector Representation:**

$$
c_1 x_1 + c_2 x_2 + \cdots + c_n x_n = b \text{ with each } c_j = \begin{bmatrix} a_{1j} \\ a_{2j} \\ \vdots \\ a_{mj} \end{bmatrix} \text{ and } b = \begin{bmatrix} b_1 \\ b_2 \\ \vdots \\ b_m \end{bmatrix}
$$

2. **Matrix Representation:**

$$
\begin{bmatrix}
a_{11} & a_{12} & \cdots & a_{1n} \\
a_{21} & a_{22} & \cdots & a_{2n} \\
\vdots & \vdots & \ddots & \vdots \\
a_{m1} & a_{m2} & \cdots & a_{mn}
\end{bmatrix}
\begin{bmatrix} x_1 \\ x_2 \\ \vdots \\ x_n \end{bmatrix}
=
\begin{bmatrix} b_1 \\ b_2 \\ \vdots \\ b_m \end{bmatrix}
$$

or even simpler, we just write $AX = B$ for the matrices we listed in the matrix equation. The matrix A is called the **coefficient** matrix. At times we may write $Ax = b$ where we view x and b as tuples.

3. **Augmented Matrix:**

$$
\left[
\begin{array}{cccc|c}
a_{11} & a_{12} & \cdots & a_{1n} & b_1 \\
a_{21} & a_{22} & \cdots & a_{2n} & b_2 \\
\vdots & \vdots & \ddots & \vdots & \vdots \\
a_{m1} & a_{m2} & \cdots & a_{mn} & b_m
\end{array}
\right]
$$

Example: Consider the following 2×3 system of linear equations:

$$
\begin{array}{rcl}
2x_1 + x_2 - x_3 &=& 0 \\
x_1 - 3x_2 + x_3 &=& 7
\end{array},
$$

The vector representation of this system is

$$\begin{bmatrix} 2 \\ 1 \end{bmatrix} x_1 + \begin{bmatrix} 1 \\ -3 \end{bmatrix} x_2 + \begin{bmatrix} -1 \\ 1 \end{bmatrix} x_3 = \begin{bmatrix} 0 \\ 7 \end{bmatrix}.$$

The matrix representation of this system is

$$\begin{bmatrix} 2 & 1 & -1 \\ 1 & -3 & 1 \end{bmatrix} \begin{bmatrix} x_1 \\ x_2 \\ x_3 \end{bmatrix} = \begin{bmatrix} 0 \\ 7 \end{bmatrix}.$$

The augmented matrix associated with this system is

$$\left[\begin{array}{ccc|c} 2 & 1 & -1 & 0 \\ 1 & -3 & 1 & 7 \end{array} \right].$$

Definition 2.2 $X_0 = [c_1, c_2, \ldots, c_n]$, *where the c_i's are scalars, is a* **solution** *to*

$$\begin{aligned} a_{11}x_1 + a_{12}x_2 + \cdots + a_{1n}x_n &= b_1 \\ a_{21}x_1 + a_{22}x_2 + \cdots + a_{2n}x_n &= b_2 \\ &\vdots \\ a_{m1}x_1 + a_{m2}x_2 + \cdots + a_{mn}x_n &= b_m \end{aligned},$$

if when we substitute $c_1, \ldots, c_n$ in for $x_1, \ldots, x_n$, then the following equations are true:

$$\begin{aligned} a_{11}c_1 + a_{12}c_2 + \cdots + a_{1n}c_n &= b_1 \\ a_{21}c_1 + a_{22}c_2 + \cdots + a_{2n}c_n &= b_2 \\ &\vdots \\ a_{m1}c_1 + a_{m2}c_2 + \cdots + a_{mn}c_n &= b_m \end{aligned}.$$

To rephrase in our simplified notation, X_0 is a **solution** to $AX = B$ if $AX_0 = B$ as matrices.

The set of all solutions to a particular system of linear equations is called its **solution set**.

Example 2.1 $[1, -2, 0]$ *is a solution to the 3×3 linear system*

$$\begin{aligned} 2x_1 + x_2 - x_3 &= 0 \\ x_1 - 3x_2 + x_3 &= 7 \\ -3x_1 + x_2 + x_3 &= -5 \end{aligned},$$

since

$$\begin{array}{rcl}
2(1) + (-2) - (0) &=& 0 \\
(1) - 3(-2) + (0) &=& 7 \\
-3(1) + (-2) + (0) &=& -5
\end{array} \ .$$

In fact, we will see that this is the only solution to this linear system. Hence, the solution set to this linear system is $\{[1, -2, 0]\}$.

In Section 2.2, we will give an algorithmic method for finding the solution set of any linear system. We often use the variables x and y in place of x_1 and x_2, and x, y and z in place of x_1, x_2 and x_3 in the case of systems with two or three unknowns. The next theorem gives a classification of the possible solution sets to a system of linear equations.

Theorem 2.1 *Given any system of linear equations, there are only three possibilities for its solution set:*

 i. No solution.

 ii. Exactly one solution.

 iii. An infinite number of solutions.

Proof 2.1 *In the next section, we will see concrete examples of linear systems with no solution, one solution and infinite solutions. Thus, to prove this result it suffices to show that if a linear system has at least two solutions, then it has an infinite number of solutions. Suppose X_0 and X_1 are two solutions to the linear system $AX = B$. Define $X_a = X_0 + a(X_0 - X_1)$ for any scalar a. First notice that each X_a is a solution to $AX = B$. Indeed, using properties of matrices we see that*

$$AX_a = A(X_0 + a(X_0 - X_1)) = AX_0 + aAX_0 - aAX_1 = B + aB - aB = B.$$

Furthermore, If $a \neq b$ then $X_a \neq X_b$ (we leave this as an exercise). Since there are an infinite number of scalars a, this implies that we have an infinite number of solutions. □

Referring to Theorem 2.1, when case i occurs we say the system of linear equations is **inconsistent**, otherwise we call the system **consistent**.

EXERCISES

1. Decide which of the following systems of equations are linear:

a.
$$\begin{cases}
2x + y - (1/2)z &=& 0 \\
x - 3y + z &=& 7/2 \\
-3x + y + z &=& -5
\end{cases}$$

b.
$$\begin{cases}
2x - 8\sqrt{y} + 4z &=& 5 \\
x - 4y + 2z &=& 0
\end{cases}$$

c. $\begin{cases} x + y - (1/z) &= 0 \\ x - y + 2z &= -3 \end{cases}$

d. $\begin{cases} \sqrt{2}x - 8y + 4z &= 5 \\ x - 4y + 2^{.125}z &= 0 \end{cases}$

e. $\begin{cases} \pi x - 3y + z &= 7 \\ -3x + y + z &= (-5)^{2/3} \end{cases}$

f. $\begin{cases} 2x - xy + z &= 5 \\ x - y + z &= -1 \end{cases}$

g. $\begin{cases} x - y + z^{-2} &= 7 \\ x + y + z &= 8 \end{cases}$

h. $\begin{cases} xyz = 1 \\ x + y + z = 1 \end{cases}$

2. Rewrite each of the linear systems of equations below in a vector, matrix and augmented representation:

a. $\begin{cases} 2x + y - z &= 0 \\ x - 3y + z &= 7 \\ -3x + y + z &= -5 \end{cases}$

b. $\begin{cases} 2x - 8y + 4z &= 5 \\ x - 4y + 2z &= 0 \end{cases}$

c. $\begin{cases} \pi x + \sqrt{2}y &= e \\ (1/2)x - 3.159y &= 0 \end{cases}$

3. Recover the linear systems of equations from the following augmented matrices:

a. $\left[\begin{array}{cc|c} 1 & 2 & -3 \\ 4 & 1 & 5 \\ 2 & -3 & -1 \end{array}\right]$

b. $\left[\begin{array}{ccc|c} 1 & -1 & 0 & 2 \\ 0 & 0 & 1 & 1 \\ 0 & 0 & 0 & 0 \end{array}\right]$

c. $\left[\begin{array}{ccc|c} 1 & -1 & 2 & 0 \\ 0 & 3 & 0 & 1 \\ 0 & 5 & 0 & 0 \end{array}\right]$

d. $\left[\begin{array}{ccc|c} 1 & 0 & 1 & 0 \\ 0 & 0 & 1 & 0 \\ 0 & -3 & 0 & 1 \end{array}\right]$

e. $\left[\begin{array}{ccc|c} 1 & 0 & 0 \\ 0 & 2 & 0 \\ 0 & 0 & 1 \end{array}\right]$

f. $\left[\begin{array}{cccc|c} 1 & -2 & 0 & 3 & 0 \\ 0 & 0 & 1 & 5 & 0 \\ 0 & 1 & -1 & 0 & 1 \end{array}\right]$

4. Verify that $[-2, 0, -2, 1]$ is a solution to the following linear system of equations:

$$\begin{cases} x_1 + x_2 + x_3 + x_4 &= -3 \\ -2x_1 - x_2 + x_3 + 3x_4 &= 5 \\ 3x_1 + 4x_2 + 6x_3 + 8x_4 &= -10 \end{cases}$$

5. Complete the proof of Theorem 2.1, i.e. show that if $a \neq b$, then $X_a \neq X_b$ (do this by proving the contrapositive).

2.2   GAUSSIAN ELIMINATION

We are ready to present a systematic way for solving systems of linear equations. This method is simple and will be used quite regularly throughout the remainder of the book. First, recall that every system of linear equations has an associated augmented matrix:

Example 2.2 *The augmented matrix associated with the linear system*

$$\begin{cases} 2x_1 + x_2 - x_3 & = & 0 \\ x_1 - 3x_2 + x_3 & = & 7 \\ -3x_1 + x_2 + x_3 & = & -5 \end{cases}$$

is

$$\left[\begin{array}{ccc|c} 2 & 1 & -1 & 0 \\ 1 & -3 & 1 & 7 \\ -3 & 1 & 1 & -5 \end{array} \right].$$

In solving a linear system we wish to manipulate the equations without altering the solution set and arrive at a more "desirable" system of equations for which we can readily identify the solution set. The operations below achieve this goal.

Definition 2.3 *The following three operations are called* **elementary row operations** *which can be applied to a system of linear equations or the associated augmented matrix:*

1. *Multiplying the ith equation (or ith row of the augmented matrix) by a non-zero scalar a. The notation is aR_i.*

2. *Switching the ith and jth equation (or ith and jth row of the augmented matrix). The notation is $R_i \leftrightarrow R_j$.*

3. *Adding a scalar a times the ith equation to the jth equation (or adding a times the ith row to the jth row of the augmented matrix). The notation is $aR_i + R_j$.*

Example: We illustrate the three operations in an example.

$$\begin{cases} 2x - y & = & 6 \\ x + 6y & = & -1 \end{cases} \xrightarrow{-3R_2} \begin{cases} 2x - y & = & 6 \\ -3x - 18y & = & 3 \end{cases}$$

$$\xrightarrow{R_1 \leftrightarrow R_2} \begin{cases} -3x - 18y & = & 3 \\ 2x - y & = & 6 \end{cases} \xrightarrow{4R_2 + R_1} \begin{cases} 5x - 22y & = & 27 \\ 2x - y & = & 6 \end{cases}.$$

Notice that all information is retained if we apply the same operations to the corresponding augmented matrix. We use the augmented matrix to make the process slightly less tedious.

$$\begin{bmatrix} 2 & -1 & 6 \\ 1 & 6 & -1 \end{bmatrix} \xrightarrow{-3R_2} \begin{bmatrix} 2 & -1 & 6 \\ -3 & -18 & 3 \end{bmatrix}$$

$$\xrightarrow{R_1 \leftrightarrow R_2} \begin{bmatrix} -3 & -18 & 3 \\ 2 & -1 & 6 \end{bmatrix} \xrightarrow{4R_2+R_1} \begin{bmatrix} 5 & -22 & 27 \\ 2 & -1 & 6 \end{bmatrix}.$$

When we wish to refer to an arbitrary elementary row operation, we will use the notation $\xrightarrow{op}$.

Theorem 2.2 *Let $AX = B$ be a system of linear equations and suppose that the linear system $CX = D$ is obtained by applying an elementary row operation to $AX = B$. Then $AX = B$ and $CX = D$ have identical solution sets.*

Proof 2.2 *To demonstrate this theorem, we need to verify the theorem's statement for each of the three elementary row operations. Our tactic is to show that each solution to $AX = B$ is also a solution to $CX = D$ and vice versa. The theorem is clear for $R_i \leftrightarrow R_j$ since rearranging the order of the equations should not have an effect on the solution set. We will verify the theorem for the first operation and leave the proof of the third operation as an exercise.*

To do this we need the expanded notation for $AX = B$:

$$\begin{cases} a_{11}x_1 + a_{12}x_2 + \cdots + a_{1n}x_n &= b_1 \\ a_{21}x_1 + a_{22}x_2 + \cdots + a_{2n}x_n &= b_2 \\ &\vdots \\ a_{i1}x_1 + a_{i2}x_2 + \cdots + a_{in}x_n &= b_i \\ &\vdots \\ a_{m1}x_1 + a_{m2}x_2 + \cdots + a_{mn}x_n &= b_m \end{cases}$$

Applying the operation aR_i ($a \neq 0$) to the system $AX = B$, we obtain the system $CX = D$:

$$\begin{cases} a_{11}x_1 + a_{12}x_2 + \cdots + a_{1n}x_n &= b_1 \\ a_{21}x_1 + a_{22}x_2 + \cdots + a_{2n}x_n &= b_2 \\ &\vdots \\ aa_{i1}x_1 + aa_{i2}x_2 + \cdots + aa_{in}x_n &= ab_i \\ &\vdots \\ a_{m1}x_1 + a_{m2}x_2 + \cdots + a_{mn}x_n &= b_m \end{cases}$$

Now, suppose that $[c_1, c_2, \ldots, c_n]$ is a solution to $AX = B$. This means that

$$\begin{cases} a_{11}c_1 + a_{12}c_2 + \cdots + a_{1n}c_n &= b_1 \\ a_{21}c_1 + a_{22}c_2 + \cdots + a_{2n}c_n &= b_2 \\ &\vdots \\ a_{i1}c_1 + a_{i2}c_2 + \cdots + a_{in}c_n &= b_i \\ &\vdots \\ a_{m1}c_1 + a_{m2}c_2 + \cdots + a_{mn}c_n &= b_m \end{cases}.$$

By properties of equality we can multiply the ith equation by $a \neq 0$ to obtain

$$\begin{cases} a_{11}c_1 + a_{12}c_2 + \cdots + a_{1n}c_n &= b_1 \\ a_{21}c_1 + a_{22}c_2 + \cdots + a_{2n}c_n &= b_2 \\ &\vdots \\ aa_{i1}c_1 + aa_{i2}c_2 + \cdots + aa_{in}c_n &= ab_i \\ &\vdots \\ a_{m1}c_1 + a_{m2}c_2 + \cdots + a_{mn}c_n &= b_m \end{cases}.$$

This last set of equations exhibits the fact that $[c_1, c_2, \ldots, c_n]$ is also a solution to $CX = D$. On the other hand, suppose that $[c_1, c_2, \ldots, c_n]$ is a solution to $CX = D$. This means that

$$\begin{cases} a_{11}c_1 + a_{12}c_2 + \cdots + a_{1n}c_n &= b_1 \\ a_{21}c_1 + a_{22}c_2 + \cdots + a_{2n}c_n &= b_2 \\ &\vdots \\ aa_{i1}c_1 + aa_{i2}c_2 + \cdots + aa_{in}c_n &= ab_i \\ &\vdots \\ a_{m1}c_1 + a_{m2}c_2 + \cdots + a_{mn}c_n &= b_m \end{cases}.$$

Since $a \neq 0$, we can multiply the ith equation by $1/a$ to obtain

$$\begin{cases} a_{11}c_1 + a_{12}c_2 + \cdots + a_{1n}c_n &= b_1 \\ a_{21}c_1 + a_{22}c_2 + \cdots + a_{2n}c_n &= b_2 \\ &\vdots \\ a_{i1}c_1 + a_{i2}c_2 + \cdots + a_{in}c_n &= b_i \\ &\vdots \\ a_{m1}c_1 + a_{m2}c_2 + \cdots + a_{mn}c_n &= b_m \end{cases}.$$

This last set of equations exhibits the fact that $[c_1, c_2, \ldots, c_n]$ is also a solution to $AX = B$. □

Corollary 2.1 *Let $AX = B$ be a system of linear equations and suppose that the linear system $CX = D$ is obtained by applying a finite number of elementary row operation to $AX = B$. Then $AX = B$ and $CX = D$ have identical solution sets.*

Proof 2.3 *The proof follows by induction and Theorem 2.2 and is left as an exercise.*
□

Using these elementary row operations, we wish to alter the augmented matrix so that the solution to the associated linear system is evident. This altered augmented matrix is a form which we now define:

Definition 2.4 *A matrix $A = [a_{ij}]$ is in* **reduced row-echelon form** *if the following conditions on the matrix are met:*

1. *Reading from left to right in each row, the first non-zero entry (if there is one) in each row is 1. Each of these 1's is called a* **pivot**.

2. *The entries above and below each pivot are 0's.*

3. *Whenever a_{ij} and a_{kl} $(i < k)$ are pivots for A, then $j < k$. We say that the pivots form a "staircase".*

4. *Reading from top to bottom, the rows entirely filled with zeros occur last.*

Example 2.3 *The following matrix is in reduced row echelon form:*

$$\begin{bmatrix} 1 & 0 & -2 & 0 \\ 0 & 1 & 1 & 0 \\ 0 & 0 & 0 & 1 \\ 0 & 0 & 0 & 0 \end{bmatrix}.$$

An important fact about reduced row-echelon form is the following:

Theorem 2.3 *Every matrix row reduces to exactly one row-echelon form.*

We remark that the proof of Theorem 2.3 would involve showing that different reduced row-echelon forms yield different solution sets (if we view these matrices as augmented matrices) and then appeal to Theorem 2.2.

Now we are ready to state our method for solving any system of linear equations, called **Gaussian Elimination**: Using elementary row operations convert the corresponding augmented matrix for the linear system into reduced row-echelon form. We will give below an example for each of the three types of solution sets of a linear system.

Example 2.4 *Consider the linear system we introduced earlier:*

$$\begin{cases} 2x_1 + x_2 - x_3 &= 0 \\ x_1 - 3x_2 + x_3 &= 7 \\ -3x_1 + x_2 + x_3 &= -5 \end{cases}.$$

The first thing one does is switch to the associated augmented matrix

$$\left[\begin{array}{ccc|c} 2 & 1 & -1 & 0 \\ 1 & -3 & 1 & 7 \\ -3 & 1 & 1 & -5 \end{array} \right].$$

Loosely speaking, working from left to right (in columns) we attempt to put pivot 1's on the principle diagonal and zeros above and below each of these 1's:

$$\left[\begin{array}{ccc|c} 2 & 1 & -1 & 0 \\ 1 & -3 & 1 & 7 \\ -3 & 1 & 1 & -5 \end{array} \right] \xrightarrow{R_1 \leftrightarrow R_2} \left[\begin{array}{ccc|c} 1 & -3 & 1 & 7 \\ 2 & 1 & -1 & 0 \\ -3 & 1 & 1 & -5 \end{array} \right]$$

$$\xrightarrow{-2R_1 + R_2} \left[\begin{array}{ccc|c} 1 & -3 & 1 & 7 \\ 0 & 7 & -3 & -14 \\ -3 & 1 & 1 & -5 \end{array} \right] \xrightarrow{3R_1 + R_3} \left[\begin{array}{ccc|c} 1 & -3 & 1 & 7 \\ 0 & 7 & -3 & -14 \\ 0 & -8 & 4 & 16 \end{array} \right]$$

$$\xrightarrow{1R_3 + R_2} \left[\begin{array}{ccc|c} 1 & -3 & 1 & 7 \\ 0 & -1 & 1 & 2 \\ 0 & -8 & 4 & 16 \end{array} \right] \xrightarrow{(-1)R_2} \left[\begin{array}{ccc|c} 1 & -3 & 1 & 7 \\ 0 & 1 & -1 & -2 \\ 0 & -8 & 4 & 16 \end{array} \right]$$

$$\xrightarrow{3R_2 + R_1} \left[\begin{array}{ccc|c} 1 & 0 & -2 & 1 \\ 0 & 1 & -1 & -2 \\ 0 & -8 & 4 & 16 \end{array} \right] \xrightarrow{8R_2 + R_3} \left[\begin{array}{ccc|c} 1 & 0 & -2 & 1 \\ 0 & 1 & -1 & -2 \\ 0 & 0 & -4 & 0 \end{array} \right]$$

$$\xrightarrow{(-1/4)R_3} \left[\begin{array}{ccc|c} 1 & 0 & -2 & 1 \\ 0 & 1 & -1 & -2 \\ 0 & 0 & 1 & 0 \end{array} \right] \xrightarrow{2R_3 + R_1} \left[\begin{array}{ccc|c} 1 & 0 & 0 & 1 \\ 0 & 1 & -1 & -2 \\ 0 & 0 & 1 & 0 \end{array} \right]$$

$$\xrightarrow{(1)R_3 + R_2} \left[\begin{array}{ccc|c} 1 & 0 & 0 & 1 \\ 0 & 1 & 0 & -2 \\ 0 & 0 & 1 & 0 \end{array} \right]$$

Now that we have the matrix in reduced row-echelon form we convert back to the associated linear system and we have our solution:

$$\begin{cases} x_1 &=& 1 \\ x_2 &=& -2 \\ x_3 &=& 0 \end{cases}$$

Notice that we tried to avoid introducing fractions in this procedure. For instance, our first step in the process could have been $(1/2)R_1$ which would have produced a pivot 1 in the appropriate place. But the rest of the process would have been much more tedious having to work with fractions. We remark, though, that regardless of how one arrives at reduced row-echelon form Theorem 2.3 guarantees that we always obtain the same one.

Example 2.5 *Having given an example of a linear system with one solution, we now give an example of a linear system with no solution. Consider the following linear system:*

$$\begin{cases} 2x - 8y + 4z &=& 5 \\ x - 4y + 2z &=& 0 \end{cases}$$

Converting to the augmented matrix, we begin the process:

$$\begin{bmatrix} 2 & -8 & 4 & 5 \\ 1 & -4 & 2 & 0 \end{bmatrix} \xrightarrow{R_1 \leftrightarrow R_2} \begin{bmatrix} 1 & -4 & 2 & 0 \\ 2 & -8 & 4 & 5 \end{bmatrix}$$

$$\xrightarrow{(-2)R_1 + R_2} \begin{bmatrix} 1 & -4 & 2 & 0 \\ 0 & 0 & 0 & 5 \end{bmatrix}$$

At this point we can stop and conclude that there is no solution, for observe the second row of the augmented matrix. As an equation it reads as $0 = 5$ which is a clear contradiction. Arriving at some sort of contradictory statement such as $0 = 5$ tells us that the system is inconsistent and has no solution, i.e. the solution set is empty.

In general, if at any point in the Gaussian Elimination process we should generate a row of the form $[0 \ 0 \ \cdots \ 0 \mid *]$, where $*$ is a non-zero number, then we can stop and conclude the system has no solution.

Example 2.6 *Finally, we give an example of a linear system with infinite solutions. Take special care to note how we eventually present the solution set. Consider the following linear system:*

$$\begin{cases} 2x - 8y + 4z &=& 5 \\ x - 4y + 2z &=& 0 \end{cases}$$

Converting to the augmented matrix we begin the process:

$$\begin{bmatrix} 2 & -8 & 1 & | & 5 \\ 1 & -4 & 2 & | & 0 \end{bmatrix} \xrightarrow{R_1 \leftrightarrow R_2} \begin{bmatrix} 1 & -4 & 2 & | & 0 \\ 2 & -8 & 1 & | & 5 \end{bmatrix}$$

$$\xrightarrow{(-2)R_1 + R_2} \begin{bmatrix} 1 & -4 & 2 & | & 0 \\ 0 & 0 & -3 & | & 5 \end{bmatrix} \xrightarrow{(-1/3)R_2} \begin{bmatrix} 1 & -4 & 2 & | & 0 \\ 0 & 0 & 1 & | & -5/3 \end{bmatrix}$$

$$\xrightarrow{(-2)R_2 + R_2} \begin{bmatrix} 1 & -4 & 0 & | & 10/3 \\ 0 & 0 & 1 & | & -5/3 \end{bmatrix}$$

Now that the augmented matrix is in reduced row-echelon form we revert back to the linear system:

$$\begin{cases} x - 4y & = & 10/3 \\ z & = & -5/3 \end{cases}$$

The variables corresponding to the pivots will be called **pivot** *variables. The pivot variables will be the dependent variables and the non-pivot variables will be the independent variables. The next step is to solve for the pivot variables:*

$$\begin{cases} x & = & 4y + 10/3 \\ z & = & -5/3 \end{cases}$$

The solution set to the linear system is then

$$\{[4y + 10/3, y, -5/3] \ : \ y \in \mathbb{R}\}.$$

We have an infinite number of solutions to the linear system since we have an infinite number of choices for y. For instance, setting $y = -1$ gives one solution to the linear system, namely, $[-2/3, -1, -5/3]$. Each choice of y yields another solution to the linear system. Oftentimes one replaces the independent variable y by a new parameter t and exhibit the solution set as

$$\{[4t + 10/3, t, -5/3] \ : \ t \in \mathbb{R}\}.$$

Example 2.7 *Consider the following linear system:*

$$\begin{cases} 2x_1 - x_2 + x_3 - x_5 & = & 1 \\ 5x_1 - 2x_2 + 2x_3 + x_4 - x_5 & = & 0 \\ -2x_1 + x_2 - x_3 - 3x_4 - 2x_5 & = & 2 \end{cases}$$

The corresponding augmented matrix is

$$\left[\begin{array}{ccccc|c} 2 & -1 & 1 & 0 & -1 & 1 \\ 5 & -2 & 2 & 1 & -1 & 0 \\ -2 & 1 & -1 & -3 & -2 & 2 \end{array}\right]$$

One can compute that this augmented matrix has the following reduced row-echelon form:

$$\left[\begin{array}{ccccc|c} 1 & 0 & 0 & 0 & 0 & -1 \\ 0 & 1 & -1 & 0 & 1 & -3 \\ 0 & 0 & 0 & 1 & 1 & -1 \end{array}\right]$$

Now that the augmented matrix is in reduced row-echelon form we revert back to the linear system:

$$\begin{cases} x_1 & = & -1 \\ x_2 - x_3 + x_5 & = & -3 \\ x_4 + x_5 & = & -1 \end{cases}$$

Solving for the pivot variables yields

$$\begin{cases} x_1 & = & -1 \\ x_2 & = & x_3 - x_5 - 3 \\ x_4 & = & -x_5 - 1 \end{cases}$$

The solution set to the linear system is then

$$\{ \, [-1, x_3 - x_5 - 3, x_3, -x_5 - 1, x_5] \; : \; x_3, x_5 \in \mathbb{R} \, \}.$$

As in the previous example, the solution set is infinite since we have an infinite number of choices for x_3 (and/or x_5).

For better readability, as in the previous example, one can replace the independent variables x_3 and x_5 by new parameters s and t to express the solution set as

$$\{ \, [-1, s - t - 3, s, -t - 1, t] \; : \; s, t \in \mathbb{F} \}.$$

EXERCISES

1. Decide which of the following matrices is in reduced row-echelon form:

a. $\left[\begin{array}{cccc} 1 & -1 & 0 & 2 \\ 0 & 0 & 1 & 1 \\ 0 & 0 & 0 & 0 \end{array}\right]$

b. $\left[\begin{array}{ccc} 1 & 0 & -3 \\ 0 & 1 & 5 \\ 0 & 0 & 0 \end{array}\right]$

c. $\left[\begin{array}{cccc} 1 & -1 & 2 & 0 \\ 0 & 0 & 0 & 1 \\ 0 & 0 & 0 & 0 \end{array}\right]$

$$
\text{d.} \begin{bmatrix} 1 & 0 & 1 & 0 \\ 0 & 0 & 1 & 0 \\ 0 & 0 & 0 & 1 \end{bmatrix}
\qquad
\text{e.} \begin{bmatrix} 1 & 0 & 0 \\ 0 & 2 & 0 \\ 0 & 0 & 1 \end{bmatrix}
\qquad
\text{f.} \begin{bmatrix} 1 & -2 & 0 & 3 & 0 \\ 0 & 0 & 1 & 5 & 0 \\ 0 & 0 & 0 & 0 & 1 \end{bmatrix}
$$

2. Solve each of the following linear systems using Gaussian Elimination and give a representation of each solution set:

a.
$$
\begin{cases} 2x - y &= -3 \\ x + 3y &= 0 \end{cases}
$$

b.
$$
\begin{cases} x - 2y + z &= 1 \\ 2x - 4y &= -2 \\ x + 2z &= 0 \end{cases}
$$

c.
$$
\begin{cases} 4x - 2y + 2z &= -3 \\ 2x + 5y + z &= 1 \\ -2x + y - z &= 2 \end{cases}
$$

d.
$$
\begin{cases} 2x_1 + 3x_2 - x_3 + x_4 &= 6 \\ 3x_1 + 2x_2 + 2x_3 + 7x_4 &= 2 \end{cases}
$$

e.
$$
\begin{cases} 6x_1 - 12x_2 - 5x_3 + 16x_4 - 2x_5 &= -53 \\ -3x_1 + 6x_2 + 3x_3 - 9x_4 + x_5 &= 29 \\ -4x_1 + 8x_2 + 3x_3 - 10x_4 + x_5 &= 33 \end{cases}
$$

f.
$$
\begin{cases} x_1 + x_2 + x_3 + x_4 &= -3 \\ -2x_1 - x_2 + x_3 + 3x_4 &= 5 \\ 3x_1 + 4x_2 + 6x_3 + 8x_4 &= -10 \end{cases}
$$

g.
$$
\begin{cases} -7x_1 - 28x_2 - 4x_3 + 2x_4 - 10x_5 &= 3 \\ -9x_1 - 36x_2 - 5x_3 + 3x_4 - 15x_5 &= 10 \\ 6x_1 + 24x_2 + 3x_3 - 3x_4 + 10x_5 &= 4 \end{cases}
$$

h.
$$
\begin{cases} 2x_1 - x_2 + x_3 - x_5 &= 1 \\ 5x_1 - 2x_2 + 2x_3 + x_4 - x_5 &= 0 \\ -2x_1 + x_2 - x_3 - 3x_4 - 2x_5 &= 2 \end{cases}
$$

i.
$$
\begin{cases} 2x - 2y + 2z &= 3 \\ -3x + 3y - 3z &= 2 \\ 2x - y + z &= 0 \end{cases}
$$

j.
$$
\begin{cases} 2x - 2y + z &= 1 \\ -3x + y - 3z &= 2 \\ 2x - y + z &= 0 \end{cases}
$$

k.
$$
\begin{cases} x + y + z &= 2 \\ 2x - y - 4z &= -1 \\ -3x - 2y - z &= 0 \end{cases}
$$

l.
$$
\begin{cases} -2x_1 - 4x_2 + x_3 + 2x_5 &= 4 \\ x_1 + 2x_2 - 2x_4 - 7x_5 &= 2 \\ x_1 + 2x_2 - x_3 + 4x_4 + 11x_5 &= -8 \end{cases}
$$

m.
$$
\begin{cases} 3x + y - z &= 5 \\ 2x - y + z &= 0 \\ 2x + y + 2z &= -2 \end{cases}
$$

n.
$$\begin{cases} x - y + z = 1 \\ 2x + y - z = -1 \\ -x - 2y + 2z = 0 \end{cases}$$
o.
$$\begin{cases} 2x_1 - x_2 + x_3 - x_5 = 1 \\ 5x_1 - 2x_2 + 2x_3 + x_4 - x_5 = 0 \\ -2x_1 + x_2 - x_3 - 3x_4 - 2x_5 = 2 \end{cases}$$

p.
$$\begin{cases} x_1 - 2x_2 + 2x_4 = 0 \\ 2x_1 - 5x_2 - 2x_3 + 4x_4 - 3x_5 = -1 \\ x_1 + 4x_3 + 2x_4 + 9x_5 = -1 \end{cases}$$

3. The general equation of a circle is $x^2 + y^2 + ax + by = c$. Use Gaussian Elimination to find the equation of the circle passing through the points $(-2, 1), (1, 1)$ and $(0, -1)$.

4. Consider the following matrices:

$$A = \begin{bmatrix} 3 & -1 & -2 \\ 2 & -2 & 1 \\ 1 & -3 & 0 \end{bmatrix} \qquad B = \begin{bmatrix} 1 \\ 1 \\ 2 \end{bmatrix}.$$

Use Gaussian Elimination to solve the linear system $AX = B$.

5. Use Gaussian Elimination to solve the following application problems:

a. I have 27 coins in my pocket made up of nickles and dimes which add up to $2.15. How many of each coin do I have?

b. Buffalo Bob makes his home in a cabin by the Columbia river. The nearest general store is 48 miles along the river by canoe. Traveling at a constant speed, it takes him three hours to get to the store and four hours to return home. Determine the speed of the river current.

c. I have 21 coins in my pocket made up of pennies, dimes and quarters whose total value is $ 1.23. The number of pennies I have is one more than twice the number of dimes. How many dimes do I have?

d. A grocer mixes three types of dried fruits together: Apricots valued at $ 2 per pound, bananas at $ 1 per pound, and papaya at $ 3 per pound. The result is a 10 pound mixture valued at $ 2 per pound. The number of pounds of apricots is one pound more than the number of pounds of bananas used. How many pounds of each were used?

6. Verify Theorem 2.2 for the elementary row operations aR_i (for $a \neq 0$) and $aR_i + R_j$.

7. Prove by induction Corollary 2.1.

2.3 APPLICATION: MARKOV CHAINS

In this section, we will consider dynamic systems which can at any moment be in exactly one of a finite number of states. Perhaps a simple example might be the weather with states "sunny", "cloudy" or "precipitating". Suppose, in addition, that we observe this system at discrete intervals (maybe once per hour, or day, or year, etc.). We are concerned with the probabilities of changing from one of the states to another and ultimately concerned with the probability of being in any of the finite number of states in the long run.

Definition 2.5 *Suppose a system can be in exactly one of the following m states: $s_1, s_2, \ldots, s_m$. The* **transition probability***, p_{ij}, represents the probability of changing from state s_j to state s_i. Set $P = [p_{ij}] \in M_{mm}$ which is called the* **transition matrix***.*

Note the following observations:

1. For all i and j, we have $0 \le p_{ij} \le 1$.

2. The sum of the entries in any column of P equals 1.

3. The number p_{ij} represents a conditional probability, namely,

$$p_{ij} = p(\text{state is now } s_i \mid \text{state was } s_j).$$

Example 2.8 *A survey is done on adults and smoking. It was found that if someone was smoking during one year, then there was a 70% chance that they would be smoking next year. If someone was not smoking during one year, there was a 10% chance they would be smoking next year.*

Our system has two states,

$$s_1 = \ smoking \qquad\qquad s_2 = \ not\ smoking.$$

Our discrete interval for observation is one year. The survey says that $p_{11} = 0.7$ and $p_{12} = 0.1$. The other probabilities are inferred, i.e. $p_{21} = 0.3$ and $p_{22} = 0.9$ and so we form the transition matrix

$$P = \begin{bmatrix} 0.7 & 0.1 \\ 0.3 & 0.9 \end{bmatrix}.$$

What we are interested in is how the system will look in the long run, that is, after an extended amount of time what is probability of being in each of the states. In our example, we would like to know what percentage of the population will be smokers and non-smokers in the long run.

Suppose initially that 30% of our population smoke. We can represent this information in a column vector

$$X_0 = \begin{bmatrix} 0.3 \\ 0.7 \end{bmatrix} \begin{matrix} \longleftarrow & Smokers \\ \longleftarrow & Non\text{-}smokers \end{matrix}.$$

Using the information in P, let's compute what the percentages will be like next year. Of the 30% who smoke 70% will remain smokers, i.e. $(0.7)(0.3) = 0.21$ or 21%. Of the 70% who do not smoke 10% will become smokers, i.e. $(0.1)(0.7) = 0.07$ or 7%. Hence, $21\% + 7\% = 28\%$ of the population will smoke on the following year, and therefore 72% will not smoke. We can represent these results as a vector.

$$X_1 = \begin{bmatrix} 0.28 \\ 0.72 \end{bmatrix} \begin{matrix} \longleftarrow & Smokers \\ \longleftarrow & Non\text{-}smokers \end{matrix}.$$

Notice that we can use matrix multiplication to obtain X_1:

$$X_1 = \begin{bmatrix} 0.28 \\ 0.72 \end{bmatrix} = \begin{bmatrix} 0.7 & 0.1 \\ 0.3 & 0.9 \end{bmatrix} \begin{bmatrix} 0.3 \\ 0.7 \end{bmatrix} = PX_0.$$

This is certainly true for the successive observations as well.

$$X_2 = PX_1 = \begin{bmatrix} 0.7 & 0.1 \\ 0.3 & 0.9 \end{bmatrix} \begin{bmatrix} 0.28 \\ 0.72 \end{bmatrix} = \begin{bmatrix} 0.268 \\ 0.732 \end{bmatrix}.$$

$$X_3 = PX_2 = \begin{bmatrix} 0.2608 \\ 0.7392 \end{bmatrix}, \quad X_4 = PX_3 = \begin{bmatrix} 0.2565 \\ 0.7435 \end{bmatrix}, \quad X_5 = PX_4 = \begin{bmatrix} 0.2539 \\ 0.7461 \end{bmatrix}, etc.$$

Thus, the question is whether or not these vectors tend towards a fixed vector. Before we get ahead of ourselves let's formalize what we have defined and observed thus far.

Definition 2.6

*1. The column vector $X_n = \begin{bmatrix} a_{1n} \\ a_{2n} \\ \vdots \\ a_{mn} \end{bmatrix}$ is called the n-th **state vector** if for $1 \leq$*

$i \leq m, a_{in}$ is the probability of being in state i after n discrete intervals of observation of the system.

2. A **steady state** vector, X_∞, for a system with state vectors $X_1, X_2, \ldots$ is defined as

$$X_\infty = \lim_{n \to \infty} X_n,$$

when the limit exists. We will forego the formal definition of a limit and simply say that the limiting steady state vector is a constant vector which the state vectors are tending towards.

3. Any vector $v \in \mathbb{R}^m$ is called a **probability vector** if its entries are non-negative and sum up to 1 (thus X_n and X_∞ are particular probability vectors).

Theorem 2.4 *Given a system with transition matrix P and state vectors X_n,*

i. $X_n = PX_{n-1}$

ii. $X_n = P^n X_0$.

Proof 2.4 *Part i is evident and part ii is proved by induction (and left as an easy exercise).* □

We would like to focus on systems where the state vectors are tending towards a fixed steady state vector. Our hope is to find certain transition matrices which guarantee a steady state vector. As we shall see below, our hopes are fulfilled and furthermore, we present a simple method for computing the steady state vector. The following matrices are the ones we consider:

Definition 2.7 *A transition matrix P is **regular** if for some positive integer k, P^k has no zero entries.*

Example 2.9 *In the previous example on smoking, certainly P is regular since P^1 has no zero entries.*

Example 2.10 *The following transition matrix is regular:*

$$P = \begin{bmatrix} 0 & 1/2 & 1/2 & 1/3 \\ 1/3 & 0 & 0 & 1/3 \\ 1/3 & 0 & 0 & 1/3 \\ 1/3 & 1/2 & 1/2 & 0 \end{bmatrix}.$$

Indeed,

$$P^2 = \begin{bmatrix} 4/9 & 1/6 & 1/6 & 1/3 \\ 1/9 & 1/3 & 1/3 & 1/9 \\ 1/9 & 1/3 & 1/3 & 1/9 \\ 1/3 & 1/6 & 1/6 & 4/9 \end{bmatrix},$$

which has no zero entries.

We shall see other examples below, but first we require the foundational result, namely that a regular transition matrix guarantees a steady state vector. The proof is somewhat technical and requires a basic knowledge of real analysis and limits. For this reason we delegate the proof to an appendix at the back of the text and simply state the result here.

Theorem 2.5 *If P is a regular transition matrix, then P^n converges to a matrix Q as $n \to \infty$ and has the following properties:*

i. The columns of Q are identical probability vectors.

ii. Each entry of Q is positive.

Let's see why this theorem guarantees a steady state vector for the system.

Corollary 2.2 *Let P and Q be as in Theorem 2.5 and set $Q = [q \ q \ \cdots \ q]$ with its identical columns. Then the following hold:*

i. For any probability vector X, we have $QX = q$.

ii. $X_\infty = q$.

Proof 2.5 *For any probability vector $X = \begin{bmatrix} a_1 \\ \vdots \\ a_m \end{bmatrix}$, by Lemma 1.2,*

$$QX = a_1 q + \cdots + a_m q = (a_1 + \cdots + a_m)q = 1q = q.$$

Therefore, by Lemma 2.4,

$$X_\infty = \lim_{n \to \infty} X_n = \lim_{n \to \infty} P^n X_0 = QX_0 = q.$$

□

Notice that Corollary 2.2 proves something stronger than just the existence of a steady state vector. By observing the proof, we see that we have shown that regardless of how the system is initially started it will always tends towards the same steady state vector.

Now we give a simple method for finding the steady state vector in the case when we know it exists (as in the case of a regular transition matrix).

Theorem 2.6 *If P is a transition matrix which has a steady state vector, then that steady state vector is the unique (probability vector) solution to $PX = X$ or equivalently, the system $(I - P)X = 0$.*

Proof 2.6 *Notice that*

$$X_\infty = \lim_{n\to\infty} X_n = \lim_{n\to\infty} PX_{n-1} = P \lim_{n\to\infty} X_{n-1} = PX_\infty.$$

Therefore, $PX_\infty = X_\infty$ or $(I-P)X_\infty = 0$, and so X_∞ is a solution to $(I-P)X = 0$. Now suppose that we have another probability vector Y which is a solution to $(I-P)X = 0$. Then $PY = Y$ and certainly $P^nY = Y$ for all $n = 1,2,3,\ldots$. But then by Theorem 2.2,

$$Y = \lim_{n\to\infty} P^nY = QY = q = X_\infty.$$

$\square$

Example 2.11 *Let's revisit our earlier example on smoking adults in which we found the (regular) transition matrix to be $\begin{bmatrix} 0.7 & 0.1 \\ 0.3 & 0.9 \end{bmatrix}$. Then*

$$I - P = \begin{bmatrix} 0.3 & -0.1 \\ -0.3 & 0.1 \end{bmatrix} \quad \text{which reduces to} \quad \begin{bmatrix} 1 & -(1/3) \\ 0 & 0 \end{bmatrix}.$$

Hence, the solution set is $\left\{ \begin{bmatrix} (1/3)b \\ b \end{bmatrix} : b \in \mathbb{R} \right\}$. The unique probability vector solution has the property that $(1/3)b + b = 1$ and so $b = 3/4$. Hence, $X_\infty = \begin{bmatrix} 1/4 \\ 3/4 \end{bmatrix}$.
Therefore, in the long run, 25% of the population will be smokers.
Of course, we can simply add the equation $a + b = 1$ to the system $(I - P)X = 0$ to immediately get this final result.

Example 2.12 *Consider a particle moving along the graph below:*

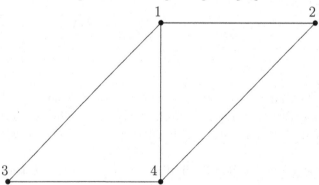

Suppose it travels from one vertex of the graph to another and observations are made after each transition is completed. In addition, assume that the particle has an equal probability of traveling to any of the other vertices (and must transition). We wish to know in the long run what are the chances the particle will be at each

of the vertices when we observe the system. Let $s_i =$ the particle is at vertex i (for $i = 1, 2, 3, 4$). Then

$$
\begin{bmatrix}
0 & 1/2 & 1/2 & 1/3 \\
1/3 & 0 & 0 & 1/3 \\
1/3 & 0 & 0 & 1/3 \\
1/3 & 1/2 & 1/2 & 0
\end{bmatrix},
$$

and as we saw earlier P is regular, since P^2 has no zero entries. Then

$$
I - P = \begin{bmatrix}
1 & -1/2 & -1/2 & -1/3 \\
-1/3 & 1 & 0 & -1/3 \\
-1/3 & 0 & 1 & -1/3 \\
-1/3 & -1/2 & -1/2 & 1
\end{bmatrix}
\quad \text{which reduces to} \quad
\begin{bmatrix}
1 & 0 & 0 & -1 \\
0 & 1 & 0 & -2/3 \\
0 & 0 & 1 & -2/3 \\
0 & 0 & 0 & 0
\end{bmatrix}.
$$

Hence, the solution set is

$$
\left\{ \begin{bmatrix} d \\ (2/3)d \\ (2/3)d \\ d \end{bmatrix} : d \in \mathbb{R} \right\}.
$$

The unique probability vector solution has the property that $d + (2/3)d + (2/3)d + d = 1$ and so $d = 3/10$. Hence,

$$
X_\infty = \begin{bmatrix} 3/10 \\ 1/5 \\ 1/5 \\ 3/10 \end{bmatrix}.
$$

Observe that vertices 1 and 4 should have the higher probabilities of 20%, since there are more connections to vertices 1 and 4 and thus more ways to get to them.

EXERCISES

1. Emotional behavior is observed in a psychiatric patient and the following data is collected:

   a. Given that the patient is happy one day there is a 30 % chance the patient will remain happy the next day and a 40 % chance the patient will become angry the next day.

   b. Given that the patient is angry one day there is an 80 % chance the patient will become depressed the next day and a 10 % chance the patient will become happy the next day.

c. Given that the patient is depressed one day there is a 90 % chance the patient will remain depressed the next day and an 8 % chance the patient will become angry the next day.

Compute the long term probabilities for the patient's three emotional states.

2. There are three spotlights flickering in a parking lot. Observations of the lights are made in regular increments. Given that one of the lights is on, there is a 50–50 chance either two or three lights will be on in the next time increment. Given two lights are on, there is a 2 in 3 chance two lights will remain on in the next time increment, otherwise one light will be on. If there are three lights on, in the next time increment there is always one light on. Find the probability that one, two or three lights are on in any given time increment.

3. Repeat the work done in the section for the Example 2.12 concerning a particle and a graph, but this time allow for the possibility that the particle remain at the vertex it is presently at for the next observation (with equal probability).

4. Repeat the work done in the section concerning a particle and graph, but now use the following graph:

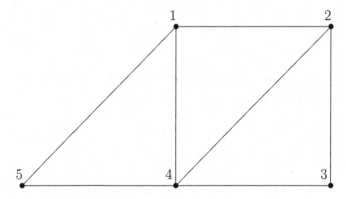

5. Consider the following floor plan with five numbered rooms which will serve as a maze for a mouse:

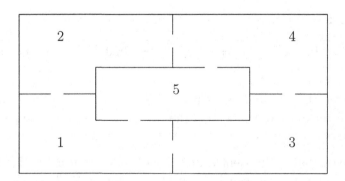

The gaps in the walls represent passageways from one room to the other. Now suppose we place a mouse in the maze and we observe his movement. At regular intervals we observe what room the mouse is in (even if he remains in the same room). Suppose the mouse has an equal probability of either remaining in the room or transitioning to each of the connected rooms.

Compute the probabilities for the mouse being in each of the rooms in the long run (hint: first, view the rooms as vertices and convert the floor plan into a graph).

6. Prove Theorem 2.4.ii

Project for Section 2.3:

Consider the following multilevel floor plan with numbered rooms which will serve as a maze for a mouse:

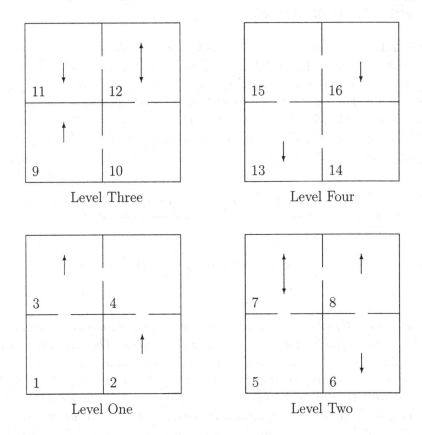

Level Three   Level Four

Level One   Level Two

The gaps in the walls represent passageways from one room to the other. An up arrow indicates the mouse can traverse up to the next level into the room directly above and a down arrow indicates the mouse can traverse down to the previous level into the room directly below. Now, suppose, we place a mouse in the maze and assume at some fixed increment of time that the mouse has an equal probability of transitioning to each of the connected rooms.

1. Over an extended period of time, compute in which room the mouse spends most of his time and where he spends the least of his time.

2. Answer the same question as part a, however assume now that the mouse can remain in the same room with equal probability to transitioning to each of the connected rooms.

2.4 APPLICATION: THE SIMPLEX METHOD

An important application of the use of elementary row operations and pivoting is the Simplex Method for solving linear programming problems. The topic of linear programming (or linear optimization) requires an entire text of its own and it makes use of linear algebra in a big way. This section is meant as a glimpse into this field of study and its connections to linear algebra. To begin, we introduce the terminology for this setting by way of an example.

Example 2.13 *The college cafeteria is offering a lunch consisting of two entrees. The first entree contains 16g of fat, 20g of carbohydrates and 15g of protein per unit serving, while the second contains 10g of fat, 30g of carbohydrates and 17g of protein per unit serving. For lunch, Harry must have at least 100g of protein, but at most 50g of fat and exactly 75g of carbohydrates. The first entree costs $ 0.45 per serving while the second costs $ 0.65 per serving. How many servings of each entree should Harry take so as to meet his nutritional needs and spend the least amount of money.*

Item	Fat	Carb	Protein	Cost per Serving
1	16	20	15	$ 0.45
2	10	30	17	$ 0.65

There are two unknowns that we wish to determine in this problem, namely the number of servings of each of the two entrees. Let's call these unknowns x and y. The **objective function** *represents the quantity z that is being optimized (maximized or minimized). In our example, it is the cost and we wish to minimize it. Mathematically, cost (in dollars) for Harry's meal is represented by $z = 0.45x + 0.65y$. The* **constraints** *of a linear programming problem are the conditions imposed on the unknowns for the particular problem. For instance, in our example, Harry must have at least 100 grams of protein, i.e. the amount of protein must be ≥ 100. Each of the two entrees will contribute to* *the total protein depending on how many servings of each are eaten and mathematically this condition translates into $15x + 17y \geq 100$. Harry cannot have more than 50 grams of fat becomes $16x + 10y \leq 50$ and exactly 75 grams of carbohydrates becomes $20x + 30y = 75$. There is also implicit in this problem a **positivity** constraint, namely that the number of servings must be positive numbers (and perhaps even integers, but we won't concern ourselves with this for the sake of simplicity), i.e. $x, y \geq 0$. Now we can state this linear programming problem in mathematical form as follows:*

$$\text{Minimize } z = 0.45x + 0.65y, \qquad \text{Subject to} \begin{cases} 15x + 17y & \geq & 100 \\ 16x + 10y & \leq & 50 \\ 20x + 30y & = & 75 \\ x, y & \geq & 0 \end{cases}$$

*A linear programming problem is in **standard form** if the objective function is being maximized, the constraints are all $\leq$ and the unknowns are all positive. Most linear programming problems can be rewritten to fit these conditions. Just consider the example above. The values of x and y which maximize $z = 0.45x + 0.65y$ will be the same values of x and y which minimize $z' = -0.45x - 0.65y$ (of course, the values of z and z' will be of opposite signs). Hence, a minimization problem can always be rewritten as a maximization problem. A constraint involving $\geq$ can always be rewritten as a constraint involving $\leq$ simply by multiplying the constraint by -1 and reversing the inequality. Again, in the example above, $15x + 17y \geq 100$ can be rewritten as $-15x - 17y \leq -100$. An equation can be rewritten as two inequalities of the form $\leq$. In the example above $20x + 30y = 75$ is equivalent to $20x + 30y \leq 75$ and $20x + 30y \geq 75$ which is equivalent to $20x + 30y \leq 75$ and $-20x - 30y \leq -75$. Hence, our example will have the following standard form:*

$$\text{Maximize } z' = -0.45x - 0.65y, \qquad \text{Subject to} \begin{cases} -15x - 17y & \leq & -100 \\ 16x + 10y & \leq & 50 \\ 20x + 30y & \leq & 75 \\ -20x - 30y & \leq & -75 \\ x, y & \geq & 0 \end{cases}$$

Finally, if an unknown can take on any value (both positive or negative), we can always replace it by a difference of two new positive unknowns. For example, if it were the case in our example that x was unrestricted, we could set $x = u - v$ with $u, v \geq 0$ and then make this substitution into our linear programming problem. For example, we would now be maximizing the objective function $z' = -0.45(u - v) - 0.65y = -0.45u + 0.45v - 0.65y$ and the constraints would also require the same substitution; for instance, the first constraint becomes $-15u + 15v - 17y \leq -100$.

The method we present in this section works precisely on those linear programming problems in standard form for which the constants on the righthand side of the constraints are all positive. Therefore, the method we will develop in this section will not work on the example we have been discussing thus far, because of the −100 or the −75 appearing in the problem when put in standard form.

Definition 2.8 *A* **feasible solution** *to a standard linear programming problem is a set of values for the unknowns which satisfy all the constraints. The collection of all such solutions is called the* **feasible region***. If, in addition to being feasible, the set of values of the unknowns also maximizes the objective function, then we call this set of values an* **optimal solution***.*

Optimal solutions are what we are striving for in this section for these are the solutions to linear programming problems. In solving a linear programming problem there are three possible outcomes.

1. EMPTY: The feasible region is empty.

2. UNBOUNDED: The feasible region is non-empty and the objective function is unbounded on the feasible region.

3. FINITE: The feasible region is non-empty and the objective function is bounded on the feasible region.

In the first case there is certainly no optimal solution since such a solution can only be found among the feasible ones. In the second case there is also no optimal solution, since the objective function can be as large as we please on the feasible region and hence there is no maximal value. It is only the last case for which there exists an optimal solution.

Let's look at a particular example of a linear programming problem which we will eventually be able to solve in this section.

Example 2.14

$$\text{Maximize } z = 2x + 3y, \qquad \text{Subject to } \begin{cases} x + y & \leq & 4 \\ x + 3y & \leq & 6 \\ x, y & \geq & 0 \end{cases}$$

The feasible region can be represented geometrically by graphing the solution to each inequality, which will be a half-plane, and then finding out where these half-planes overlap. Such techniques are developed in any basic algebra course so we omit the details and simply represent the feasible region.

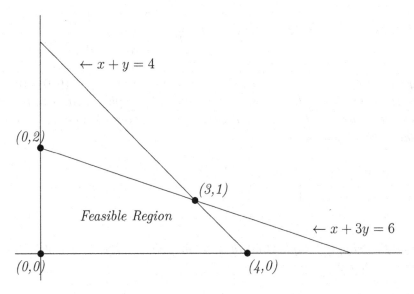

Notice that the feasible region has four corners at $(0,0)$, $(0,2)$, $(3,1)$ and $(4,0)$. We call such corners extreme points. More formally, an **extreme point** *of a feasible region is a point in the region which does not lie on the interior of a line segment entirely contained in the region.*

However, for our purposes, extreme points will be the sharp edges of a feasible region (note that the extreme points of a disk consists of all the points on its boundary circle and a circle surely has no corners! However, such regions do not occur in a linear setting). The extreme points turn out to be the key in determining an optimal solution as the following theorem attests to this fact:

Theorem 2.7 *(Extreme Value Theorem) Let F be the feasible region for some linear programming problem.*

1. *If $F \neq \emptyset$ and is bounded, then an optimal solution exists and occurs at the extreme points of F.*

2. *If $F \neq \emptyset$ and is unbounded, then* **if** *an optimal solution exists, then it occurs at the extreme points of F.*

3. *If there is no optimal solution, then either $F = \emptyset$ or F is unbounded.*

Informally, the use of the word unbounded in the above theorem means basically that the region cannot fit within any circle (or sphere/hyper-sphere for higher dimensions). The proof of this theorem is much too far afield and therefore will be omitted. However, we can instantly use it to solve the problem at hand in our example.

Example 2.15 *For the linear programming problem*

$$\text{Maximize } z = 2x + 3y, \qquad \text{Subject to } \begin{cases} x + y & \leq & 4 \\ x + 3y & \leq & 6 \\ x, y & \geq & 0 \end{cases},$$

the feasible region is bounded and we found the extreme points to be $(0,0)$, $(0,2)$, $(3,1)$ and $(4,0)$. Hence, by the Extreme Value Theorem, it's simply a matter of testing the four points in the objective function and whichever yields the largest value must correspond to the optimal solution.

(x,y)	$(0,0)$	$(0,2)$	$(3,1)$	$(4,0)$
$z = 2x + 3y$	0	6	9	8

Thus, we see from this simple computation that the maximum value of objective function is 9 and occurs when $x = 3$ and $y = 1$.

One can guess the deficiencies in using a picture to solve a linear programming problem. It could be rather complicated to visualize if there are many constraints or it could have three unknowns which would have us visualizing 3-dimensional planes. Beyond three unknowns, we cannot even draw a picture and the method completely breaks down! Therefore, one sees the need to develop a strictly algebraic algorithm that does not rely on a picture. Such an algorithm, which we shall develop in what follows, is called the **simplex method**.

As a first step, we need to convert our linear programming problem into canonical form which requires that all the constraints be equations. This is easy to do by introducing additional variables. For instance, an inequality like $x + y \leq 4$ can be transformed into an equation as $x + y + u = 4$ with $u \geq 0$. If it is reversed, like $x + y \geq 4$, we get $x + y - u = 4$ with $u \geq 0$. We wish the new variables to also have the positivity constraint for the purposes of this method. These additional variables are called **slack** variables for in a sense they *take up the slack* of the inequality.

Example 2.16 *In the earlier example, it was already in standard form.*

$$\text{Maximize } z = 2x + 3y, \qquad \text{Subject to } \begin{cases} x + y & \leq & 4 \\ x + 3y & \leq & 6 \\ x, y & \geq & 0 \end{cases}$$

In canonical form it looks like this:

$$\text{Maximize } z = 2x + 3y, \qquad \text{Subject to } \begin{cases} x + y + u & = & 4 \\ x + 3y + v & = & 6 \\ x, y, u, v & \geq & 0 \end{cases}$$

Our next step is to put all the information concerning the linear programming problem into a matrix. We need to rewrite the objective function as $-2x - 3y + z = 0$

and treat it as an additional equation. This equation together with the two constraint can be put into what is called a **tableau** *which is basically an annotated matrix. Here is what this example's initial tableau looks like.*

$$
\begin{array}{c}
\\
u \\
\\
v \\
\\
\\
\end{array}
\left[
\begin{array}{ccccc|c}
x & y & u & v & z & \\
1 & 1 & 1 & 0 & 0 & 4 \\
1 & 3 & 0 & 1 & 0 & 6 \\
\hline
-2 & -3 & 0 & 0 & 1 & 0
\end{array}
\right]
$$

Notice that the columns corresponding to the variables u and v have pivoting 1's and we keep track of this fact on the left-hand side of the matrix. The numbers 4 and 6 in the right-hand columns correspond to the values of u and v, respectively, for a feasible solution to the problem. Let's call these variables u and v the **pivoting variables**. *Looking back at the canonical form we see then that x and y equal zero and the feasible solution to the canonical form of the problem is $(x, y, u, v) = (0, 0, 4, 6)$. Note that it will always be the case that the non-pivots of the tableau will be zero. Furthermore, dropping the slack variable, $(x, y) = (0, 0)$ is an extreme point of the feasible region corresponding to the standard form of the problem. Notice also the bottom right corner of the tableau is 0 which corresponds to the value of the objective function at $(x, y) = (0, 0)$. All of these observations can be stated more formally and proven with mathematical rigor, however for the sake of brevity, suffice it to say that the Simplex Method produces a series of Tableaus which move from one extreme point to another of the feasible region of the standard form in such a way that the objective function increases to its maximal value (if the problem has one). Hence, we need not find the extreme points of the feasible region by creating a graph. Furthermore, the Simplex Method will pass through only some of the extreme points on its way to the optimal solution.*

The crucial step is how to get to the next tableau (on our way to the optimal solution). We will step through one transition carefully and give a justification as to why it makes sense, but the reader should take note that in the long run this computation will be done without considering the justification at every turn. We first decide which of the non-slack variables x and y more effectively increases $z = 2x + 3y$ as it increases. Clearly, it will be the variable which has the larger coefficient in the objective function, which in our example is y which has a coefficient of 3 in the objective function. In the tableau, this corresponds to looking at the bottom row (called the **objective row***) to the left of the vertical bar and finding the smallest negative number, namely -3.*

$$\begin{array}{c} \\ u \\ \\ v \\ \\ \\ \end{array} \left[\begin{array}{ccccc|c} x & y & u & v & z & \\ 1 & 1 & 1 & 0 & 0 & 4 \\ 1 & 3 & 0 & 1 & 0 & 6 \\ \hline -2 & \underline{\underline{-3}} & 0 & 0 & 1 & 0 \end{array} \right]$$

If there are no negative numbers in the objective row, then increasing x or y will not increase z and hence we must already be at the optimal solution. This analysis is called the **objective criterion**. *Having chosen y which we label as the* **entering variable**, *we now look at the two constraints and solve for the pivoting variables:*

$$u = 4 - x - y \qquad\qquad v = 6 - x - 3y.$$

As we increase y, we have to insure that u and v remain positive so that we remain in the feasible region of the canonical form of the problem, i.e.

$$u = 4 - x - y \geq 0 \qquad\qquad v = 6 - x - 3y \geq 0,$$

or

$$y \leq 4 - x \qquad\qquad y \leq \frac{6 - x}{3}.$$

Since we have opted to increase y, the value of x remains 0 and so we have the inequalities

$$y \leq 4 \qquad and \qquad y \leq 2.$$

Hence, y can be increased by no more than 2, since we must choose the smaller of the two positive numbers in order to satisfy both inequalities. By increasing y by 2, v now becomes zero, since $u = 4 - x - y = 4 - 0 - 2 = 2$. The variable v is called the **departing variable**.

There is a way to obtain this same information from our tableau. Simply form the ratios of the entries in the last column and the column corresponding to our entering variable y.

$$\begin{array}{c} \\ u \\ \\ v \\ \\ \\ \end{array} \left[\begin{array}{ccccc|c} x & y & u & v & z & \\ 1 & \underline{\underline{1}} & 1 & 0 & 0 & \underline{4} \\ 1 & \underline{3} & 0 & 1 & 0 & \underline{\underline{6}} \\ \hline -2 & -3 & 0 & 0 & 1 & 0 \end{array} \right] \begin{array}{l} \\ \leftarrow 4/1 = 4 \\ \\ \leftarrow 6/3 = 2 \\ \\ \\ \end{array}$$

We then choose the smaller of the two ratios to decide how much to increase the entering variable. We only choose from the positive ratios and if all the ratios turn out to be non-positive (including an undefined ratio in which we are dividing by zero), then we must conclude that the linear programming problem has no solution. For instance, take the case where all the ratios were negative. Suppose the first constraint were $x - y + u = 4$ and as before we had selected y as the entering variable. The ratio in this case would be $4/-1 = -4 < 0$. Solving again for the slack variable and setting $x = 0$, $u = 4 - x + y = 4 + y \geq 0$ and so we get the condition that $y \geq -4$. Now this puts no restriction on y at all, since we are increasing its value which began at zero. This in turn would put no restriction on how large z could be. Hence, there would be no finite maximal value and therefore the linear programming problem would have no optimal solution.

We now want a new tableau where u and the new entering variable y are the pivoting columns. We do this using elementary row operations just as we did in Gaussian Elimination (except, of course, we are not putting the matrix in reduced row-echelon form). We want the pivoting 1 at the intersection of the entering column and the departing row.

$$
\begin{array}{c|ccccc|c}
 & x & \underline{\underline{y}} & u & v & z & \\
\hline
u & 1 & 1 & 1 & 0 & 0 & 4 \\
\underline{\underline{v}} & 1 & 3 & 0 & 1 & 0 & 6 \\
\hline
 & -2 & -3 & 0 & 0 & 1 & 0
\end{array}
$$

We put a 1 in place of the 3 via the operation $\frac{1}{3}R_2$ and get

$$
\begin{array}{ccccc|c}
x & y & u & v & z & \\
1 & 1 & 1 & 0 & 0 & 4 \\
1/3 & 1 & 0 & 1/3 & 0 & 2 \\
\hline
-2 & -3 & 0 & 0 & 1 & 0
\end{array}.
$$

We then put a zero above and below the pivoting 1 via the operations $-R_2 + R_1$ and $3R_2 + R_3$ to get

$$
\begin{array}{c|ccccc|c}
 & x & y & u & v & z & \\
\hline
u & 2/3 & 0 & 1 & -1/3 & 0 & 2 \\
y & 1/3 & 1 & 0 & 1/3 & 0 & 2 \\
\hline
 & -1 & 0 & 0 & 1 & 1 & 6
\end{array}.
$$

At this point, reading off the information from the tableau as we did previously,
$(x, y, u, v) = (0, 2, 2, 0)$ *with* $z = 6$. *Take note that any non pivot variables must
be zero, just like in the initial tableau. Notice that* y *has taken the place of* v *in the
left-most column as the new pivoting variable (hence, the terminology entering and
departing from pivoting status). Notice also that we are now at the extreme point*
$(0, 2)$ *of the feasible region of the standard problem.*

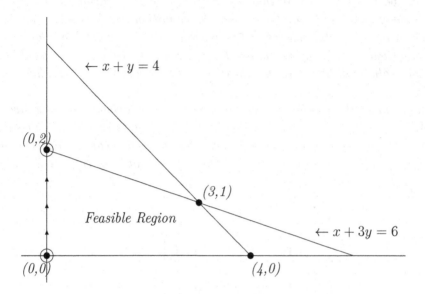

*We are not yet at the optimal solution, since there are still negative values in the
objective row. We will now obtain the next tableau without any more fanfare. There
is a* -1 *in the* x *column, and being the only negative number* x *must be the next
entering variable. The resulting ratios are then* $2/(2/3) = 3$ *and* $2/(1/3) = 6$ *with
the first being the smaller, and so* u *is now the next departing variable. The* $2/3$ *will
become the next pivoting 1 via the operation* $(3/2)R_1$ *to get*

$$
\begin{bmatrix}
x & y & u & v & z & \\
1 & 0 & 3/2 & -1/2 & 0 & 3 \\
1/3 & 1 & 0 & 1/3 & 0 & 2 \\
\hline
-1 & 0 & 0 & 1 & 1 & 6
\end{bmatrix}.
$$

We then put zeros below the pivoting 1 via the operations $-(1/3)R_1 + R_2$ *and* $R_1 + R_3$
to get

$$\begin{bmatrix} & x & y & u & v & z & \\ x & 1 & 0 & 3/2 & -1/2 & 0 & 3 \\ y & 0 & 1 & -1/2 & 1/2 & 0 & 1 \\ & 0 & 0 & 3/2 & 1/2 & 1 & 9 \end{bmatrix}.$$

Reading off the information from the tableau, $(x, y, u, v) = (3, 1, 0, 0)$ with $z = 9$. Notice that x has now taken the place of u in the left-most column as the new pivoting variable. Again, notice that we are now at the extreme point $(3, 1)$ of the feasible region of the standard problem.

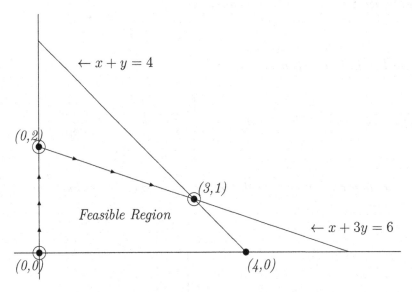

We are now at the optimal solution, since the values in the objective row are all non-negative. Hence, the optimal solution to the linear programming problem is $x = 3$ and $y = 1$ with corresponding maximal value of $z = 9$, just as we had found before using the geometric method of looking at the picture.

There is one last comment to make about the Simplex Method. Notice that the z column never changes throughout the whole process and in fact adds no additional information about the optimal solution. Hence, from now on we shall omit the z column. Let's solve another problem from start to finish by applying the algorithm that was just gone over carefully.

Example 2.17 *A toy manufacturer is producing two kinds of toys. The first toy will net a profit of $ 2, while the second toy nets him a profit of $ 3. Both toys weigh one pound. The first toy takes up a cubic foot of space while the second takes up two cubic feet. These toys have to be packed into crate to be delivered. The crate can carry no more than 12 pounds and can hold at most 20 cubic feet of toys. The first toy* *requires two workers to produce it, while the second requires only one. The company has 20 workers at its disposal, however each worker will only work on a single toy. How many toys of each type should be produced in order to maximize the company's profits?*

If we let x represent the number of the first toy produced and y the number of the second toy produced, then our profit objective function becomes z = 2x + 3y. The information about the weight of the toys leads to the constraint x + y ≤ 12. The information about the space taken by the toys leads to the constraint x + 2y ≤ 20. The information about the number of workers needed to produce the toys leads to the constraint 2x + y ≤ 20. Therefore, our linear programming problem (already in standard form) is

$$\text{Maximize } z = 2x + 3y, \qquad \text{Subject to} \begin{cases} x + y & \leq & 12 \\ x + 2y & \leq & 20 \\ 2x + y & \leq & 20 \\ x, y & \geq & 0 \end{cases}$$

In canonical form the problem becomes

$$\text{Maximize } z = 2x + 3y, \qquad \text{Subject to} \begin{cases} x + y + u & = & 12 \\ x + 2y + v & = & 20 \\ 2x + y + w & = & 20 \\ x, y, u, v, w & \geq & 0 \end{cases}$$

Therefore, the initial tableau (with the z column now omitted as per earlier comments) is

$$\begin{bmatrix} & x & y & u & v & w & \\ u & 1 & 1 & 1 & 0 & 0 & 12 \\ v & 1 & 2 & 0 & 1 & 0 & 20 \\ w & 2 & 1 & 0 & 0 & 1 & 20 \\ \hline & -2 & -3 & 0 & 0 & 0 & 0 \end{bmatrix}.$$

Since −3 is the smallest negative number in the objective row, we select y as the entering variable and compute the three corresponding ratios:

$$
\begin{array}{c c}
\begin{array}{c}
\\
u \\
\\
v \\
\\
w \\
\\
\\
\end{array}
&
\left[
\begin{array}{ccccc|c}
x & y & u & v & w & \\
1 & 1 & 1 & 0 & 0 & 12 \\
1 & 2 & 0 & 1 & 0 & 20 \\
2 & 1 & 0 & 0 & 1 & 20 \\
\hline
-2 & \underline{\underline{-3}} & 0 & 0 & 0 & 0
\end{array}
\right]
\end{array}
\begin{array}{l}
\leftarrow 12/1 = 12 \\
\\
\leftarrow 20/2 = 10 \ . \\
\\
\leftarrow 20/1 = 20
\end{array}
$$

The smallest positive ratio is 10 and so we select v as the departing variable. So we must put a pivoting 1 in the second row, second column via the operation $(1/2)R_2$:

$$
\left[
\begin{array}{ccccc|c}
x & y & u & v & w & \\
1 & 1 & 1 & 0 & 0 & 12 \\
1/2 & \underline{1} & 0 & 1/2 & 0 & 10 \\
2 & 1 & 0 & 0 & 1 & 20 \\
\hline
-2 & -3 & 0 & 0 & 0 & 0
\end{array}
\right]
$$

Now we put zeros in the rest of the column via the operations $-R_2 + R_1$, $-R_2 + R_3$ and $3R_2 + R_4$ to get

$$
\begin{array}{c c}
\begin{array}{c}
\\
u \\
\\
y \\
\\
w \\
\\
\\
\end{array}
&
\left[
\begin{array}{ccccc|c}
x & y & u & v & w & \\
1/2 & 0 & 1 & -1/2 & 0 & 2 \\
1/2 & 1 & 0 & 1/2 & 0 & 10 \\
3/2 & 0 & 0 & -1/2 & 1 & 10 \\
\hline
-1/2 & 0 & 0 & 3/2 & 0 & 30
\end{array}
\right]
\end{array} \ .
$$

Notice that the optimal criterion is not yet satisfied, since there is still a negative number in the objective row, so we must repeat the steps in the process. The entering variable corresponds to the smallest negative value (in this case the only negative value) in the objective row, namely $-1/2$, and so x is now the entering variable. Again, we compute the corresponding ratios:

$$
\begin{array}{c|ccccc|c}
 & x & y & u & v & w & \\
\hline
u & 1/2 & 0 & 1 & -1/2 & 0 & 2 \\
y & 1/2 & 1 & 0 & 1/2 & 0 & 10 \\
w & 3/2 & 0 & 0 & -1/2 & 1 & 10 \\
\hline
 & -1/2 & 0 & 0 & 3/2 & 0 & 30
\end{array}
\quad
\begin{array}{l}
\leftarrow 2/(1/2) = 4 \\[1em]
\leftarrow 10/(1/2) = 20 \\[1em]
\leftarrow 10/(3/2) = 20/3
\end{array}
$$

The smallest positive ratio is 4 and so we select u as the departing variable. So we must put a pivoting 1 in the first row, first column via the operation $2R_1$:

$$
\begin{array}{c|ccccc|c}
 & x & y & u & v & w & \\
\hline
 & \underline{1} & 0 & 2 & -1 & 0 & 4 \\
 & 1/2 & 1 & 0 & 1/2 & 0 & 10 \\
 & 3/2 & 0 & 0 & -1/2 & 1 & 10 \\
\hline
 & -1/2 & 0 & 0 & 3/2 & 0 & 30
\end{array}
\;\; .
$$

Now we put zeros in the rest of the column via the operations $-(1/2)R_1 + R_2$, $-(3/2)R_1 + R_3$ *and* $(1/2)R_1 + R_4$ *to get*

$$
\begin{array}{c|ccccc|c}
 & x & y & u & v & w & \\
\hline
x & 1 & 0 & 2 & -1 & 0 & 4 \\
y & 0 & 1 & -1 & 1 & 0 & 8 \\
w & 0 & 0 & -3 & 1 & 1 & 4 \\
\hline
 & 0 & 0 & 1 & 1 & 0 & 32
\end{array}
$$

Now the optimal criterion is satisfied with $(x, y, u, v, w) = (4, 8, 0, 0, 4)$. *Hence, the optimal solution is* $z = 32$ *with* $x = 4$ *and* $y = 8$. *In other words, the company should have the workers produce four of first toy and eight of the second in order to achieve a maximal profit of $ 32.*

We could have also solved this problem geometrically, the graph of the feasible solution derived from the constraints together with the extreme points would look like this:

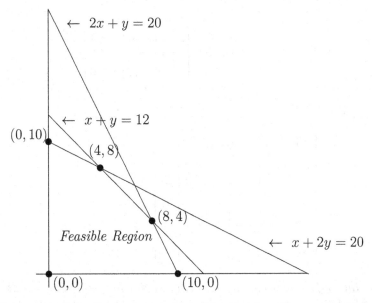

We now test the five points in the objective function and whichever yields the largest value must correspond to the optimal solution.

(x, y)	$(0, 0)$	$(0, 10)$	$(10, 0)$	$(4, 8)$	$(8, 4)$
$z = 2x + 3y$	0	30	20	$\underline{\underline{32}}$	28

So we reconfirm that the maximum value of objective function is 32 and occurs when $x = 4$ and $y = 8$.

Example 2.18 We give one last example to illustrate what the case of no solution can look like. The linear programming problem is the following:

$$\text{Maximize } z = x_1 + 2x_2 + 3x_3, \quad \text{Subject to } \begin{cases} x_1 - x_2 - x_3 & \leq & 2 \\ x_1 + x_2 - x_3 & \geq & -1 \\ x_1, x_2, x_3 & \geq & 0 \end{cases}$$

The linear programming problem as it stands is not in standard form. This can easily be amended by multiplying the second constraint by -1 to get

$$\text{Maximize } z = x_1 + 2x_2 + 3x_3, \quad \text{Subject to } \begin{cases} x_1 - x_2 - x_3 & \leq & 2 \\ -x_1 - x_2 + x_3 & \leq & 1 \\ x_1, x_2, x_3 & \geq & 0 \end{cases}$$

The canonical form is then

$$\text{Maximize } z = x_1 + 2x_2 + 3x_3, \qquad \text{Subject to } \begin{cases} x_1 - x_2 - x_3 + u & = 2 \\ -x_1 - x_2 + x_3 + v & = 1 \\ x_1, x_2, x_3, u, v & \geq 0 \end{cases}$$

The initial tableau is then

$$
\begin{bmatrix}
 & x_1 & x_2 & x_3 & u & v & \\
u & 1 & -1 & -1 & 1 & 0 & 2 \\
v & -1 & -1 & 1 & 0 & 1 & 1 \\
\hline
 & -1 & -2 & -3 & 0 & 1 & 0
\end{bmatrix}.
$$

Because of the -3 in the objective row, the entering variable is x_3. The resulting ratios yield only one which is positive, so the departing variable is v which puts the pivot in the second row, third column.

$$
\begin{bmatrix}
 & x_1 & x_2 & x_3 & u & v & \\
u & 1 & -1 & -1 & 1 & 0 & 2 \\
v & -1 & -1 & 1 & 0 & 1 & 1 \\
\hline
 & -1 & -2 & \underline{-3} & 0 & 1 & 0
\end{bmatrix}
\begin{array}{l} \leftarrow 2/-1 = -2 \\[1.2em] \leftarrow 1/1 = 1 \end{array}
$$

The pivot is already 1, so we proceed to put zeros in the rest of the column via the operations $R_2 + R_1$ and $3R_2 + R_3$ and get

$$
\begin{bmatrix}
 & x_1 & x_2 & x_3 & u & v & \\
u & 0 & -2 & 0 & 1 & 1 & 3 \\
v & -1 & -1 & 1 & 0 & 1 & 1 \\
\hline
 & -4 & -5 & 0 & 0 & 4 & 3
\end{bmatrix}
$$

Because of the -5 in the objective row, the next entering variable is x_2. Notice now that the resulting ratios yield only negative ratios and it is at this point that we can conclude that the linear programming problem has no solution.

$$
\begin{array}{c}
\quad\begin{array}{ccccc} x_1 & x_2 & x_3 & u & v \end{array}\\
\left[\begin{array}{ccccc|c}
 & & & & & \\
0 & -2 & 0 & 1 & 1 & 3 \\
 & & & & & \\
-1 & -1 & 1 & 0 & 1 & 1 \\
\hline
-4 & \underset{=}{-5} & 0 & 0 & 4 & 3
\end{array}\right]
\begin{array}{l}
\\
\leftarrow 3/-2 = -(3/2)\\
\\
\leftarrow 1/-1 = -1\\
\\
\end{array}
\end{array}
$$

with row labels u and v on the left of the two middle rows.

EXERCISES

1. Put each of the following problems in first standard form and then canonical form and determine which of them can be solved by the method given in this section:

   a.

   $$\text{Minimize } z = x - y, \qquad \text{Subject to } \begin{cases} x - 2y & \geq & 24 \\ 2x - y & = & 18 \\ x, y & \geq & 0 \end{cases}$$

   b.

   $$\text{Minimize } z = x_1 - x_2 - x_3 + 2x_4$$

   $$\text{Subject to } \begin{cases} x_1 - x_2 + 2x_3 + x_4 & = & 4 \\ x_1 - x_2 - 3x_3 - x_4 & \geq & 2 \\ 4x_1 + 6x_2 + 8x_3 + 3x_4 & \leq & -1 \\ x_1, x_2, x_3, x_4 & \geq & 0 \end{cases}$$

2. Solve the following linear programming problems using both the geometric method and the simplex method:

   a.

   $$\text{Maximize } z = x + y, \qquad \text{Subject to } \begin{cases} 4x + 3y & \leq & 24 \\ 2x + 3y & \leq & 18 \\ x, y & \geq & 0 \end{cases}$$

   b.

   $$\text{Maximize } z = 2x + 4y, \qquad \text{Subject to } \begin{cases} x + y & \leq & 15 \\ 2x + y & \leq & 24 \\ x + 3y & \leq & 36 \\ x, y & \geq & 0 \end{cases}$$

3. Solve the following linear programming problems using the simplex method:

   a.

   $$\text{Minimize } z = x - 2y - 3z, \qquad \text{Subject to } \begin{cases} x + y + z & \leq & 10 \\ x + y - z & \leq & 30 \\ x - y + z & \leq & 20 \\ x, y, z & \geq & 0 \end{cases}$$

b.

$$\text{Maximize } z = x_1 + 2x_2 + 3x_3 + x_4$$

$$\text{Subject to } \begin{cases} 2x_1 + 3x_2 + 6x_3 + 4x_4 & \leq & 4 \\ -10x_1 - 4x_2 - 3x_3 - 2x_4 & \geq & -6 \\ 4x_1 + 6x_2 + 8x_3 + 3x_4 & \leq & 8 \\ x_1, x_2, x_3, x_4 & \geq & 0 \end{cases}$$

c. The college cafeteria is offering a lunch consisting of two entrees. The first entree contains 10g of fat, 20g of carbohydrates and 12g of protein per unit serving, while the second contains 8g of fat, 12g of carbohydrates and 16g of protein per unit serving. For lunch, Harry must have no more than 100g of protein, at most 50g of fat and at most 75g of carbohydrates. The first entree costs $ 0.45 per serving while the second costs $ 0.65 per serving. How many servings of each entree should Harry take so as to meet his nutritional needs and spend the least amount of money.

Item	Fat	Carb	Protein	Cost per Serving
1	10	20	12	$ 0.45
2	8	12	16	$ 0.65

Project for 2.4:

The following is an example of a well known linear programming problem called the **Transportation Problem**:

Two manufacturing plants located in New York City and Los Angeles produce a certain synthetic polymer. The plant in N.Y.C. can produce 100 tons of the polymer, while the one in L.A. produces 120 tons. This product is to be shipped to three cities: San Francisco, Dallas and Chicago. The demand in each of these cities are 60 tons, 70 tons and 90 tons, respectively.

The following table lists the relative shipping costs to transport one ton of the product from each origin city to each destination city:

	SF	DAL	CHI
NYC	10	7	6
LA	6	7	8

Let x_{ij} represent the amount of polymer shipped from origin city $i = 1, 2$ to destination city $j = 1, 2, 3$.

1. What is the cost function which we wish to minimize?

2. Assuming that the amount shipped exactly meets the demand and that all that is being produced is in fact shipped, write out the constraints for this linear programming problem and solve using a computer software package.

2.5 ELEMENTARY MATRICES AND MATRIX EQUIVALENCE

At this point, we have attained the first goal of the chapter, namely we have presented a systematic way of solving linear systems. Indeed, Gaussian elimination is an algorithm which can be coded up in a computer and performed without the need of doing so by hand. Many computer algebra systems have a command which puts a matrix in reduced row echelon form. With the second goal in mind, namely to develop a theorem which lists statements which are equivalent to the statement that a matrix has an inverse, we define a special class of matrices:

Definition 2.9 *An* **elementary** *matrix is a matrix obtained by applying exactly one elementary row operation to the identity matrix (note that elementary matrices are necessarily square).*

Example 2.19 *a. An example of a 2×2 elementary matrix is* $\begin{bmatrix} 1 & -2 \\ 0 & 1 \end{bmatrix}$ *since*

$$I_2 = \begin{bmatrix} 1 & 0 \\ 0 & 1 \end{bmatrix} \xrightarrow{-2R_2 + R_1} \begin{bmatrix} 1 & -2 \\ 0 & 1 \end{bmatrix}.$$

b. *An example of a 3 × 3 elementary matrix is* $\begin{bmatrix} 1 & 0 & 0 \\ 0 & 0 & 1 \\ 0 & 1 & 0 \end{bmatrix}$ *since*

$$I_3 = \begin{bmatrix} 1 & 0 & 0 \\ 0 & 1 & 0 \\ 0 & 0 & 1 \end{bmatrix} \xrightarrow{R_2 \leftrightarrow R_3} \begin{bmatrix} 1 & 0 & 0 \\ 0 & 0 & 1 \\ 0 & 1 & 0 \end{bmatrix}.$$

The next important fact shows that there is a direct connection between elementary row operations and matrix multiplication.

Theorem 2.8 *Every elementary row operation corresponds to multiplication on the left by the corresponding elementary matrix. Stated mathematically, if $A \xrightarrow{op} B$ and $I \xrightarrow{op} E$, then $B = EA$.*

Example 2.20 *Let's illustrate this fact.*

a. *Consider the matrix* $A = \begin{bmatrix} 1 & 2 \\ 3 & 4 \end{bmatrix}$ *and the elementary row operation*

$$A = \begin{bmatrix} 1 & 2 \\ 3 & 4 \end{bmatrix} \xrightarrow{-2R_2 + R_1} \begin{bmatrix} -5 & -6 \\ 3 & 4 \end{bmatrix} = B.$$

We have just seen above that the elementary matrix corresponding to this elementary row operation is $E = \begin{bmatrix} 1 & -2 \\ 0 & 1 \end{bmatrix}.$

To illustrate Theorem 2.8 notice that the product

$$EA = \begin{bmatrix} 1 & -2 \\ 0 & 1 \end{bmatrix} \begin{bmatrix} 1 & 2 \\ 3 & 4 \end{bmatrix} = \begin{bmatrix} -5 & -6 \\ 3 & 4 \end{bmatrix} = B$$

yields the same result as applying the elementary row operation to A.

b. *Consider the matrix* $A = \begin{bmatrix} 1 & 2 & 3 \\ 4 & 5 & 6 \\ 7 & 8 & 9 \end{bmatrix}$ *and the elementary row operation*

$$A = \begin{bmatrix} 1 & 2 & 3 \\ 4 & 5 & 6 \\ 7 & 8 & 9 \end{bmatrix} \xrightarrow{R_2 \leftrightarrow R_3} \begin{bmatrix} 1 & 2 & 3 \\ 7 & 8 & 9 \\ 4 & 5 & 6 \end{bmatrix} = B.$$

We have also just seen above that the elementary matrix corresponding to this elementary row operation is $E = \begin{bmatrix} 1 & 0 & 0 \\ 0 & 0 & 1 \\ 0 & 1 & 0 \end{bmatrix}$.

To illustrate Theorem 2.8 again, notice that the product

$$EA = \begin{bmatrix} 1 & 0 & 0 \\ 0 & 0 & 1 \\ 0 & 1 & 0 \end{bmatrix} \begin{bmatrix} 1 & 2 & 3 \\ 4 & 5 & 6 \\ 7 & 8 & 9 \end{bmatrix} = \begin{bmatrix} 1 & 2 & 3 \\ 7 & 8 & 9 \\ 4 & 5 & 6 \end{bmatrix} = B$$

yields the same result as applying the elementary row operation to A.

Definition 2.10 *A matrix B is said to be **equivalent** to another matrix A if B can be obtained by applying a finite number of elementary row operations to A.*

We can represent this equivalence by

$$A \xrightarrow{op_1} A_1 \xrightarrow{op_2} A_2 \xrightarrow{op_3} \cdots \xrightarrow{op_k} B.$$

Example 2.21 *a.* $B = \begin{bmatrix} 3 & 4 \\ -5 & -6 \end{bmatrix}$ *is equivalent to* $A = \begin{bmatrix} 1 & 2 \\ 3 & 4 \end{bmatrix}$ *since*

$$A = \begin{bmatrix} 1 & 2 \\ 3 & 4 \end{bmatrix} \xrightarrow{-2R_2+R_1} \begin{bmatrix} -5 & -6 \\ 3 & 4 \end{bmatrix} \xrightarrow{R_1 \leftrightarrow R_2} \begin{bmatrix} 3 & 4 \\ -5 & -6 \end{bmatrix} = B.$$

b. $B = \begin{bmatrix} 3 & 6 & 9 \\ 18 & 21 & 24 \\ 7 & 8 & 9 \end{bmatrix}$ *is equivalent to* $A = \begin{bmatrix} 1 & 2 & 3 \\ 4 & 5 & 6 \\ 7 & 8 & 9 \end{bmatrix}$ *since*

$$A = \begin{bmatrix} 1 & 2 & 3 \\ 4 & 5 & 6 \\ 7 & 8 & 9 \end{bmatrix} \xrightarrow{3R_1} \begin{bmatrix} 3 & 6 & 9 \\ 4 & 5 & 6 \\ 7 & 8 & 9 \end{bmatrix} \xrightarrow{2R_3+R_2} \begin{bmatrix} 3 & 6 & 9 \\ 18 & 21 & 24 \\ 7 & 8 & 9 \end{bmatrix} = B.$$

We combine all that we have covered so far in this section in the following fact:

Theorem 2.9 *If B is equivalent to A, then there exist elementary matrices $E_1, E_2, \ldots E_n$ such that $B = E_n \cdots E_2 E_1 A$.*

Proof 2.7 *This result follows from Theorem 2.8 and a simple induction proof and is left as an exercise.* □

We illustrate this result with the following examples:

Example 2.22 *a. Consider the previous example where op_1 is $-2R_2 + R_1$ and op_2 is $R_1 \leftrightarrow R_2$.*

The corresponding elementary matrices are

$$E_1 = \begin{bmatrix} 1 & -2 \\ 0 & 1 \end{bmatrix} \text{ and } E_2 = \begin{bmatrix} 0 & 1 \\ 1 & 0 \end{bmatrix}.$$

So we have $A \xrightarrow{op_1} A_1 \xrightarrow{op_2} B$ and

$$E_2 E_1 A = \begin{bmatrix} 0 & 1 \\ 1 & 0 \end{bmatrix} \begin{bmatrix} 1 & -2 \\ 0 & 1 \end{bmatrix} \begin{bmatrix} 1 & 2 \\ 3 & 4 \end{bmatrix} = \begin{bmatrix} -5 & -6 \\ 3 & 4 \end{bmatrix} = B.$$

b. Consider the previous example where op_1 is $3R_1$ and op_2 is $2R_3 + R_2$.

The corresponding elementary matrices are

$$E_1 = \begin{bmatrix} 3 & 0 & 0 \\ 0 & 1 & 0 \\ 0 & 0 & 1 \end{bmatrix} \text{ and } E_2 = \begin{bmatrix} 1 & 0 & 0 \\ 0 & 1 & 2 \\ 0 & 0 & 1 \end{bmatrix}.$$

So we have $A \xrightarrow{op_1} A_1 \xrightarrow{op_2} B$ and

$$E_2 E_1 A = \begin{bmatrix} 1 & 0 & 0 \\ 0 & 1 & 2 \\ 0 & 0 & 1 \end{bmatrix} \begin{bmatrix} 3 & 0 & 0 \\ 0 & 1 & 0 \\ 0 & 0 & 1 \end{bmatrix} \begin{bmatrix} 1 & 2 & 3 \\ 4 & 5 & 6 \\ 7 & 8 & 9 \end{bmatrix} = \begin{bmatrix} 3 & 6 & 9 \\ 18 & 21 & 24 \\ 7 & 8 & 9 \end{bmatrix} = B.$$

Definition 2.11 *Let op be an arbitrary elementary row operation. We say that op^{-1} is the **inverse operation** of op if whenever $A \xrightarrow{op} B$, we also have $B \xrightarrow{op^{-1}} A$.*

In a sense, op^{-1} undoes whatever op does to a matrix.

$$A \xrightarrow{op} B \xrightarrow{op^{-1}} A.$$

Lemma 2.1

1. *Every elementary row operation has an inverse operation.*

2. *The inverse operation is an elementary operation of the same type.*

Proof 2.8 *The inverse of $R_i \leftrightarrow R_j$ is $R_i \leftrightarrow R_j$. The inverse of aR_i is $(1/a)R_i$. The inverse of $aR_i + R_j$ is $-aR_i + R_j$.* □

Theorem 2.10 *Let A and B be two matrices. Then*

i. A is equivalent to A

ii. If B is equivalent to A then A is equivalent to B.

iii. If A is equivalent to B and B is equivalent to C, then A is equivalent to C.

Proof 2.9 *To prove i, apply the operation* $(1)R_1$ *to A. To prove ii, since B is equivalent to A, we have*

$$A \xrightarrow{op_1} A_1 \xrightarrow{op_2} A_2 \xrightarrow{op_3} \cdots \xrightarrow{op_k} B.$$

By Lemma 2.1, $op_1^{-1}, \ldots, op_k^{-1}$ *exist and we have*

$$B \xrightarrow{op_k^{-1}} A_1 \xrightarrow{op_{k-1}^{-1}} A_2 \xrightarrow{op_{k-2}^{-1}} \cdots \xrightarrow{op_1^{-1}} A,$$

and hence A is equivalent to B. We leave the proof of iii as an exercise. □

As a result of part ii of the previous theorem, we can simply state that two matrices are equivalent. In other words, we can say A and B are equivalent instead of A is equivalent to B or B is equivalent to A.

A more practical way to determine whether or not two matrices are equivalent relies on the following result (the proof of which is left as an exercise):

Theorem 2.11 *Two matrices are equivalent iff they have the same reduced row-echelon form.*

Example 2.23 *a.* $A = \begin{bmatrix} 1 & 1 \\ 0 & 2 \end{bmatrix}$ *and* $B = \begin{bmatrix} 3 & 0 \\ 1 & 1 \end{bmatrix}$ *are equivalent since*

$$A = \begin{bmatrix} 1 & 1 \\ 0 & 2 \end{bmatrix} \xrightarrow{\frac{1}{2}R_2} \begin{bmatrix} 1 & 1 \\ 0 & 1 \end{bmatrix} \xrightarrow{-R_2+R_1} \begin{bmatrix} 1 & 0 \\ 0 & 1 \end{bmatrix} \quad and$$

$$B = \begin{bmatrix} 3 & 0 \\ 1 & 1 \end{bmatrix} \xrightarrow{\frac{1}{3}R_1} \begin{bmatrix} 1 & 0 \\ 1 & 1 \end{bmatrix} \xrightarrow{-R_1+R_2} \begin{bmatrix} 1 & 0 \\ 0 & 1 \end{bmatrix}.$$

Thus, since A and B have the same reduced row-echelon form by Theorem 2.11, A and B are equivalent. Furthermore, using inverse operations, we can exhibit a set of elementary row operations which make them equivalent. This is achieved by linking the two sequences listed above. One needs to reverse and invert the second sequence to get the following:

$$A = \begin{bmatrix} 1 & 1 \\ 0 & 2 \end{bmatrix} \xrightarrow{\frac{1}{2}R_2} \begin{bmatrix} 1 & 1 \\ 0 & 1 \end{bmatrix} \xrightarrow{-R_2+R_1} \begin{bmatrix} 1 & 0 \\ 0 & 1 \end{bmatrix} \xrightarrow{R_1+R_2} \begin{bmatrix} 1 & 0 \\ 1 & 1 \end{bmatrix} \xrightarrow{3R_1} \begin{bmatrix} 3 & 0 \\ 1 & 1 \end{bmatrix} = B.$$

b. $A = \begin{bmatrix} -3 & 0 & 0 \\ 2 & 0 & 1 \\ 0 & 0 & 0 \end{bmatrix}$ *and* $B = \begin{bmatrix} 1 & 0 & 0 \\ 0 & 3 & 0 \\ 0 & 3 & 0 \end{bmatrix}$ *are **not** equivalent since*

$$A = \begin{bmatrix} -3 & 0 & 0 \\ 2 & 0 & 1 \\ 0 & 0 & 0 \end{bmatrix} \xrightarrow{-\frac{1}{3}R_1} \begin{bmatrix} 1 & 0 & 0 \\ 2 & 0 & 1 \\ 0 & 0 & 0 \end{bmatrix} \xrightarrow{-2R_1+R_2} \begin{bmatrix} 1 & 0 & 0 \\ 0 & 0 & 1 \\ 0 & 0 & 0 \end{bmatrix} \quad while$$

$$B = \begin{bmatrix} 1 & 0 & 0 \\ 0 & 3 & 0 \\ 0 & 3 & 0 \end{bmatrix} \xrightarrow{-R_2+R_3} \begin{bmatrix} 1 & 0 & 0 \\ 0 & 3 & 0 \\ 0 & 0 & 0 \end{bmatrix} \xrightarrow{\frac{1}{3}R_2} \begin{bmatrix} 1 & 0 & 0 \\ 0 & 1 & 0 \\ 0 & 0 & 0 \end{bmatrix}.$$

Thus, by Theorem 2.11, A and B are not equivalent since they do **not** have the same reduced row-echelon form.

EXERCISES

1. Compute the elementary matrix corresponding to each of the following elementary operations: $R_1 \leftrightarrow R_3$, $-3R_2$, $-2R_2 + R_3$. Then illustrate Theorem 2.8 for each of these elementary operations using the matrix

$$A = \begin{bmatrix} 1 & 2 & 3 \\ 2 & 3 & 1 \\ 3 & 2 & 1 \end{bmatrix}$$

2. Using Theorem 2.11, decide whether or not each of the following two matrices are elementary equivalent:

a.
$$A = \begin{bmatrix} 1 & 1 \\ 0 & 1 \end{bmatrix}, \qquad B = \begin{bmatrix} 1 & 0 \\ 0 & 2 \end{bmatrix}$$

b.
$$A = \begin{bmatrix} 1 & 1 \\ 2 & 2 \end{bmatrix}, \qquad B = \begin{bmatrix} 1 & 2 \\ 3 & 4 \end{bmatrix}$$

c.
$$A = \begin{bmatrix} 1 & 0 & 0 \\ 0 & 1 & 0 \\ 1 & 0 & 0 \end{bmatrix}, \qquad B = \begin{bmatrix} 1 & 0 & 0 \\ 0 & -1/2 & 0 \\ 0 & 0 & 0 \end{bmatrix}$$

d.
$$A = \begin{bmatrix} 1 & 0 & -1 \\ 2 & 3 & 1 \\ 1 & 1 & 2 \end{bmatrix}, \qquad B = \begin{bmatrix} 2 & 4 & 4 \\ 0 & 3 & 3 \\ 1 & 4 & 5 \end{bmatrix}$$

e.
$$A = \begin{bmatrix} 2 & 3 & -1 \\ 3 & 5 & -2 \\ -2 & 4 & -1 \end{bmatrix}, \qquad B = \begin{bmatrix} 6 & -1 & 2 \\ -4 & 2 & -1 \\ 2 & -2 & 3 \end{bmatrix}$$

3. In the previous problem, for the matrices which were elementary equivalent, illustrate Theorem 2.9.

4. Consider the matrices listed below:

$$A = \begin{bmatrix} 2 & 1 \\ 1 & 0 \end{bmatrix}, \qquad B = \begin{bmatrix} 2 & 2 \\ 0 & 1 \end{bmatrix}.$$

   a. Prove that A and B are equivalent matrices.

   b. Illustrate the equivalence in part a, by a series of elementary row operations.

   c. Illustrate how B can be obtained from A by multiplication on the left by elementary matrices.

5. Consider the matrices listed below:

$$A = \begin{bmatrix} 1 & 0 & 0 \\ 2 & 1 & 0 \\ 0 & -3 & 0 \end{bmatrix}, \qquad B = \begin{bmatrix} 0 & 0 & 0 \\ 0 & 1 & 0 \\ 2 & 0 & 0 \end{bmatrix}.$$

   a. Prove that A and B are equivalent matrices.

   b. Illustrate the equivalence in part a, by a series of elementary row operations.

   c. Illustrate how B can be obtained from A by multiplication on the left by elementary matrices.

6. Consider the matrices listed below:

$$A = \begin{bmatrix} 1 & 0 & 0 \\ 1 & 1 & 0 \\ 0 & 0 & 2 \end{bmatrix} \qquad B = \begin{bmatrix} 0 & 1 & 0 \\ 1 & 0 & 0 \\ 0 & 2 & 3 \end{bmatrix}$$

   a. Prove that A and B are equivalent matrices.

   b. Illustrate the equivalence in part a, by a series of elementary row operations.

   c. Illustrate how B can be obtained from A by multiplication on the left by elementary matrices.

7. Show that if E is an elementary matrix, then so is E^T

   (hint: consider each of the three elementary row operations separately)

8. Prove Theorem 2.11.

9. Prove part iii of Theorem 2.10.

10. Prove Theorem 2.9.

2.6 INVERSE OF A MATRIX

If a is a non-zero real number, then the multiplicative inverse of a is $1/a$ since $a(1/a) = 1 = (1/a)a$. Note for a real number to have a multiplicative inverse it must be non-zero. We now investigate the existence of multiplicative inverses for matrices using matrix multiplication. We will see that they do not always exist, indeed for more than just the zero matrix. However, in the case that the inverse does exist, we can conclude a number of seemingly unrelated equivalent conditions for its existence. This theorem which we will derive slowly for the remainder of the chapter is the second goal of this chapter. We also give a systematic way to find the inverse of a matrix when it exists.

Definition 2.12 *Let A be a square matrix. B is the **inverse** of A if $AB = I = BA$. When A has an inverse we say that A is **invertible** (or **non-singular**). Otherwise, we say A is **non-invertible** (or **singular**).*

Note that A must be a square matrix in order for both products AB and BA to be possible.

Example 2.24 *The inverse of* $\begin{bmatrix} 2 & 1 \\ 1 & 1 \end{bmatrix}$ *is* $\begin{bmatrix} 1 & -1 \\ -1 & 2 \end{bmatrix}$ *since*

$$\begin{bmatrix} 2 & 1 \\ 1 & 1 \end{bmatrix} \begin{bmatrix} 1 & -1 \\ -1 & 2 \end{bmatrix} = \begin{bmatrix} 1 & 0 \\ 0 & 1 \end{bmatrix} = \begin{bmatrix} 1 & -1 \\ -1 & 2 \end{bmatrix} \begin{bmatrix} 2 & 1 \\ 1 & 1 \end{bmatrix}.$$

A number of remarks are in order here.

- The inverse of A is necessarily square and of the same dimensions as A.

- The inverse of a matrix does not always exists. Take the case of $A = 0_{nn}$; it has no inverse because for all matrices B, $AB = 0_{nn} \neq I_n$. In addition certain non-zero matrices have no inverse. For instance, $A = \begin{bmatrix} 1 & 0 \\ 0 & 0 \end{bmatrix}$ (similar argument). In fact, one of our goals is to determine which matrices do have an inverse.

- When A has an inverse, it has exactly one. In other words a matrix has at most one inverse. For suppose B_1 and B_2 are both inverses of A, i.e. $AB_1 = I = B_1 A$ and $AB_2 = I = B_2 A$. Then

$$B_1 = B_1 I = B_1(AB_2) = (B_1 A)B_2 = I B_2 = B_2.$$

- Since inverses are unique we will denote the inverse of A (when it exists) by A^{-1}.

Some basic properties of inverse are the following:

Theorem 2.12 *Let A and B be invertible matrices of the same dimensions and C and D any matrices of the appropriate dimensions. Then*

i. $(A^{-1})^{-1} = A$.

ii. AB is invertible with $(AB)^{-1} = B^{-1}A^{-1}$.

iii. If $AC = AD$ then $C = D$.

iv. A^T is invertible with $(A^T)^{-1} = (A^{-1})^T$.

Proof 2.10 *To prove i, we know that $AA^{-1} = I = A^{-1}A$. But this says that A is the inverse of A^{-1}, i.e. $A = (A^{-1})^{-1}$. To prove ii, one need only check that $(AB)(B^{-1}A^{-1}) = I = (B^{-1}A^{-1})AB$ (which is easy to compute) and so $B^{-1}A^{-1}$ is the inverse of AB, i.e. $(AB)^{-1} = B^{-1}A^{-1}$. The remainder of the statements are left as exercises.* □

A consequence of Theorem 2.12.ii (which follows by induction) is the following:

Corollary 2.3 *If $A_1, A_2, \ldots, A_n$ are invertible, then so is $A_1 A_2 \cdots A_n$ and*
$$(A_1 A_2 \cdots A_n)^{-1} = A_n^{-1} \cdots A_2^{-1} A_1^{-1}.$$

When A is invertible, matrix exponentiation can be extended to negative exponents as follows:

- $A^{-1} =$ the inverse of A,

- $A^{-2} = A^{-1}A^{-1}$,

- For $k \geq 2$, $A^{-k} = A^{-1}A^{-1}\cdots A^{-1}$ (k times).

The exponentiation rules stated earlier also extend here, namely

$$A^k A^l = A^{k+l} \qquad \text{and} \qquad (A^k)^l = A^{kl}.$$

Now we draw on material from the previous section. The lemmas that follow lead to our second goal. We motivate the next lemma by illustrating it with an example.

Example 2.25 *Consider the elementary matrix $E = \begin{bmatrix} 1 & 0 \\ -2 & 1 \end{bmatrix}$ obtained by the elementary row operation $-2R_1 + R_2$. The inverse operation is $2R_1 + R_2$ and the corresponding elementary row matrix is $E' = \begin{bmatrix} 1 & 0 \\ 2 & 1 \end{bmatrix}$. Notice that $EE' = I = E'E$. So we see that the elementary matrix corresponding to the inverse operation is the inverse of the original elementary matrix.*

Lemma 2.2 *Elementary matrices are invertible. More specifically, if $I \xrightarrow{op} E$, then $I \xrightarrow{op^{-1}} E^{-1}$.*

Proof 2.11 *Let E be an elementary matrix obtained by the elementary operation op, i.e. $I \xrightarrow{op} E$. Let op^{-1} be the inverse operation of op (which we know exists). Let E' be the elementary matrix corresponding to op^{-1}, i.e. $I \xrightarrow{op^{-1}} E'$. We claim that E' is the inverse of E (and so we are done). By the definition of inverse operation, we have*

$$I \xrightarrow{op} E \xrightarrow{op^{-1}} I.$$

By Theorem 2.8, $I = E'E$. Similarly,

$$I \xrightarrow{op^{-1}} E' \xrightarrow{op} I.$$

and $I = EE'$. Hence, $E' = E^{-1}$. $\square$

Lemma 2.3 *For a square matrix $A \in M_n(F)$, A is equivalent to I iff $AX = B$ has a solution for any B.*

Proof 2.12 *If A is equivalent to I, i.e.*

$$A \xrightarrow{op_1} A_1 \xrightarrow{op_2} A_2 \xrightarrow{op_3} \cdots \xrightarrow{op_k} I,$$

then apply the same elementary row operations to the augmented matrix $[A|B]$ to obtain

$$[A|B] \xrightarrow{op_1} [A_1|B_1] \xrightarrow{op_2} [A_2|B_2] \xrightarrow{op_3} \cdots \xrightarrow{op_k} [I|C]$$

with $C = \begin{bmatrix} c_1 \\ c_2 \\ \vdots \\ c_n \end{bmatrix}$. Then $AX = B$ has solution $x_1 = c_1$, $x_2 = c_2$, ..., $x_m = c_m$.

*We prove the reverse direction of the iff by contrapositive. Assume that A is **not** equivalent to I. Now a bit of cleverness is necessary here. We need to find a B for which $AX = B$ has no solution. If we succeed then the proof is complete. Let R be the reduced row echelon form for the matrix A with*

$$A \xrightarrow{op_1} A_1 \xrightarrow{op_2} A_2 \xrightarrow{op_3} \cdots \xrightarrow{op_k} R.$$

Observe that since R is not I it must be the case that at least the last row of R is a row filled with zeros. Let $E_1, E_2, \ldots E_k$ be the elementary matrices corresponding to $op_1, op_2, \ldots op_k$. Set $B = (E_k \cdots E_1)^{-1} e_n$ where $e_n = \begin{bmatrix} 0 \\ 0 \\ \vdots \\ 0 \\ 1 \end{bmatrix}$. Let C be such that

$$B \xrightarrow{op_1} B_1 \xrightarrow{op_2} B_2 \xrightarrow{op_3} \cdots \xrightarrow{op_k} C.$$

By Theorem 2.9, $C = E_k \cdots E_1 B = E_k \cdots E_1 (E_k \cdots E_1)^{-1} e_n = e_n$. Applying Gaussian Elimination,

$$[A|B] \xrightarrow{op_1} [A_1|B_1] \xrightarrow{op_2} [A_2|B_2] \xrightarrow{op_3} \cdots \xrightarrow{op_k} [R|C].$$

The last row translates into $0 = 1$ which is a contradiction. Hence, $AX = B$ has no solution. □

Lemma 2.4 *If A is equivalent to I then A is invertible and a product of elementary matrices.*

Proof 2.13 *If A is equivalent to I then I is equivalent to A and*

$$A = E_k \cdots E_2 E_1 I = E_k \cdots E_2 E_1,$$

by Theorem 2.9. Hence, A is a product of elementary matrices. Since $E_1, E_2, \ldots, E_n$ are invertible and the product of invertible matrices is invertible (Corollary 2.3), A is therefore invertible. □

We illustrate how one can express an invertible matrix as a product of elementary matrices.

Example 2.26 *Consider the following matrix:*

$$A = \begin{bmatrix} 1 & 0 & 0 \\ 0 & 2 & 3 \\ 0 & 1 & 0 \end{bmatrix}.$$

A is invertible, since A is equivalent to I_3. Indeed,

$$A = \begin{bmatrix} 1 & 0 & 0 \\ 0 & 2 & 3 \\ 0 & 1 & 0 \end{bmatrix} \xrightarrow{R_2 \leftrightarrow R_3} \begin{bmatrix} 1 & 0 & 0 \\ 0 & 1 & 0 \\ 0 & 2 & 3 \end{bmatrix} \xrightarrow{-2R_2 + R_3}$$

$$\begin{bmatrix} 1 & 0 & 0 \\ 0 & 1 & 0 \\ 0 & 0 & 3 \end{bmatrix} \xrightarrow{\frac{1}{3}R_3} \begin{bmatrix} 1 & 0 & 0 \\ 0 & 1 & 0 \\ 0 & 0 & 1 \end{bmatrix} = I_3.$$

Now observe the corresponding inverse elementary row operations:

$$I_3 = \begin{bmatrix} 1 & 0 & 0 \\ 0 & 1 & 0 \\ 0 & 0 & 1 \end{bmatrix} \xrightarrow{3R_3} \begin{bmatrix} 1 & 0 & 0 \\ 0 & 1 & 0 \\ 0 & 0 & 3 \end{bmatrix} \xrightarrow{2R_2 + R_3}$$

$$\begin{bmatrix} 1 & 0 & 0 \\ 0 & 1 & 0 \\ 0 & 2 & 3 \end{bmatrix} \xrightarrow{R_2 \leftrightarrow R_3} \begin{bmatrix} 1 & 0 & 0 \\ 0 & 2 & 3 \\ 0 & 1 & 0 \end{bmatrix} = A.$$

For these inverse operations we compute their corresponding elementary matrices:

$$E_1 = \begin{bmatrix} 1 & 0 & 0 \\ 0 & 1 & 0 \\ 0 & 0 & 3 \end{bmatrix}, \quad E_2 = \begin{bmatrix} 1 & 0 & 0 \\ 0 & 1 & 0 \\ 0 & 2 & 1 \end{bmatrix}, \quad E_3 = \begin{bmatrix} 1 & 0 & 0 \\ 0 & 0 & 1 \\ 0 & 1 & 0 \end{bmatrix}.$$

By Theorem 2.9, $A = E_3 E_2 E_1 I = E_3 E_2 E_1$, *i.e.*

$$A = \begin{bmatrix} 1 & 0 & 0 \\ 0 & 2 & 3 \\ 0 & 1 & 0 \end{bmatrix} = \begin{bmatrix} 1 & 0 & 0 \\ 0 & 0 & 1 \\ 0 & 1 & 0 \end{bmatrix} \begin{bmatrix} 1 & 0 & 0 \\ 0 & 1 & 0 \\ 0 & 2 & 1 \end{bmatrix} \begin{bmatrix} 1 & 0 & 0 \\ 0 & 1 & 0 \\ 0 & 0 & 3 \end{bmatrix}.$$

Hence, we have a method for expressing an invertible matrix A as a product of elementary matrices.

Lemma 2.5 *Let A and C be square matrices. If $AC = I$ or $CA = I$, then A is invertible with inverse C.*

Proof 2.14 *Assume $AC = I$. Consider the linear system $AX = B$ for any B. Then CB is a solution to $AX = B$ since $A(CB) = (AC)B = IB = B$. By Lemma 2.3, A is equivalent to I and by Lemma 2.4, A is invertible. To complete the proof in this case we need to show that $C = A^{-1}$. We know $AC = I = AA^{-1}$. By Theorem 2.12.iii, $C = A^{-1}$.*

If $CA = I$ then by the work above C is invertible with inverse A. Hence $CA = I = AC$ and so A is invertible with inverse C. □

There is a practical side to the previous lemma. It says that to check that B is the inverse of A it is sufficient to verify only one of the products $AB = I$ or $BA = I$.

Now we state the main theorem of this chapter. We will add more statements to it in later sections.

Theorem 2.13 *For a square matrix A the following are equivalent:*

i. A is invertible.

ii. The linear system $AX = B$ has a unique solution for any $B \in F^n$.

iii. A is equivalent to I.

iv. A is a product of elementary matrices.

Proof 2.15 *A lot of the work for this proof has already been done. To prove ii implies iii, we appeal to Lemma 2.3. To prove iii implies iv, we also appeal to Lemma 2.4. To prove iv implies i, we appeal to Lemma 2.2 and Corollary 2.3. For the final implication, i implies ii, given that A^{-1} exists, a solution to $AX = B$ is $A^{-1}B$, since*

$$A(A^{-1}B) = (AA^{-1})B = IB = B.$$

It is the only solution, for suppose X_0 is a solution to $AX = B$, i.e. $AX_0 = B$. Multiplying on the left by A^{-1} yields $A^{-1}(AX_0) = A^{-1}B$ and simplifying gives $X_0 = A^{-1}B$. Now our proof is complete. □

Now we give an algorithm for finding the inverse of A (when it exists). It is a simple, though sometimes tedious, procedure: Begin with the augmented matrix $[A|I]$. By applying elementary row operations, convert this matrix to the form $[I|B]$ for some B (recall that when A is invertible, A can be row reduced to I). Then $B = A^{-1}$. Let us see why this is so. We have

$$[A|I] \xrightarrow{op_1} \cdots \xrightarrow{op_n} [I|B].$$

Then $I = (E_n \cdots E_2 E_1)A$ and $B = (E_n \cdots E_2 E_1)I$ for the corresponding elementary matrices $E_1, E_2, \ldots, E_n$. Combining these two equations we have $I = BA$ and by Lemma 2.5, B is the inverse of A.

Example 2.27 *We compute the inverse of $A = \begin{bmatrix} 2 & 1 \\ 1 & 1 \end{bmatrix}$.*

$$\begin{bmatrix} 2 & 1 & | & 1 & 0 \\ 1 & 1 & | & 0 & 1 \end{bmatrix} \xrightarrow{-R_2+R_1} \begin{bmatrix} 1 & 0 & | & 1 & -1 \\ 1 & 1 & | & 0 & 1 \end{bmatrix} \xrightarrow{-R_1+R_2} \begin{bmatrix} 1 & 0 & | & 1 & -1 \\ 0 & 1 & | & -1 & 2 \end{bmatrix},$$

and so $A^{-1} = \begin{bmatrix} 1 & -1 \\ -1 & 2 \end{bmatrix}$. If A could not be reduced to the identity matrix in this procedure, then we can conclude that A has no inverse.

Example 2.28 *We compute the inverse of $B = \begin{bmatrix} 1 & 2 & 0 \\ 0 & 1 & 0 \\ 0 & 0 & 3 \end{bmatrix}$.*

$$\begin{bmatrix} 1 & 2 & 0 & | & 1 & 0 & 0 \\ 0 & 1 & 0 & | & 0 & 1 & 0 \\ 0 & 0 & 3 & | & 0 & 0 & 1 \end{bmatrix} \xrightarrow{-2R_2+R_1} \begin{bmatrix} 1 & 0 & 0 & | & 1 & -2 & 0 \\ 0 & 1 & 0 & | & 0 & 1 & 0 \\ 0 & 0 & 3 & | & 0 & 0 & 1 \end{bmatrix}$$

$$\xrightarrow{\frac{1}{3}R_3} \begin{bmatrix} 1 & 0 & 0 & | & 1 & -2 & 0 \\ 0 & 1 & 0 & | & 0 & 1 & 0 \\ 0 & 0 & 1 & | & 0 & 0 & 1/3 \end{bmatrix},$$

and so $B^{-1} = \begin{bmatrix} 1 & -2 & 0 \\ 0 & 1 & 0 \\ 0 & 0 & 1/3 \end{bmatrix}$.

With the introduction of the inverse we are now in a position to give an alternative way to solve a linear system of equations in the case when it has exactly one solution. In fact, the proof of Theorem 2.13 (i implies iii) give the method: If $AX = B$ has exactly one solution, then the unique solution is $A^{-1}B$. We illustrate this method in the following example:

Example 2.29 *Consider the linear system*

$$\begin{cases} 2x + y & = & 1 \\ x + y & = & 2 \end{cases} \text{ with matrix equation } \begin{bmatrix} 2 & 1 \\ 1 & 1 \end{bmatrix} \begin{bmatrix} x \\ y \end{bmatrix} = \begin{bmatrix} 1 \\ 2 \end{bmatrix}.$$

The coefficient matrix for this linear system is $A = \begin{bmatrix} 2 & 1 \\ 1 & 1 \end{bmatrix}$. *In the previous*

example we computed $A^{-1} = \begin{bmatrix} 1 & -1 \\ -1 & 2 \end{bmatrix}$. *Therefore,*

$$\begin{bmatrix} x \\ y \end{bmatrix} = X = A^{-1}B = \begin{bmatrix} 1 & -1 \\ -1 & 2 \end{bmatrix} \begin{bmatrix} 1 \\ 2 \end{bmatrix} = \begin{bmatrix} -1 \\ 3 \end{bmatrix}.$$

Thus, the solution is $x = -1$ *and* $y = 3$.

EXERCISES

1. If possible, find the inverse of each of the following matrices:

$$A = \begin{bmatrix} 2 & -1 \\ 1 & 3 \end{bmatrix} \qquad B = \begin{bmatrix} 1 & -2 & 1 \\ 2 & -4 & 0 \\ 1 & 0 & 2 \end{bmatrix} \qquad C = \begin{bmatrix} 4 & -2 & 2 \\ 2 & 5 & 1 \\ -2 & 1 & -1 \end{bmatrix}$$

2. Use your results in the first problem to solve each of the following linear systems (when there is a unique solution):

$$\begin{cases} 2x - y & = & -3 \\ x + 3y & = & 0 \end{cases} \qquad \begin{cases} x - 2y + z & = & 1 \\ 2x - 4y & = & -2 \\ x + 2z & = & 0 \end{cases} \qquad \begin{cases} 4x - 2y + 2z & = & -3 \\ 2x + 5y + z & = & 1 \\ -2x + y - z & = & 2 \end{cases}$$

3. Again, use your results in the first problem to write the following matrices as products of elementary matrices (when it is possible):

$$\begin{bmatrix} 2 & -1 \\ 1 & 3 \end{bmatrix} \qquad \begin{bmatrix} 1 & -2 & 1 \\ 2 & -4 & 0 \\ 1 & 0 & 2 \end{bmatrix} \qquad \begin{bmatrix} 4 & -2 & 2 \\ 2 & 5 & 1 \\ -2 & 1 & -1 \end{bmatrix}$$

4. Let $A = \begin{bmatrix} 2 & 1 \\ 1 & 2 \end{bmatrix}$

    a. Find the reduced row-echelon form of A.

    b. Use part a. to decide whether or not A is invertible (explain).

    c. Explain why we can write A as a product of elementary matrices and then do so.

5. Consider the matrix $A = \begin{bmatrix} 2 & 2 \\ 1 & 2 \end{bmatrix}$

    a. Find the reduced row echelon form for A.

    b. Use part a to explain why A is invertible.

    c. Find A^{-1}.

    d. Use part c to solve the following linear system:

$$\begin{cases} 2x + 2y &=& -1 \\ x + 2y &=& 2 \end{cases}$$

    e. Express A as a product of elementary matrices.

6. Consider the matrix $A = \begin{bmatrix} 3 & 2 \\ 1 & 0 \end{bmatrix}$.

    a. Find the reduced row echelon form for A using elementary row operations.

    b. Use part a to explain why A is invertible.

    c. Use part a to find A^{-1}.

    d. Without solving, what can be said about the following linear system?

$$\begin{cases} 3x + 2y &=& -1 \\ x &=& 2 \end{cases}$$

    e. Now use part c to find the solution set of the linear system in part d.

    f. Express A as a product of elementary matrices.

7. We wish to find an equation of the form $x^2 + axy + y^2 = b$ which contains the points $(-1, 1)$ and $(2, 1/2)$.

    a. Set up the matrix representation of a linear system needed for finding such an equation.

    b. Solve part a. using the inverse of the coefficient matrix.

8. Using the inverse of a matrix, find the equation of the circle $x^2 + y^2 + ax + by = c$ passing through the points $(0, 2), (3, 1)$ and $(4, 0)$.

9. Let A and B be $n \times n$ matrices. Prove that if AB is invertible, then so are A and B invertible.

10. Prove that if $A \in M_{nn}$ is symmetric and invertible, then so is A^{-1}.

11. Prove that if A is invertible and B is equivalent to A, then B is also invertible.

12. Prove the following statement:

    A is equivalent to B iff there is an invertible matrix P such that $B = PA$.

13. Show that if $A \in M_{21}$ and $B \in M_{12}$, then AB is not invertible.

14. Show that if $A \in M_{m1}$ and $B \in M_{1m}$, then AB is not invertible.

    (note: you can in fact replace 1 by n with $n < m$ and the result still holds)

15. Prove parts iii and iv of Theorem 2.12.

2.7 APPLICATION: THE SIMPLEX METHOD REVISITED

In this section we introduce a way of implementing the Simplex Method without the use of elementary row operations. Just as in Gaussian Elimination where we could replace elementary row operations by multiplication on the left by an elementary matrices, so too in the Simplex Method can we replace pivoting by multiplication on the left by an appropriate matrix in order to reach the next tableau.

First, it would make our discussion easier if we introduce matrix notation for representing a linear programming problem. Assume for this section that matrices with either one of their dimension being 1 are column matrices.

Let $A = [a_{ij}]$ and $B = [b_{ij}]$ be two matrices of the same dimensions. Then the notation $A < B$ means for each i and j that $a_{ij} < b_{ij}$. The notation $>, \leq$ and $\geq$ are defined in a similar manner.

Definition 2.13 *If our linear programming problem has n unknowns and m contraints, then the matrix representation of a* **standard** *linear programming problem has the form*

$$\text{Maximize } z = C^T X \text{ Subject to } AX \leq B, \quad X \geq 0$$

where A is an $m \times n$ matrix, C is an $n \times 1$ matrix, B is an $m \times 1$ matrix, and X is an $n \times 1$ matrix of unknowns $x_1, x_2, \ldots, x_n$.

Definition 2.14 *If our linear programming problem has n unknowns and m contraints, then the matrix representation of a* **canonical** *linear programming problem has the form*

$$\text{Maximize } z = [C|0_m]^T X' \quad \text{Subject to } [A|I_m]X = B, \quad X \geq 0$$

where A is an $m \times n$ matrix, C is an $n \times 1$ matrix, B is an $m \times 1$ matrix, and X' is an $(m + n) \times 1$ matrix of unknowns $x_1, x_2, \ldots, x_n$ together with slack variables $u_1, u_2, \ldots, u_m$. Note that adding m slack variables corresponds to replacing A by $[A|I_m]$ and does nothing to effect the objective function, hence the 0_m.

Example 2.30 Consider the following linear programming problem in standard form.

$$\text{Maximize } z = 2x + 3y, \qquad \text{Subject to } \begin{cases} x + y & \leq & 4 \\ x + 3y & \leq & 6 \\ 3x + y & \leq & 2 \\ x, y & \geq & 0 \end{cases}$$

The standard matrix form will be

$$\text{Maximize } z = \begin{bmatrix} 2 & 3 \end{bmatrix} \begin{bmatrix} x \\ y \end{bmatrix}$$

$$\text{Subject to } \begin{bmatrix} 1 & 1 \\ 1 & 3 \\ 3 & 1 \end{bmatrix} \begin{bmatrix} x \\ y \end{bmatrix} \leq \begin{bmatrix} 4 \\ 6 \\ 2 \end{bmatrix}, \qquad \begin{bmatrix} x \\ y \end{bmatrix} \geq \begin{bmatrix} 0 \\ 0 \end{bmatrix}$$

The canonical matrix form will be

$$\text{Maximize } z = \begin{bmatrix} 2 & 3 & 0 & 0 & 0 \end{bmatrix} \begin{bmatrix} x \\ y \\ u \\ v \\ w \end{bmatrix}$$

$$\text{Subject to } \begin{bmatrix} 1 & 1 & 1 & 0 & 0 \\ 1 & 3 & 0 & 1 & 0 \\ 3 & 1 & 0 & 0 & 1 \end{bmatrix} \begin{bmatrix} x \\ y \\ u \\ v \\ w \end{bmatrix} = \begin{bmatrix} 4 \\ 6 \\ 2 \end{bmatrix}, \qquad \begin{bmatrix} x \\ y \\ u \\ v \\ w \end{bmatrix} \geq \begin{bmatrix} 0 \\ 0 \\ 0 \\ 0 \\ 0 \end{bmatrix}$$

Example 2.31 Consider the example we presented in the earlier section on linear programming.

$$\text{Maximize } z = 2x + 3y, \qquad \text{Subject to } \begin{cases} x + y & \leq & 4 \\ x + 3y & \leq & 6 \\ x, y & \geq & 0 \end{cases}$$

We will verify our method by means of this example and omit the task of justifying that the technique works in general, since this would take us too far afield and out of the scope of this text. Let's list the tableaus produced by the Simplex Method.

$$
\begin{array}{c}
 \\ u \\ \\ v \\ \\
\end{array}
\left[
\begin{array}{cccc|c}
x & y & u & v & \\
1 & 1 & 1 & 0 & 4 \\
1 & 3 & 0 & 1 & 6 \\
\hline
-2 & -3 & 0 & 0 & 0 \\
\end{array}
\right]
\rightarrow
\left[
\begin{array}{cccc|c}
x & y & u & v & \\
2/3 & 0 & 1 & -1/3 & 2 \\
1/3 & 1 & 0 & 1/3 & 2 \\
\hline
-1 & 0 & 0 & 1 & 6 \\
\end{array}
\right]
\rightarrow
$$

$$
\begin{array}{c}
 \\ x \\ \\ y \\ \\
\end{array}
\left[
\begin{array}{cccc|c}
x & y & u & v & \\
1 & 0 & 3/2 & -1/2 & 3 \\
0 & 1 & -1/2 & 1/2 & 1 \\
\hline
0 & 0 & 3/2 & 1/2 & 9 \\
\end{array}
\right].
$$

*We will illustrate how one can get from one tableau to the next without the need of elementary row operations. For the sake of illustration, we will go from the **second** tableau to the **final** tableau via this new method.*

*Let T denote the body of the **initial** tableau above the horizontal bar, i.e.,*

$$
T = \left[
\begin{array}{cccc|c}
1 & 1 & 1 & 0 & 4 \\
1 & 3 & 0 & 1 & 6 \\
\end{array}
\right].
$$

*Next, as usual, determine what the pivot variables will be in the **next** (which in our case is the **final**) tableau. Form a square matrix D consisting of the columns of the **initial** tableau corresponding to these pivot variables (in order) and then compute also D^{-1}. In our case, the next set of pivot variables (in order) will be x and y,*

$$
D = \left[
\begin{array}{cc}
1 & 1 \\
1 & 3 \\
\end{array}
\right], \quad \text{and one can compute } D^{-1} = \left[
\begin{array}{cc}
3/2 & -1/2 \\
-1/2 & 1/2 \\
\end{array}
\right].
$$

One can show that the body of the next tableau will be $D^{-1}T$, which turns out to be

$$
\left[
\begin{array}{cc}
3/2 & -1/2 \\
-1/2 & 1/2 \\
\end{array}
\right]
\left[
\begin{array}{cccc|c}
1 & 1 & 1 & 0 & 4 \\
1 & 3 & 0 & 1 & 6 \\
\end{array}
\right]
=
\left[
\begin{array}{cccc|c}
1 & 0 & 3/2 & -1/2 & 3 \\
0 & 1 & -1/2 & 1/2 & 1 \\
\end{array}
\right].
$$

Annotate the new tableau with the variables, as usual, but in addition place the coefficients of the objective function $z = 2x + 3y$ next to the corresponding variables:

$$\begin{bmatrix} & & 2 & 3 & 0 & 0 & \\ & & x & y & u & v & \\ 2 & x & 1 & 0 & 3/2 & -1/2 & 3 \\ 3 & y & 0 & 1 & -1/2 & 1/2 & 1 \end{bmatrix}.$$

Our final task is to recover the objective row of the new tableau. The column of numbers to the left of the pivot variables x and y will be denoted by C_D, i.e.

$$C_D = \begin{bmatrix} 2 \\ 3 \end{bmatrix}.$$

One can show that the values in the objective row can be obtained by taking the dot product of C_D with each of the columns of the new tableau and then subtracting the corresponding coefficient of z listed at the top of the column. In other words, the first value in the objective row will be

$$\begin{bmatrix} 2 \\ 3 \end{bmatrix} \cdot \begin{bmatrix} 1 \\ 0 \end{bmatrix} - 2 = 2 - 2 = 0.$$

The second value in the objective row will be

$$\begin{bmatrix} 2 \\ 3 \end{bmatrix} \cdot \begin{bmatrix} 0 \\ 1 \end{bmatrix} - 3 = 3 - 3 = 0.$$

The third value in the objective row will be

$$\begin{bmatrix} 2 \\ 3 \end{bmatrix} \cdot \begin{bmatrix} 3/2 \\ -1/2 \end{bmatrix} - 0 = 3/2.$$

The fourth value in the objective row will be

$$\begin{bmatrix} 2 \\ 3 \end{bmatrix} \cdot \begin{bmatrix} -1/2 \\ 1/2 \end{bmatrix} - 0 = 1/2.$$

Finally, the value of the objective function is simply the dot product

$$\begin{bmatrix} 2 \\ 3 \end{bmatrix} \cdot \begin{bmatrix} 3 \\ 1 \end{bmatrix} = 9.$$

Hence, the next tableau is

$$
\begin{array}{cc}
 & \begin{array}{cccc} 2 & 3 & 0 & 0 \\ x & y & u & v \end{array} \\
\begin{array}{cc} 2 & x \\ 3 & y \end{array} &
\left[
\begin{array}{cccc|c}
1 & 0 & 3/2 & -1/2 & 3 \\
0 & 1 & -1/2 & 1/2 & 1 \\
\hline
0 & 0 & 3/2 & 1/2 & 9
\end{array}
\right].
\end{array}
$$

As the reader can see, we were able to compute the next tableau without the use of any elementary row operations.

Example 2.32 *Let's redo the toy problem example from Section 2.4 using this new method. Recall that the linear programming problem was*

$$
\text{Maximize } z = 2x + 3y, \qquad \text{Subject to } \begin{cases} x + y & \leq & 12 \\ x + 2y & \leq & 20 \\ 2x + y & \leq & 20 \\ x, y & \geq & 0 \end{cases}
$$

In canonical form the problem is

$$
\text{Maximize } z = 2x + 3y, \qquad \text{Subject to } \begin{cases} x + y + u & = & 12 \\ x + 2y + v & = & 20 \\ 2x + y + w & = & 20 \\ x, y, u, v, w & \geq & 0 \end{cases}
$$

The initial tableau is

$$
\begin{array}{cc}
 & \begin{array}{ccccc} x & y & u & v & w \end{array} \\
\begin{array}{c} u \\ v \\ w \end{array} &
\left[
\begin{array}{ccccc|c}
1 & 1 & 1 & 0 & 0 & 12 \\
1 & 2 & 0 & 1 & 0 & 20 \\
2 & 1 & 0 & 0 & 1 & 20 \\
\hline
-2 & -3 & 0 & 0 & 0 & 0
\end{array}
\right].
\end{array}
$$

We do the same analysis as before to find the entering variable is y and departing variable is v. Hence, the next set of pivot variables will be u, y, w (in order). Then

$$
D = \begin{bmatrix} 1 & 1 & 0 \\ 0 & 2 & 0 \\ 0 & 1 & 1 \end{bmatrix}, \text{ and one can compute } D^{-1} = \begin{bmatrix} 1 & -1/2 & 0 \\ 0 & 1/2 & 0 \\ 0 & -1/2 & 1 \end{bmatrix}
$$

Hence, the body of the next tableau will be

$$
\begin{bmatrix} 1 & -1/2 & 0 \\ 0 & 1/2 & 0 \\ 0 & -1/2 & 1 \end{bmatrix}
\begin{bmatrix} 1 & 1 & 1 & 0 & 0 & 12 \\ 1 & 2 & 0 & 1 & 0 & 20 \\ 2 & 1 & 0 & 0 & 1 & 20 \end{bmatrix}
=
\begin{bmatrix} 1/2 & 0 & 1 & -1/2 & 0 & 2 \\ 1/2 & 1 & 0 & 1/2 & 0 & 10 \\ 3/2 & 0 & 0 & -1/2 & 1 & 10 \end{bmatrix}.
$$

Now add all the annotations to get

$$
\begin{bmatrix}
 & & 2 & 3 & 0 & & 0 & 0 & \\
 & & x & y & u & & v & w & \\
0 & u & 1/2 & 0 & 1 & -1/2 & 0 & 2 \\
3 & y & 1/2 & 1 & 0 & 1/2 & 0 & 10 \\
0 & w & 3/2 & 0 & 0 & -1/2 & 1 & 10
\end{bmatrix}.
$$

As in the previous example, we dot the column of values next to the pivot variables with each column in the tableau and subtract the values above the variable columns to get

$$
\begin{bmatrix}
 & & 2 & 3 & 0 & & 0 & 0 & \\
 & & x & y & u & & v & w & \\
0 & u & 1/2 & 0 & 1 & -1/2 & 0 & 2 \\
3 & y & 1/2 & 1 & 0 & 1/2 & 0 & 10 \\
0 & w & 3/2 & 0 & 0 & -1/2 & 1 & 10 \\
 & & -1/2 & 0 & 0 & 3/2 & 0 & 30
\end{bmatrix}.
$$

We are now at the next tableau, so we begin the process again. Determine entering and departing variables as usual which in this case are x and u, respectively. Hence, the next set of pivot variables will be x, y, w (in order). Then

$$
D = \begin{bmatrix} 1 & 1 & 0 \\ 1 & 2 & 0 \\ 2 & 1 & 1 \end{bmatrix}, \text{ and one can compute } D^{-1} = \begin{bmatrix} 2 & -1 & 0 \\ -1 & 1 & 0 \\ -3 & 1 & 1 \end{bmatrix}
$$

Hence, the body of the next tableau will be

$$
\begin{bmatrix} 2 & -1 & 0 \\ -1 & 1 & 0 \\ -3 & 1 & 1 \end{bmatrix}
\left[\begin{array}{ccccc|c} 1 & 1 & 1 & 0 & 0 & 12 \\ 1 & 2 & 0 & 1 & 0 & 20 \\ 2 & 1 & 0 & 0 & 1 & 20 \end{array}\right]
=
\left[\begin{array}{ccccc|c} 1 & 0 & 2 & -1 & 0 & 4 \\ 0 & 1 & -1 & 1 & 0 & 8 \\ 0 & 0 & -3 & 1 & 1 & 4 \end{array}\right].
$$

Now add all the annotations to get

		2	3	0	0	0	
		x	y	u	v	w	
2	x	1	0	2	-1	0	4
3	y	0	1	-1	1	0	8
0	w	0	0	-3	1	1	4

Now, dot the column of values next to the pivot variables with each column in the tableau and subtract the values above the variable columns to get

		2	3	0	0	0	
		x	y	u	v	w	
2	x	1	0	2	-1	0	4
3	y	0	1	-1	1	0	8
0	w	0	0	-3	1	1	4
		0	0	1	1	0	32

As before in section 2.4, the optimal criterion is now satisfied with the optimal solution being $z = 32$ with $x = 4$ and $y = 8$.

EXERCISES

1. Redo Exercise 2 in Section 2.4 using the new method presented in this section.

2. Redo Exercise 2a in Section 2.4 using the new method presented in this section.

3. Redo Exercise 2b in Section 2.4 using the new method presented in this section.

4. Redo Exercise 3 in Section 2.4 using the new method presented in this section.

5. Redo Exercise 3b in Section 2.4 using the new method presented in this section.

Project for 2.7:

The following is an example of a well known linear programming problem called the **Maximal Flow Problem**:

Consider the following diagram called a **network**:

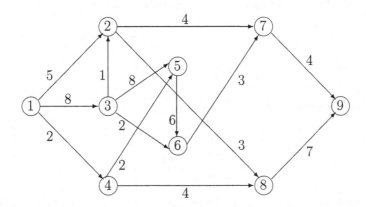

View the diagram as representing the flow of material through a grid of points; these circles are called **nodes** and the connections are called **arcs** which we will represent by $\overrightarrow{(i,j)}$. The material traversing the grid might be water, oil, electricity, etc. The node labeled by 1 is called the **source** since all material flows from it, and the node labeled by 9 is called the **sink** since all material flows towards it. Each node connection has a direction and a capacity, where the capacity represents the maximal amount of material which that line can carry.

Let x_{ij} ($i \neq j$ and $i, j = 1, \ldots, 9$) be the unknown representing the amount of material we decide to let flow across the corresponding arc $\overrightarrow{(i,j)}$ (which we will make to be zero if there should be no connection between i and j). Let c_{ij} ($i \neq j$ and $i, j = 1, \ldots, 9$) represent the capacity of the arc $\overrightarrow{(i,j)}$ (which again we will make to be zero if there should be no connection between i and j). We wish to maximize the flow of material across the grid from the source to the sink subject to certain constraints. One contraint is implicit in the problem, namely that the amount of flow across an arc cannot exceed the capacity of the arc. The second constraint will be that the amount of flow of material into a node must equal the amount flowing out of that node; in other words material is not allowed to collect at any node.

1. What is the flow function which we wish to maximize?

2. Write out the constraints for this linear programming problem and solve using a computer software package.

2.8 HOMOGENEOUS/NON-HOMOGENEOUS SYSTEMS AND RANK

In this section, we distinguish two kinds on linear systems of equations. There is a natural division of linear systems of equations into two types: homogeneous and non-homogeneous.

Definition 2.15 *A* **homogeneous** *linear system of equations has the form $AX = 0$, where 0 is a column of zeros. A* **non-homogeneous** *linear system of equations has the form $AX = B$, where $B \neq 0$, i.e. B is a column with nonzero entries.*

Example 2.33 *The linear system below is homogeneous.*

$$\begin{cases} 2x_1 + x_2 - x_3 & = & 0 \\ x_1 - 3x_2 + x_3 & = & 0 \\ -3x_1 + x_2 + x_3 & = & 0 \end{cases}$$

The linear system below is non-homogeneous.

$$\begin{cases} 2x_1 + x_2 - x_3 & = & -1 \\ x_1 - 3x_2 + x_3 & = & 0 \\ -3x_1 + x_2 + x_3 & = & 6 \end{cases}$$

We remark that any homogeneous linear system has at least one solution, namely $X = 0$ called the **trivial** solution. Some results can be proved at this point.

Theorem 2.14 *For a square matrix A the following are equivalent:*

  i. A is invertible.

  ii. A is equivalent to I.

  iii. The linear system $AX = B$ has a unique solution for any $B \in \mathbb{R}^n$.

  iv. A is a product of elementary matrices.

  v. The linear system $AX = 0$ has only the trivial solution.

Proof 2.16 *Part iii certainly implies part v. To complete the proof, we show that part v implies part ii. If $AX = 0$ has only the trivial solution, then it must be the case that the augmented matrix $[A|0]$ reduces to $[I|0]$, and so A is equivalent to I.* □

Theorem 2.15 *Given a linear system $AX = B$ which has at least one solution, the solution set has the form*

$$\{X_p + X_h \; : \; \text{for any } X_h \text{ a solution to the homogeneous system } AX = 0\}.$$

where X_p is one of the solutions to $AX = B$ (called a **particular** *solution).*

Proof 2.17 *First note that for any X_h a solution to $AX = 0$, $X_p + X_h$ is indeed a solution to $AX = B$, since*

$$A(X_p + X_h) = AX_p + AX_h = B + 0 = B.$$

Second, if X_0 is a solution to $AX = B$, then $X_0 - X_p$ is a solution to $AX = 0$, since

$$A(X_0 - X_p) = AX_0 - AX_p = B - B = 0.$$

Hence, $X_0 = (X_0 - X_p) + X_p$ where $X_h = X_0 - X_p$ is a solution to $AX = 0$. □

Example 2.34 *In a previous example, we found the solution set to*

$$\begin{cases} 2x - 8y + z = 5 \\ x - 4y + 2z = 0 \end{cases} \quad is \quad \{[4y + 10/3,\ y,\ -5/3] : y \in F\}.$$

Now notice that $[4y + 10/3, y, -5/3] = [10/3, 0, -5/3] + [4y, y, 0] = X_p + X_h$ as in the theorem above.

A consequence of the theorem above is the following:

Theorem 2.16 *For $A \in M_{mn}$, the following are equivalent:*

   i. $AX = 0$ has only the trivial solution.

   ii. $AX = B$ has at most one solution, for any $B \in \mathbb{R}^m$.

Proof 2.18 *First, assuming that $AX = 0$ has only the trivial solution, we show that $AX = B$ has at most one solution, for any $B \in \mathbb{R}^m$. Suppose that $AX = B$ has a solution, say X_p. Then, by Theorem 2.15, the solution set of $AX = B$ has the form $\{X_p + 0\} = \{X_p\}$, where X_p is a solution to $AX = B$. Hence, $AX = B$ has exactly one solution.*

Now we assume $AX = B$ has at most one solution, for any $B \in \mathbb{R}^m$ and show that $AX = 0$ has only the trivial solution. Since $AX = B$ has at most one solution, for any $B \in \mathbb{R}^m$, in particular, $AX = 0$ has at most one solution. Now, since $X = 0$ is always a solution to $AX = 0$, we conclude that $AX = 0$ has only the trivial solution. □

Definition 2.16 *Let A be any matrix and R its corresponding reduced row-echelon form. The **rank** A, written $rk(A)$, is the number of non-zero rows in R.*

Example 2.35 *One can show that if*

$$A = \begin{bmatrix} 1 & 1 & -1 & 0 \\ 2 & -4 & 3 & 1 \\ 3 & 15 & -13 & -2 \end{bmatrix}, \ then \ R = \begin{bmatrix} 1 & 0 & -1/6 & 1/6 \\ 0 & 1 & -5/6 & -1/6 \\ 0 & 0 & 0 & 0 \end{bmatrix}.$$

*Hence $rk(A) = 2$. We remark that to compute the rank of a matrix one does not need to totally reduce the matrix A to R. It is sufficient to produce non-zero pivots and 0's below these pivots (this form of a matrix is sometimes called **row-echelon form**). For instance, in this example, we could have stopped at the following matrix in order to compute the rank:*

$$\begin{bmatrix} 1 & 1 & -1 & 0 \\ 0 & -6 & 5 & 1 \\ 0 & 0 & 0 & 0 \end{bmatrix}.$$

The next result extends Theorem 2.14 by yet another statement.

Theorem 2.17 *For an $n \times n$ square matrix A the following are equivalent:*

i. A is invertible.

ii. A is equivalent to I.

iii. The linear system $AX = B$ has a unique solution for any $B \in \mathbb{R}^n$.

iv. A is a product of elementary matrices.

v. The linear system $AX = 0$ has only the trivial solution.

vi. $rk(A) = n$.

Proof 2.19 *We show that ii to equivalent to vi. If A is equivalent to I, then the reduced row-echelon form of A is I, and so by definition, $rk(A) = n$. If $rk(A) = n$ then, since A is square, the reduced row-echelon form of A must be I.* □

The next two results show that rank of a matrix can be used to describe the solution set of a homogeneous linear system.

Theorem 2.18 *Let $A \in M_{mn}$ be any matrix.*

i. If $rk(A) = n$ then $AX = 0$ has only the trivial solution.

ii. If $rk(A) < n$ then $AX = 0$ has infinitely many solutions.

Proof 2.20 *If $rk(A) = n$, then all the variables in $AX = 0$ will be pivot variables. Hence, there will be no independent variables, and so $AX = 0$ cannot have infinitely many solutions. The only other possibility for $AX = 0$ is that it have only the trivial solution. By the same token, if $rk(A) < n$, then there will be non-pivot variables, and so independent variables, and so infinitely many solutions for $AX = 0$.* □

Corollary 2.4 *If $A \in M_{mn}$ with $m < n$, then $AX = 0$ has infinitely many solutions.*

Proof 2.21 *Since $rk(A) \leq m < n$, by Theorem 2.18.ii, the result follows.* □

Example 2.36 *Using Corollary 2.4, without any computation we can assert that the following homogeneous system has infinite solutions (since $m = 2 < 3 = n$):*

$$\begin{cases} 2x + y - 2z &= 0 \\ x - y + 6z &= 0 \end{cases}.$$

EXERCISES

1. Referring to Section 2.2 Exercise 2, express the solution set of each system of equations which resulted in infinitely many solutions in the form $X_p + X_h$ as in Theorem 2.15.

2. Consider the following matrix:

$$A = \begin{bmatrix} 2 & 1 & 2 & 3 \\ -2 & 0 & -1 & 1 \\ 4 & 1 & 1 & 1 \\ 0 & 5 & 3 & 19 \end{bmatrix}.$$

   a. Compute $rk(A)$.

   b. Use part a. to decide whether or not A has an inverse. (explain)

   c. Use part a. to describe the solution set of $AX = 0$. (explain)

3. Consider the following matrix:

$$A = \begin{bmatrix} 1 & 2 & 3 \\ 3 & 2 & 1 \\ 2 & 1 & 3 \\ 3 & 1 & 2 \end{bmatrix}.$$

   a. Compute the $rk(A)$.

   b. Use part a. describe the solution set to $AX = 0$.

4. Consider the following matrix:

$$A = \begin{bmatrix} 2 & 0 & 1 & -1 \\ 0 & -3 & -2 & 1 \\ -1 & 0 & -2 & 1 \\ 1 & -3 & 0 & 0 \end{bmatrix}.$$

   a. Compute $rk(A)$.

   b. Use part a. to describe the solution set of the following system of equations:

$$\begin{cases} 2x_1 + x_3 - x_4 & = & 0 \\ -3x_2 - 2x_3 + x_4 & = & 0 \\ -x_1 - 2x_3 + x_4 & = & 0 \\ x_1 - 3x_2 & = & 0 \end{cases}$$

5. Consider the following matrix:

$$B = \begin{bmatrix} 1 & -2 & 3 \\ 3 & -3 & 0 \\ 2 & -1 & -3 \\ -2 & 4 & -6 \end{bmatrix}.$$

a. Compute the $rk(B)$.

b. Use part a. describe the solution set to $BX = 0$.

6. Consider the following matrices:

$$B = \begin{bmatrix} 3 & -6 & 0 & 3 \\ -2 & 4 & 2 & -2 \\ 4 & -8 & 6 & 7 \end{bmatrix} \qquad C = \begin{bmatrix} 0 \\ 3 \\ 9 \end{bmatrix}.$$

a. Without actually computing it, what can we say about $rk(B)$?

b. Using part a, can we conclude anything about the solution set of $BX = C$?

c. Use Gaussian Elimination to solve the system in part b.

7. Prove that if $a \in \mathbb{R}$ and X_0, X_1 are solutions to the homogeneous linear system $AX = 0$, then $aX_0 + X_1$ is also a solution to $AX = 0$.

8. Our goal in this exercise is to prove that for $A \in M_{mn}$ and $B \in M_{nr}$, we have $rk(AB) \leq rk(A)$.

a. Show that if A is equivalent to C, then $rk(A) = rk(C)$.

b. Show that for an elementary matrix $E \in M_{mn}$, we have $rk(EB) = rk(B)$ (use part a.).

c. If A has k zero rows, what can be said about $rk(A)$?

d. Show that if $R \in M_{mn}$ is a matrix in reduced row-echelon form that $rk(RB) \leq rk(R)$ (use part c.).

e. Use parts a. through d. to achieve the goal of this exercise.

2.9 DETERMINANT

The determinant is a function whose inputs are square matrices and whose outputs are scalars. We will eventually give a general definition of the determinant, but first we introduce it in some special cases.

Let us first establish our notation. For a square matrix A, the determinant of A will be denoted by $|A|$ or $det(A)$ (keep in mind that $|A|$ is a scalar).

- For a 1×1 matrix $A = [a]$, $|A| = a$.

  Example 2.37 *If* $A = [-6]$, *then* $|A| = -6$.

- For a 2×2 matrix $A = \begin{bmatrix} a & b \\ c & d \end{bmatrix}$, $|A| = \begin{vmatrix} a & b \\ c & d \end{vmatrix} = ad - bc$.

  Example 2.38 $\begin{vmatrix} 1 & -3 \\ 2 & 4 \end{vmatrix} = (1)(4) - (-3)(2) = 10.$

- For a 3×3 matrix $A = \begin{bmatrix} a & b & c \\ d & e & f \\ g & h & i \end{bmatrix}$,

$$|A| = \begin{vmatrix} a & b & c \\ d & e & f \\ g & h & i \end{vmatrix} = (a) \begin{vmatrix} e & f \\ h & i \end{vmatrix} - (b) \begin{vmatrix} d & f \\ g & i \end{vmatrix} + (c) \begin{vmatrix} d & e \\ g & h \end{vmatrix}.$$

Notice that the 2×2 determinants in the definition of a 3×3 determinant are easy to remember. For instance, the 2×2 determinant next to a is obtained by crossing out in the 3×3 determinant the row and column in which a appears. For b cross out the row and column in which b appears, and the same for c.

Example 2.39

$$\begin{vmatrix} 1 & 2 & -1 \\ 2 & 0 & 1 \\ 1 & 1 & -1 \end{vmatrix} = (1) \begin{vmatrix} 0 & 1 \\ 1 & -1 \end{vmatrix} - (2) \begin{vmatrix} 2 & 1 \\ 1 & -1 \end{vmatrix} + (-1) \begin{vmatrix} 2 & 0 \\ 1 & 1 \end{vmatrix}$$

$$= (0 - 1) - 2(-2 - 1) - (2 - 0) = -1 + 6 - 2 = 3.$$

Now, we give a general definition of the determinant of an $n \times n$ matrix which agrees with the definition we gave above:

Definition 2.17 *Let $A = [a_{ij}]$ be a square $n \times n$ matrix. Set A_{ij} by the matrix obtained by removing the ith row and jth column from A. The* **determinant** *of A, written $\det(A)$ or*

$$|A| = \sum_{j=1}^{n} (-1)^{1+j} a_{1j} |A_{1j}|.$$

We call this way of computing the determinant *expanding on row 1.*

The next result shows that in computing the determinant we can expand on any row or column. The proof of this result, for the sake of continuity, will be delegated to the last section of this chapter and basically follows from the fact that the determinant has certain properties which are unique to it.

Theorem 2.19 *Let A be any square matrix.*

   i. (Expanding on a row) Fix any row i in A. Then

$$|A| = \sum_{j=1}^{n} (-1)^{i+j} a_{ij} |A_{ij}|.$$

   ii. (Expanding on a column) Fix any column j in A. Then

$$|A| = \sum_{i=1}^{n} (-1)^{i+j} a_{ij} |A_{1j}|.$$

Example 2.40 *We compute the determinant by expanding on column 2 (notice that we get the same answer as we did before when we computed it):*

$$\begin{vmatrix} 1 & 2 & -1 \\ 2 & 0 & 1 \\ 1 & 1 & -1 \end{vmatrix}$$

$$= (-1)^{1+2}(2) \begin{vmatrix} 2 & 1 \\ 1 & -1 \end{vmatrix} + (-1)^{2+2}(0) \begin{vmatrix} 1 & -1 \\ 1 & -1 \end{vmatrix} + (-1)^{3+2}(1) \begin{vmatrix} 1 & -1 \\ 2 & 1 \end{vmatrix}$$

$$= -2(-2-1) + 0 - 1(1+2) = 6 - 3 = 3.$$

We remark that one need not be so careful about how the signs in front of each 2×2 determinant is computed. Imagine a checkerboard of $+$'s and $-$'s where we begin with a "$+$" in the upper left-hand corner of the matrix:

$$\begin{vmatrix} + & - & + \\ - & + & - \\ + & - & + \end{vmatrix}$$

These $+$'s and $-$'s correspond to the $(-1)^{i+j}$ in the definition of $|A|$ (This observation extends to square matrices of any dimension).

Notice that this more general definition of determinant gives us more freedom on how we can compute $|A|$. No matter what row or column we expand on to compute $|A|$ we always get the same result. The example below illustrates how advantageous this can be.

Example 2.41 *For the determinant below, we certainly want to expand on row three, since it contains the most zeros and so most of the 3×3 determinants will have a 0 coefficient in front of them. Then the resulting 3×3 determinant is easiest to compute by expanding on row 3:*

$$\begin{vmatrix} 1 & 2 & 3 & 4 \\ 1 & 1 & 2 & -1 \\ 0 & 1 & 0 & 0 \\ 0 & -2 & 0 & 4 \end{vmatrix} = +(0) - (1) \begin{vmatrix} 1 & 3 & 4 \\ 1 & 2 & -1 \\ 0 & 0 & 4 \end{vmatrix} + (0) - (0)$$

$$= - \left(+(0) - (0) + (4) \begin{vmatrix} 1 & 3 \\ 1 & 2 \end{vmatrix} \right) = -4(2 - 3) = 4.$$

We present some basic properties about determinant in the theorem below:

Theorem 2.20 *Let A be a square matrix. Then*

i. $|A^T| = |A|$.

ii. If B is obtained from A by $A \xrightarrow{aR_i} B$ $(a \neq 0)$, then $|B| = a|A|$. Consequently, one can factor a scalar out of any row (or by i, column) of a determinant.

iii. If B is obtained from A by $A \xrightarrow{R_i \leftrightarrow R_j} B$, then $|B| = -|A|$. Consequently, switching any two rows (or by i, columns) introduces a minus sign in front of the determinant.

iv. If B is obtained from A by $A \xrightarrow{aR_i + R_j} B$, then $|B| = |A|$.

v. If A is upper (or by i, lower)-triangular, then $|A| = a_{11} a_{22} \cdots a_{nn}$, i.e. the determinant of A is just the product of its diagonal entries.

Proof 2.22 *We prove i by induction on the dimension n of A. If $n = 1$, then $A = A^T$ so the result follows immediately. Now assume the statement is true for $n = k$ and prove the statement is true $n = k + 1$. Therefore, consider a $(k + 1) \times (k + 1)$ matrix,*

A. Set $B = A^T$ with $B = [b_{ij}]$ and $b_{ij} = a_{ji}$. Observe that $B_{ij} = A_{ji}^T$. Computing the determinant of A^T by expanding on column j and using induction yields

$$|A^T| = |B| = \sum_{i=1}^{n}(-1)^{i+j}b_{ij}|B_{ij}| = \sum_{i=1}^{n}(-1)^{i+j}a_{ji}|A_{ji}^T| =$$

$$\sum_{i=1}^{n}(-1)^{j+i}a_{ji}|A_{ji}| = |A|.$$

Notice that the last equality follows from expansion on row j.

For the proof of ii (in the case of a row), expand the determinant on row i to get

$$|B| = \sum_{j=1}^{n}(-1)^{i+j}(aa_{ij})|A_{ij}| = a\sum_{j=1}^{n}(-1)^{i+j}a_{ij}|A_{ij}| = a|A|.$$

The proofs of iii and iv are similar to ii. We prove v by induction on the dimension n of A. If $n = 1$, then $A = [a]$ and $|A| = a$ which is the product of its diagonal entries. Now assume the statement true for $n = k$ and prove the statement for a $(k+1)\times(k+1)$ matrix A. Computing the determinant by expanding on column 1 yields only one $k \times k$ determinant with a non-zero constant in front of it, i.e. $|A| = a_{11}|A_{11}|$. Now A_{11} is a $k \times k$ upper-triangular matrix, so by induction, $|A_{11}| = a_{22}a_{33}\cdots a_{k+1,k+1}$. Putting the two computations above together gives us the result. □

This theorem gives us another way to compute determinant and is especially useful on determinants of larger dimension (i.e. ≥ 4). The algorithm proceeds as follows: Use elementary row operations to put the determinant in upper-triangular form. Then use Theorem 2.20.v to complete the computation. The example below illustrates the method.

Example 2.42

$$\begin{vmatrix} 2 & 3 & 4 & 5 \\ 1 & 2 & -1 & 3 \\ 0 & -1 & 14 & -13 \\ 2 & 2 & -2 & 0 \end{vmatrix} \underset{R_1 \leftrightarrow R_2}{=} - \begin{vmatrix} 1 & 2 & -1 & 3 \\ 2 & 3 & 4 & 5 \\ 0 & -1 & 14 & -13 \\ 2 & 2 & -2 & 0 \end{vmatrix}$$

$$\underset{-2R_1+R_4}{\overset{-2R_1+R_2}{=}} - \begin{vmatrix} 1 & 2 & -1 & 3 \\ 0 & -1 & 6 & -1 \\ 0 & -1 & 14 & -13 \\ 0 & -2 & 0 & -6 \end{vmatrix} \underset{-2R_2+R_4}{\overset{-R_2+R_3}{=}} - \begin{vmatrix} 1 & 2 & -1 & 3 \\ 0 & -1 & 6 & -1 \\ 0 & 0 & 8 & -12 \\ 0 & 0 & -12 & -4 \end{vmatrix}$$

$$= -(2)\begin{vmatrix} 1 & 2 & -1 & 3 \\ 0 & -1 & 6 & -1 \\ 0 & 0 & 4 & -6 \\ 0 & 0 & -12 & -4 \end{vmatrix} \underset{3R_3+R_4}{=} (-2)\begin{vmatrix} 1 & 2 & -1 & 3 \\ 0 & -1 & 6 & -1 \\ 0 & 0 & 4 & -6 \\ 0 & 0 & 0 & -22 \end{vmatrix}$$

$$= (-2)(1)(-1)(4)(-22) = -176.$$

The next few results allow us to add one more statement to Theorem 2.17.

Lemma 2.6 *Let A and B be two square matrices of the same dimensions.*

   i. If A is equivalent to B, then $|A| = a|B|$ for some scalar $a \neq 0$.

   ii. If A has a row (or column) of 0's, then $|A| = 0$.

   iii. $|I| = 1$.

   iv. If A has two identical rows or columns, then $|A| = 0$.

Proof 2.23 *To prove i, the best proof is a proof by induction on m, where*

$$A \xrightarrow{op_1} A_1 \xrightarrow{op_2} \cdots \xrightarrow{op_{m-1}} A_m \xrightarrow{op_m} B.$$

To prove the result for $m = 1$ we simply appeal to Theorem 2.20.ii–iv. Now assume the statement is true for $m = k$ and prove for $m = k + 1$. So we have

$$A \xrightarrow{op_1} A_1 \xrightarrow{op_2} \cdots \xrightarrow{op_k} A_k \xrightarrow{op_{k+1}} B.$$

By induction, $|A| = b|A_k|$ for $b \neq 0$. By Theorem 2.20.ii–iv again, $|A_k| = c|B|$ for $c \neq 0$. Hence, $|A| = (bc)|B|$ where $bc \neq 0$ as desired.

To prove ii one merely expands the determinant on the row (or column) of 0's in A to get the desired result.

To prove iii simply appeal to Theorem 2.20.v.

The proof of iv is left as an exercise. □

Theorem 2.21 *For an $n \times n$ square matrix A the following are equivalent:*

   i. A is invertible.

   ii. A is equivalent to I.

   iii. The linear system $AX = B$ has a unique solution for any $B \in \mathbb{R}^n$.

   iv. A is a product of elementary matrices.

   v. The linear system $AX = 0$ has only the trivial solution.

   vi. $rk(A) = n$.

   vii. $|A| \neq 0$.

Proof 2.24 *We show that ii is equivalent to vii. If A is equivalent to I, then by Lemma 2.6.i and iii, $|A| = a|I| = a \neq 0$. By contrapositive, if A is not equivalent to I, then the reduced row-echelon form R of A must have a row of 0's. By Lemma 2.6.i and iv, $|A| = a|R| = a \cdot 0 = 0$.* □

Finally, to complete this section we prove an important property of determinant, namely $|AB| = |A||B|$ called the **multiplicative** property. The next few results will achieve this goal.

Lemma 2.7 *If B and an elementary matrix E are square matrices of the same dimensions, then $|EB| = |E||B|$.*

Proof 2.25 *In order to prove this result one needs to consider the three types of elementary matrices corresponding to the three types of elementary row operations. We will consider one type and leave the rest as exercises. Let E be the elementary matrix obtained by $I \xrightarrow{aR_i} E$. By Theorem 2.20.ii, $|E| = a|I| = a$. Recall that $B \xrightarrow{aR_i} EB$, by Theorem 2.8. Then by Theorem 2.20.ii, $|EB| = a|B| = |E||B|$.* □

A simple induction proof yields the following consequence of the result above:

Corollary 2.5 *If B and elementary matrices $E_1, E_2, \ldots, E_n$ are square matrices of the same dimensions, then*

$$|E_1 E_2 \cdots E_n B| = |E_1||E_2| \cdots |E_n||B|.$$

Now we can prove the multiplicative property of determinant.

Theorem 2.22 *If A and B are square matrices of the same dimension, then $|AB| = |A||B|$.*

Proof 2.26 *We prove this in two cases. First suppose that A is invertible. By Theorem 2.21, $A = E_1 E_2 \cdots E_n$ where $E_1, E_2, \ldots E_n$ are elementary matrices. Now by Corollary 2.5,*

$$|AB| = |E_1 E_2 \cdots E_n B| = |E_1||E_2| \cdots |E_n||B| = |E_1 E_2 \cdots E_n||B| = |A||B|.$$

Now suppose A is not invertible. By Theorem 2.21, $|A| = 0$. Now AB is not invertible as well, for if it were then $(AB)^{-1}$ would exist. But then $A[B(AB)^{-1}] = (AB)(AB)^{-1} = I$ and A would have an inverse, namely $B(AB)^{-1}$. But we are assuming A is not invertible. Hence AB cannot be invertible and by Theorem 2.21 again, $|AB| = 0$. Hence,

$$|AB| = 0 = 0 \cdot |B| = |A||B|.$$

□

EXERCISES

1. Consider the following matrix:

$$A = \begin{bmatrix} 3 & 3 & 0 & 1 \\ 2 & 7 & -1 & 1 \\ 1 & 2 & 1 & -1 \\ -2 & 1 & 1 & 0 \end{bmatrix}.$$

   a. Calculate $|A|$ using elementary row operations.

   b. Use part a. to describe the solution set of $AX = 0$.

2. Consider the following matrix:

$$B = \begin{bmatrix} 4 & -2 & 2 \\ 2 & 5 & 1 \\ -2 & 1 & -1 \end{bmatrix}.$$

   a. Find the rank of B.

   b. Use part a. to determine $|B|$.

3. Consider the following matrix:

$$C = \begin{bmatrix} 1 & -2 & 1 \\ 2 & -4 & 0 \\ 1 & 0 & 2 \end{bmatrix}.$$

   a. Calculate $|C|$ by expanding on a row or column.

   b. Explain why part a. guarantees the existence of C^{-1}.

4. Use elementary row operations to compute the determinant of the following matrix:

$$\begin{bmatrix} 2 & 3 & 1 & 6 \\ 4 & 2 & 3 & -1 \\ -6 & -1 & 0 & 2 \\ -2 & 1 & 8 & 1 \end{bmatrix}.$$

5. Consider the following matrix:

$$A = \begin{bmatrix} 2 & 1 & 1 & 1 \\ -3 & 2 & 1 & 0 \\ 2 & 2 & -1 & -1 \\ 3 & 7 & 0 & -1 \end{bmatrix}.$$

a. Calculate $rk(A)$.

b. Without calculation, what is the value of $|A|$?

c. Without calculation, describe the solution set to the following linear system of equations:

$$\begin{aligned}
2x_1 + x_2 + x_3 + x_4 &= 1 \\
-3x_1 + 2x_2 + x_3 &= 6 \\
2x_1 + 2x_2 - x_3 - x_4 &= 0 \\
3x_1 + 7x_2 - x_4 &= -9
\end{aligned}$$

d. Without calculation, is A invertible?

6. Consider the following matrix:

$$B = \begin{bmatrix} 3 & 2 & 0 & 0 \\ -2 & 3 & 2 & 1 \\ 1 & -1 & 0 & 2 \\ 2 & 1 & 0 & 1 \end{bmatrix}.$$

a. Compute $|B|$ by row-reducing to an upper-triangular matrix.

b. Compute $|B|$ by expanding on rows or columns.

c. Using part a. or b., what can be said about the solution set of the following linear system of equations?

$$\begin{aligned}
3x_1 + 2x_2 &= 0 \\
-2x_1 + 3x_2 + 2x_3 + x_4 &= 0 \\
x_1 - x_2 + 2x_4 &= 0 \\
2x_1 + x_2 + x_4 &= 0
\end{aligned}$$

7. Consider the following matrix:

$$C = \begin{bmatrix} -3 & 0 & 2 & 1 \\ 1 & 0 & 0 & 3 \\ 2 & 1 & -1 & 1 \\ -2 & 0 & 4 & 2 \end{bmatrix}.$$

a. Compute $|C|$ by row-reducing to an upper-triangular matrix.

b. Compute $|C|$ by expanding on a row or a column.

8. Consider the following matrix:

$$D = \begin{bmatrix} 2 & -1 & 2 & 0 \\ 3 & 1 & 1 & -2 \\ 0 & 2 & 0 & 0 \\ 4 & -1 & 1 & 0 \end{bmatrix}$$

a. Compute $|D|$ by expanding on a row or a column.

b. Compute $|D|$ by row reducing to an upper-triangular matrix.

c. Use part a. or b. to decide whether or not D is invertible (explain).

9. Consider the following matrix:

$$A = \begin{bmatrix} 1 & 0 & -2 & 3 \\ 2 & 0 & -1 & 0 \\ -3 & 2 & 0 & 4 \\ -1 & 0 & -6 & -2 \end{bmatrix}.$$

a. Compute $|A|$ by expanding on a row or column.

b. Compute $|A|$ by using elementary row operations.

c. Is A equivalent to the identity matrix? (explain)

10. Consider the following matrices:

$$A = \begin{bmatrix} 2 & -4 & -4 \\ -1 & 2 & 2 \\ 3 & -6 & -6 \end{bmatrix} \qquad B = \begin{bmatrix} -2 \\ 1 \\ -3 \end{bmatrix}.$$

a. Calculate $rk(A)$.

b. Using part a., what can be said about $|A|$?

c. Using part a., decide whether or not A is invertible.

d. Express the solution set of $AX = B$ as $X_p + X_h$ as in Theorem 2.15.

11. Repeat the previous exercise with the following matrices:

$$A = \begin{bmatrix} 2 & 0 & 6 \\ 1 & 1 & -1 \\ 0 & -1 & -1 \end{bmatrix} \qquad B = \begin{bmatrix} 0 \\ -2 \\ 1 \end{bmatrix}.$$

12. Consider the following matrix:

$$A = \begin{bmatrix} 2 & 1 & 1 & 1 \\ 0 & 5 & 10 & -5 \\ 1 & 2 & 1 & 2 \\ -1 & -2 & 9 & -10 \end{bmatrix}.$$

a. Compute $|A|$ by using elementary rows operations to put A in upper-triangular form.

b. Using part a., what can be said about the solution set to the following linear system (explain):

$$2x_1 + x_2 + x_3 + x_4 = 0$$
$$5x_2 + 10x_3 - 5x_5 = 0$$
$$x_1 + 2x_2 + x_3 + 2x_4 = 0$$
$$-x_1 - 2x_2 + 9x_3 - 10x_4 = 0$$

c. Using part a., what can be said about the rank of A (explain).

13. Consider the following matrix:

$$A = \begin{bmatrix} 1 & 0 & -2 & 3 \\ 2 & 0 & -1 & 0 \\ -3 & 2 & 0 & 4 \\ -1 & 0 & -6 & 0 \end{bmatrix}.$$

a. Compute $|A|$ by expanding on a row or column (choose wisely).

b. Using part a., what can be said about the rank of A (explain).

c. Compute $|A|$ by using elementary rows operations to put A in upper-triangular form.

14. Prove that $|A^{-1}| = |A|^{-1}$.

15. Let $A \in M_{nn}$ with the property that $A^T A = I$. Prove that $|A| = \pm 1$.

16. Show that $|AA^T| \geq 0$ for any square matrix A.

17. Let $A, B \in M_{nn}$ and suppose there is an invertible $C \in M_{nn}$ such that $B = C^{-1}AC$. Prove that $|A| = |B|$.

18. Let $A, B \in M_{nn}$. Prove that if $rk(A) < n$ or $rk(B) < n$, then $rk(AB) < n$.

19. Prove parts iii and iv of Theorem 2.20.

20. Prove part iv of Lemma 2.6.

21. Complete the proof of Lemma 2.7.

2.10 APPLICATIONS OF THE DETERMINANT

With the introduction of the determinant in the previous section we are now in a position to give alternative algorithms for solving systems of linear equations (assuming that there is a unique solution) and computing the inverse of a matrix (assuming that it exists). We will also introduce the cross product for elements of $\mathbb{R}^n$, since it can be defined and remembered using a mnemonic relating to the determinant. The following result, called Cramer's Rule, achieves the first goal of this section. It is a method for finding the solution to a linear system having exactly one solution and is defined completely in terms of determinants.

Theorem 2.23 *Let $AX = B$ be a linear system in unknowns $x_1, x_2, \ldots, x_n$ with A invertible. Set $A = [c_1 \ c_2 \ \ldots \ c_i \ \ldots \ c_n]$, a column representation of A. Let $A_i = [c_1 \ c_2 \ \ldots \ B \ \ldots \ c_n]$, where the ith column of A has been replaced by B. Then the unique solution $[a_1 \ a_2 \ \ldots \ a_n]$ to $AX = B$ has the form $a_i = |A_i|/|A|$ for $i = 1, 2, \ldots, n$.*

Proof 2.27 *By Theorem 2.21, we know that $AX = B$ has a unique solution, say $X_0 = [a_1, \ldots, a_n]$ (so $AX_0 = B$). Define*

$$C_i = [e_1, \ldots, e_{i-1}, X_0, e_{i+1}, \ldots, e_n] = \begin{bmatrix} 1 & 0 & & a_1 & & 0 \\ 0 & 1 & & a_2 & & 0 \\ 0 & 0 & \cdots & a_3 & \cdots & 0 \\ \vdots & \vdots & & \vdots & & \vdots \\ 0 & 0 & & a_n & & 1 \end{bmatrix}.$$

Observe that $|C_i| = a_i$ (expand on row i). In the calculation which follows we appeal to Lemma 1.1.

Notice that

$$AC_i = [Ae_1, \ldots, Ae_{i-1}, AX_0, Ae_{i+1}, \ldots, Ae_n]$$

$$= [c_1, \ldots, c_{i-1}, B, c_{i+1}, \ldots, c_n] = A_i.$$

Therefore, $|A_i| = |AC_i| = |A||C_i| = |A|a_i$ and solving for a_i yields

$$a_i = \frac{|A_i|}{|A|}.$$

□

Example 2.43 *Consider the following system of equations:*

$$\begin{cases} 2x - y = -5 \\ x + y = 2 \end{cases} \quad or \quad \begin{bmatrix} 2 & -1 \\ 1 & 1 \end{bmatrix} \begin{bmatrix} x \\ y \end{bmatrix} = \begin{bmatrix} -5 \\ 2 \end{bmatrix}.$$

The coefficient matrix for this linear system is $A = \begin{bmatrix} 2 & -1 \\ 1 & 1 \end{bmatrix}$ and $|A| = 3$. Then $A_1 = \begin{bmatrix} -5 & -1 \\ 2 & 1 \end{bmatrix}$ and $|A_1| = -3$. Also $A_2 = \begin{bmatrix} 2 & -5 \\ 1 & 2 \end{bmatrix}$ and $|A_2| = 9$. Hence,

$$x = \frac{|A_1|}{|A|} = \frac{-3}{3} = -1 \quad and \quad y = \frac{|A_2|}{|A|} = \frac{9}{3} = 3.$$

Now we give a method for finding A^{-1} whose formula is defined completely in terms of determinants.

Definition 2.18 *Let $A = [a_{ij}]$ and define $a'_{ij} = (-1)^{i+j}|A_{ij}|$. Set $A' = [a'_{ij}]$. The* **adjoint** *of A, written $\operatorname{adj}(A) = (A')^T$.*

Example 2.44 *Consider the matrix $A = \begin{bmatrix} 1 & 2 & 3 \\ 3 & 2 & 1 \\ 1 & 3 & 2 \end{bmatrix}$. Then*

$$a'_{11} = (-1)^{1+1} \begin{vmatrix} 2 & 1 \\ 3 & 2 \end{vmatrix} = 1, \; a'_{12} = (-1)^{1+2} \begin{vmatrix} 3 & 1 \\ 1 & 2 \end{vmatrix} = -5, \; etc.$$

These computations above yield

$$A' = \begin{bmatrix} 1 & -5 & 7 \\ 5 & -1 & -1 \\ -4 & 8 & -4 \end{bmatrix} \; and \; so \; \operatorname{adj}(A) = \begin{bmatrix} 1 & 5 & -4 \\ -5 & -1 & 8 \\ 7 & -1 & -4 \end{bmatrix}.$$

Now, we prove the result which justifies the method we seek for finding the inverse of a matrix.

Theorem 2.24 *For any square matrix A, $\operatorname{adj}(A) \cdot A = |A|I = A \cdot \operatorname{adj}(A)$.*

Proof 2.28 *Set $\operatorname{adj}(A) = [c_{ij}]$ and fix an i and an j with $1 \leq i, j \leq n$. Let B be the matrix obtained by replacing the jth row of A by its ith row. Let's name the entries of $B = [b_{ij}]$. Notice that when $i \neq j$, B has two identical rows and $A = B$ when $i = j$. Hence,*

$$|B| = \begin{cases} |A|, & if \; i = j \\ 0, & if \; i \neq j \end{cases}.$$

Let's compute $|B|$ in another way, by expanding on row j:

$$|B| = \sum_{k=1}^{n}(-1)^{j+k}b_{jk}|B_{jk}| = \sum_{j=1}^{n}(-1)^{j+k}a_{ik}|A_{jk}| = \sum_{j=1}^{n}a_{ik}a'_{jk}.$$

Equating our two computations of $|B|$ yields

$$\sum_{j=1}^{n}a_{ik}a'_{jk} = \begin{cases} |A|, & if \; i = j \\ 0, & if \; i \neq j \end{cases}.$$

Using this last statement gives

$$A(\mathrm{adj}(A)) = [a_{ij}][c_{ij}] = \left[\sum_{k=1}^{n} a_{ik}c_{kj}\right] = \left[\sum_{k=1}^{n} a_{ik}a'_{jk}\right] =$$

$$\begin{bmatrix} |A| & 0 & \cdots & 0 \\ 0 & |A| & \cdots & 0 \\ \vdots & \vdots & \ddots & \vdots \\ 0 & 0 & \cdots & |A| \end{bmatrix} = |A|I.$$

□

Corollary 2.6 *If A is invertible, then $A^{-1} = (1/|A|)\mathrm{adj}(A)$.*

Proof 2.29 *This follows quite quickly from Theorem 2.24. Since A is assumed invertible, by Theorem 2.21, $|A| \neq 0$. Therefore, we can multiply the equation in Theorem 2.24 by $1/|A|$ to obtain*

$$\frac{1}{|A|}\mathrm{adj}(A) \cdot A = I = A \cdot \frac{1}{|A|}\mathrm{adj}(A).$$

Therefore, by definition, $\frac{1}{|A|}\mathrm{adj}(A)$ is the inverse of A. □

Example 2.45 *Refer to the previous example. One can compute that $|A| = 12$ and so by Corollary 2.6,*

$$A^{-1} = \begin{bmatrix} 1/12 & 5/12 & -4/12 \\ -5/12 & -1/12 & 8/12 \\ 7/12 & -1/12 & -4/12 \end{bmatrix}.$$

One final comment before leaving the notion of the adjoint of a matrix. Corollary 2.6 yields a nice and easily remembered formula for computing the inverse of a 2×2 matrix: If $A = \begin{bmatrix} a & b \\ c & d \end{bmatrix}$ then

$$A^{-1} = \begin{bmatrix} \frac{d}{|A|} & -\frac{b}{|A|} \\ -\frac{c}{|A|} & \frac{a}{|A|} \end{bmatrix}.$$

In other words, reverse the entries on the diagonal, negate the entries off the diagonal and divide all entries by $|A|$.

Example 2.46 *For $A = \begin{bmatrix} 1 & 2 \\ 3 & 4 \end{bmatrix}$, $|A| = -2$ and*

$$A^{-1} = \begin{bmatrix} \frac{4}{-2} & \frac{-2}{-2} \\ \frac{-3}{-2} & \frac{1}{-2} \end{bmatrix} = \begin{bmatrix} -2 & 1 \\ 3/2 & -1/2 \end{bmatrix}.$$

The last topic of this section is a new operation for elements of $\mathbb{R}^n$ (only for $n = 2$ or 3) called the *cross* (or *outer*, or *wedge*, or *vector*) product.

Definition 2.19 *Let $u = [d, e, f]$ and $v = [g, h, i]$ be two vectors in $\mathbb{R}^3$. The **cross product** of u and v, written $u \times v$ equals*

$$\begin{vmatrix} e & f \\ h & i \end{vmatrix} \hat{\imath} - \begin{vmatrix} d & f \\ g & i \end{vmatrix} \hat{\jmath} + \begin{vmatrix} d & e \\ g & h \end{vmatrix} \hat{k}.$$

A simple mnemonic for remembering this formula is

$$u \times v = \begin{vmatrix} \hat{\imath} & \hat{\jmath} & \hat{k} \\ d & e & f \\ g & h & i \end{vmatrix}.$$

Example 2.47 *Let $u = [1, 2, 3]$ and $v = [-1, 0, 2]$. Then*

$$u \times v = \begin{vmatrix} \hat{\imath} & \hat{\jmath} & \hat{k} \\ 1 & 2 & 3 \\ -1 & 0 & 2 \end{vmatrix} = \begin{vmatrix} 2 & 3 \\ 0 & 2 \end{vmatrix} \hat{\imath} - \begin{vmatrix} 1 & 3 \\ -1 & 2 \end{vmatrix} \hat{\jmath} + \begin{vmatrix} 1 & 2 \\ -1 & 0 \end{vmatrix} \hat{k} =$$

$$4\hat{\imath} - 5\hat{\jmath} + 2\hat{k} = [4, -5, 2].$$

The cross product can also be performed on vector in $\mathbb{R}^2$ by viewing these vectors as lying in the xy-plane of $\mathbb{R}^3$, i.e. for $u = [a, b]$ and $[c, d]$, $u \times v = [a, b, 0] \times [c, d, 0]$.

Below are some basic properties of cross product, the proofs of which we leave as exercises.

Theorem 2.25 *Let $u, v, w \in \mathbb{R}^3$ and $a \in \mathbb{R}$. Then*

i. $v \times u = -(u \times v)$.

ii. $a(u \times v) = (au) \times v = u \times (av)$.

iii. $u \times (v + w) = (u \times v) + (u \times w)$.

iv. $(u + v) \times w = (u \times w) + (v \times w)$.

v. $u \cdot (v \times w) = (u \times v) \cdot w$.

vi. $u \times (v \times w) = (u \cdot w)v - (u \cdot v)w$.

vii. $|u \times v| = |u||v| \sin \theta$ where θ is as it was defined in relation to the dot product.

There is a geometric interpretation of the cross product. One can show that $u \times v$ is a vector perpendicular to the plane containing the vectors u and v (see Figure 2.1).

Using Theorem 2.25.vii, one can prove two additional geometric interpretations of the cross product.

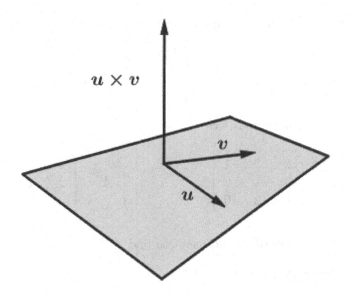

Figure 2.1 Geometric interpretation of cross product.

Corollary 2.7 *If* $u, v \in \mathbb{R}^3$, *then*

i. $|u \times v|$ *represents the area of the parallelogram with sides* u *and* v.

ii. u *and* v *are parallel iff* $u \times v = 0$.

EXERCISES

1. Consider the following linear system:

$$\begin{cases} 3x - y & = & -5 \\ 2x + 3y & = & 4 \end{cases}.$$

a. Solve the system using Cramer's Rule.

b. Solve the system using the inverse of the coefficient matrix.

2. Use Cramer's Rule to find the equation of the parabola $y = x^2 + ax + b$ containing the points $(1, 2)$ and $(-1, 4)$.

3. Consider the following matrix:

$$A = \begin{bmatrix} -7 & 5 & 3 \\ 3 & -2 & -2 \\ 3 & -2 & -1 \end{bmatrix}.$$

a. Find the A^{-1} using the Gaussian Elimination method.

b. Find the A^{-1} using the adjoint matrix given that the cofactor matrix is

$$A' = \begin{bmatrix} -2 & -3 & 0 \\ -1 & & 1 \\ -4 & & -1 \end{bmatrix}.$$

4. Consider the following matrices:

$$A = \begin{bmatrix} 1 & 2 & -1 \\ 3 & -1 & 2 \\ 1 & -2 & 0 \end{bmatrix} \qquad A' = \begin{bmatrix} 4 & 2 & -5 \\ 2 & 1 & \\ -5 & -7 \end{bmatrix}$$

Complete the entries of A' and use it to find A^{-1}.

5. Consider the following matrix:

$$A = \begin{bmatrix} 2 & 1 & 4 \\ 3 & 2 & 5 \\ 0 & -1 & 1 \end{bmatrix}.$$

a. Calculate $|A|$ by expanding on a row or column.

b. Use part a. to describe the solution set of the following linear system:

$$\begin{cases} 2x + y + 4z & = & 3 \\ 3x + 2y + 5z & = & 4 \\ -y + z & = & 1 \end{cases}.$$

c. Use Cramer's Rule to solve the system in part b.

d. Use part a. to explain why A^{-1} exists.

e. Find A^{-1} using the adjoint matrix.

f. Use part e. to solve the system in part b.

6. Consider the following matrices:

$$A = \begin{bmatrix} 2 & 1 \\ -3 & -2 \end{bmatrix} \qquad B = \begin{bmatrix} 1 \\ 0 \end{bmatrix}.$$

a. Calculate $|A|$ and explain why $AX = B$ has a unique solution.

b. Use Cramer's Rule to find the solution to $AX = B$.

c. Explain why A^{-1} exists and find A^{-1} using the 2×2 formula given after Corollary 2.6.

d. Use part c. to find the solution to $AX = B$.

7. Apply Cramer's Rule to find the equation of a parabola of the form $y = x^2 + ax + b$ passing through the points $(2, 11/4)$ and $(-1, 7/4)$.

8. Repeat the previous exercise for the points $(-3, 47/6)$ and $(1, 7/6)$.

9. Consider the equation of the circle $x^2 + y^2 + ay + b = 0$ where a and b are real numbers. We wish to find the equation of such a circle passing through the points $(-2, 3)$ and $(4, -1)$.

    a. Set up the system of equations for finding the circle.

    b. Solve the system in part a. using the inverse of the coefficient matrix.

    c. Solve the system in part a. using Cramer's Rule.

10. Consider the following matrices:

$$A = \begin{bmatrix} -4 & 7 & 6 \\ 3 & -5 & -4 \\ -2 & 4 & 3 \end{bmatrix} \qquad B = \begin{bmatrix} -2 \\ 0 \\ 1 \end{bmatrix}.$$

    a. Compute $rk(A)$.

    b. Use part a. to decide whether or not A is invertible.

    c. What is the 1st row, 3rd column entry of adj(A)?

    d. Express A as a product of elementary matrices.

    e. Compute $|A|$ by expanding on column 2.

    f. Compute $|A|$ by row-reducing to an upper-triangular matrix.

    g. From any previous part, what can be said about the solution set of $AX = B$?

    h. Solve $AX = B$ using the inverse of A.

11. Repeat the previous exercise for the following matrices:

$$A = \begin{bmatrix} -7 & 5 & 3 \\ 3 & -2 & -2 \\ 3 & -2 & -1 \end{bmatrix} \qquad B = \begin{bmatrix} -2 \\ 0 \\ 1 \end{bmatrix}.$$

12. Consider the following matrix:

$$A = \begin{bmatrix} -1 & 2 & 2 \\ -2 & 4 & 2 \\ 4 & -9 & -6 \end{bmatrix}.$$

    a. Calculate $|A|$ by expanding on a row or column.

b. Use part a. to decide whether or not A is invertible.

c. Complete the adjoint matrix below:

$$\text{adj}(A) = \begin{bmatrix} -6 & -6 & * \\ -4 & -2 & -2 \\ 2 & * & 0 \end{bmatrix}.$$

d. Use parts a. and c. to find A^{-1}.

e. Now find A^{-1} using Gaussian Elimination.

f. Use part d. to solve the following linear system:

$$\begin{cases} -x + 2y + 2z &= 1 \\ -2x + 4y + 2z &= -2 \\ 4x - 9y - 6z &= 1 \end{cases}.$$

g. Now use Cramer's Rule to solve the system in part f.

13. Consider the following matrix:

$$A = \begin{bmatrix} -3 & 6 & 4 \\ -2 & 4 & 2 \\ 4 & -7 & -8 \end{bmatrix}.$$

a. Find $rk(A)$.

b. Use part a. to decide whether or not A is invertible.

c. Express A as a product of elementary matrices.

d. What is the 3rd row, 2nd column entry of $\text{adj}(A)$.

e. Compute $|A|$ by expanding on column 2.

f. Compute $|A|$ using elementary row operations.

14. Consider the following system of linear equations:

$$\begin{cases} 2x + y - z &= 3 \\ x - 2y + z &= -6 \\ -x + 2y - 2z &= 11 \end{cases}.$$

a. Compute the determinant of the coefficient matrix.

b. From part a, what can be said about the solution set of the system? (explain)

c. Use Cramer's Rule to solve the system.

d. Use Gaussian Elimination to find A^{-1}.

e. Given that you know A' below, use the adjoint matrix to find A^{-1}.

$$A' = \begin{bmatrix} 2 & 1 & 0 \\ 0 & -5 & \\ -3 & -5 & \end{bmatrix}.$$

f. Use part d. or e. to solve the system.

15. Consider the following system of linear equations:

$$\begin{cases} 2x + y - z = 3 \\ x - 2y + z = -6 \\ -x + 2y - 2z = 11 \end{cases}$$

a. Use Cramer's Rule to find the value of z in the solution to the system.

b. Use A' partially computed below to find A^{-1}.

$$A' = \begin{bmatrix} 2 & 1 & 0 \\ 0 & -5 & \\ -3 & -5 & \end{bmatrix}.$$

16. Consider the following system of equations:

$$\begin{cases} 5x - 3y - 10z = -9 \\ 2x + 2y - 3z = 4 \\ -3x - y + 5z = -1 \end{cases}$$

a. Solve the system by the Gaussian Elimination Method.

b. Solve the system by using Cramer's Rule.

c. Solve the system by using the inverse of the coefficient matrix.

17. Compute $u \times v$ for each of the following pair of vectors:

$$u = [1, -2, 3], \; v = [3, 0, -1] \in \mathbb{R}^3 \qquad u = [2, -3], v = [-3, 1] \in \mathbb{R}^2$$

18. Let $A \in M_{nn}$. Prove the following facts related to adjoint:

a. $A \, \mathrm{adj}(A) = 0_{nn}$ iff A is not invertible.

b. $\mathrm{adj}(A^T) = \mathrm{adj}(A)^T$.

c. For $a \in \mathbb{R}$, we have $\mathrm{adj}(aA) = a^{n-1}\mathrm{adj}(A)$.

d. $|\mathrm{adj}(A)| = |A|^{n-1}$

e. If $|A| = 1$, then $\mathrm{adj}(\mathrm{adj}(A)) = A$.

f. If $|A| \neq 0$, then $rk(\mathrm{adj}(A)) = n$.

    g. If $|A| = 0$, then $rk(A) + rk(\text{adj}(A)) \leq n$.

19. Prove Theorem 2.25.

20. Prove Corollary 2.7.

2.11 APPLICATION: LU FACTORIZATION

Matrix factorizations or decompositions play an important role in numerical methods of computational linear algebra. They help in speeding up algorithms used in linear algebra such as solving linear systems, inverting a matrix or computing its determinant. One such decomposition is the LU factorization. In this section, we will describe this factorization as well as when and to what it applies. The LU factorization expresses a square matrix as a product of a unit lower triangular matrix times an upper triangular matrix, where a unit triangular matrix has ones on the diagonal.

Example 2.48 *Below we express a matrix $A = LU$, where L is unit lower triangular and U is upper triangular. Shortly, we will see the algorithm for performing this factorization.*

$$\begin{bmatrix} 1 & -3 & 1 \\ 2 & -8 & -1 \\ -3 & 1 & 1 \end{bmatrix} = \begin{bmatrix} 1 & 0 & 0 \\ 2 & 1 & 0 \\ -3 & 4 & 1 \end{bmatrix} \begin{bmatrix} 1 & -3 & 1 \\ 0 & -2 & -3 \\ 0 & 0 & 16 \end{bmatrix}$$

    We will show that such a factorization is possible when one can row reduce a square matrix to an upper triangular matrix using only Type 3 elementary row operations, $aR_i + R_j$. To prove this we need a lemma the proof of which is left as an exercise. We also point out that this is not always possible. For example, for the matrix $\begin{bmatrix} 0 & 1 \\ 1 & 0 \end{bmatrix}$ it is not possible (exercise).

Lemma 2.8 *The following statements about unit triangular matrices hold.*

1. *A finite product of unit lower (upper) triangular matrices is unit lower (upper) triangular.*

2. *The inverse of a unit lower (upper) triangular matrix is unit lower (upper) triangular.*

Note first that the elementary matrix corresponding to an elementary row operation of the form $aR_i + R_j$ with $i < j$ is a unit lower triangular matrix (exercise). We can now prove the main result in this section.

Theorem 2.26 *Given a square matrix A, suppose we can put A in upper triangular form U using only Type 3 elementary row operations, $aR_i + R_j$. Then there exists a unit lower triangular matrix L such that $A = LU$*

Proof 2.30 *By assumption A row reduces to an upper triangular matrix U using only Type 3 elementary row operations, say $op_1, op_2, \cdots, op_k$. Let $E_1, E_2, \cdots, E_k$ be the corresponding elementary matrices. By Theorem 2.9, $U = E_k \cdots E_2 E_1 A$ and so $A = (E_k \cdots E_2 E_1)^{-1} U$. Since each elementary matrix is unit lower triangular and by Lemma 2.8, $L = (E_k \cdots E_2 E_1)^{-1}$ is unit lower triangular.* □

Although Theorem 2.26 is a constructive proof and gives an algorithm for performing the LU factorization, one more observation is needed to create a more efficient algorithm for LU factorization.

Recall that the inverse operation for $aR_i + R_j$ is $-aR_i + R_j$ and the corresponding elementary matrices will have the value a for its jith entry and $-a$ in its jith entry. Furthermore, multiplying two such elementary matrices of the form $aR_i + R_j$ and $bR_{i'} + R_{j'}$ with $\{i, j\} \neq \{i', j\}$ as sets results in a matrix with a in the jith entry and b in the $j'i'$th entry. Therefore, it is easy to determine L by simply negating. An example will make this clear.

Example 2.49 *Let's revisit Example 2.48 with*

$$A = \begin{bmatrix} 1 & -3 & 1 \\ 2 & -8 & -1 \\ -3 & 1 & 1 \end{bmatrix} \xrightarrow{-2R_1 + R_2} \begin{bmatrix} 1 & -3 & 1 \\ 0 & -2 & -3 \\ -3 & 1 & 1 \end{bmatrix}$$

$$\xrightarrow{3R_1 + R_3} \begin{bmatrix} 1 & -3 & 1 \\ 0 & -2 & -3 \\ 0 & -8 & 4 \end{bmatrix} \xrightarrow{-4R_2 + R_3} \begin{bmatrix} 1 & -3 & 1 \\ 0 & -2 & -3 \\ 0 & 0 & 16 \end{bmatrix} = U.$$

Notice, we achieved the upper triangular matrix U performing only elementary row operations of Type 3. Furthermore,

$$L = \begin{bmatrix} 1 & 0 & 0 \\ 2 & 1 & 0 \\ -3 & 4 & 1 \end{bmatrix}.$$

Having introduced the LU factorization we now show how it can be useful in several numerical methods. First, we use it to solve square linear systems with a unique solution. Here is the general algorithm. Given a linear system $AX = B$ with LU factorization $A = LU$,

1. Solve $LX = B$ by forward substitution. Call that solution B_0.

2. Solve $UX = B_0$ by back substitution. Call that solution X_0.

3. Then X_0 is the solution to $AX = B$, since

$$AX_0 = (LU)X_0 = L(UX_0) = LB_0 = B.$$

Example 2.50 *Consider the linear system*

$$\begin{cases} x - 3y + z = 1 \\ 2x - 8y - z = -2 \\ -3x + y + z = 0 \end{cases}$$

The coefficient matrix for this system is the matrix A from Example 2.48. We have already found the LU factorization with

$$L = \begin{bmatrix} 1 & 0 & 0 \\ 2 & 1 & 0 \\ -3 & 4 & 1 \end{bmatrix} \quad and \quad U = \begin{bmatrix} 1 & -3 & 1 \\ 0 & -2 & -3 \\ 0 & 0 & 16 \end{bmatrix}.$$

Following the steps in the algorithm we first solve by forward substitution

$$\begin{cases} x = 1 \\ 2x + y = -2 \\ -3x + 4y + z = 0. \end{cases}$$

Since $x = 1$, we have $y = -2 - 2x = -4$ and so $z = 3x - 4y = 19$. Therefore, the solution is $x = 1$, $y = -4$ and $z = 19$. Now solve by back substitution

$$\begin{cases} x - 3y + z = 1 \\ -2y - 3z = -4 \\ 16z = 19 \end{cases}$$

Since $z = 19/16$, we have $-2y = -4 + 3z = 7/16$ and so $y = 7/32$. Then $x = 1 + 3y - z = 15/32$. Therefore, the solution is $x = 15/32$, $y = 7/32$ and $z = 19/16$ which is also the solution to the original system.

We can also compute the determinant easily with the LU factorization, for if $A = LU$, then $|A| = |L||U| = |U|$ and $|U|$ is obtained by multiplying together its diagonal entries.

Example 2.51 *Consider again the matrix A from Example 2.48 with LU factorization*

$$L = \begin{bmatrix} 1 & 0 & 0 \\ 2 & 1 & 0 \\ -3 & 4 & 1 \end{bmatrix} \quad and \quad U = \begin{bmatrix} 1 & -3 & 1 \\ 0 & -2 & -3 \\ 0 & 0 & 16 \end{bmatrix}.$$

Then $|A| = |U| = (1)(-2)(16) = -32$.

Finally, we can use the LU factorization to find the inverse of a matrix. Indeed, if $A = LU$, then $A^{-1} = U^{-1}L^{-1}$ and we can immediately obtain L^{-1} simply by negating its entries below the diagonal. Therefore, we are reduced to finding the inverse of U

which is an easier problem since it is upper triangular, thus making the Gaussian elimination quicker.

Example 2.52 *Consider again the matrix A from Example 2.48 with LU factorization*

$$L = \begin{bmatrix} 1 & 0 & 0 \\ 2 & 1 & 0 \\ -3 & 4 & 1 \end{bmatrix} \quad and \quad U = \begin{bmatrix} 1 & -3 & 1 \\ 0 & -2 & -3 \\ 0 & 0 & 16 \end{bmatrix}.$$

Then

$$L^{-1} = \begin{bmatrix} 1 & 0 & 0 \\ -2 & 1 & 0 \\ 3 & -4 & 1 \end{bmatrix}.$$

To find U^{-1}, we now reduce

$$\begin{bmatrix} 1 & -3 & 1 & | & 1 & 0 & 0 \\ 0 & -2 & -3 & | & 0 & 1 & 0 \\ 0 & 0 & 16 & | & 0 & 0 & 1 \end{bmatrix}, \text{ which reduces to } \begin{bmatrix} 1 & 0 & 0 & | & 1 & -3/2 & -11/32 \\ 0 & 1 & 0 & | & 0 & -1/2 & -3/32 \\ 0 & 0 & 1 & | & 0 & 0 & 1/16 \end{bmatrix}.$$

Therefore,

$$A^{-1} = \begin{bmatrix} 1 & -3/2 & -11/32 \\ 0 & -1/2 & -3/32 \\ 0 & 0 & 1/16 \end{bmatrix} \begin{bmatrix} 1 & 0 & 0 \\ -2 & 1 & 0 \\ 3 & -4 & 1 \end{bmatrix} = \begin{bmatrix} 7/32 & -1/8 & -11/32 \\ -1/32 & -1/8 & -3/32 \\ 11/16 & -1/4 & 1/16 \end{bmatrix}.$$

EXERCISES

1. If possible, compute the LU factorization for each matrix.

a. $A = \begin{bmatrix} 1 & 2 \\ 3 & 4 \end{bmatrix}$

b. $B = \begin{bmatrix} 1 & -1 & 1 \\ 4 & 2 & 5 \\ 3 & 1 & 4 \end{bmatrix}$

c. $C = \begin{bmatrix} 2 & -4 & 0 \\ 1 & -2 & 1 \\ 1 & 0 & 2 \end{bmatrix}$

d. $D = \begin{bmatrix} 1 & 2 & 1 & -1 \\ 3 & 3 & 0 & 1 \\ 2 & 7 & -1 & 1 \\ -2 & 1 & 1 & 0 \end{bmatrix}$

2. Solve each system of equations by using the LU factorization found in Exercise 1.

a. $\begin{cases} x + 2y &= 1 \\ 3x + 4y &= -2 \end{cases}$

b. $\begin{cases} x - y + z &= 1 \\ 4x + 2y + 5z &= -1 \\ 3x + y + 4z &= 0 \end{cases}$

c. $\begin{cases} x_1 + 2x_2 + x_3 - x_4 &= 2 \\ 3x_1 + 3x_2 + x_4 &= -1 \\ 2x_1 + 7x_2 - x_3 + x_4 &= 0 \\ -2x_1 + x_2 + x_3 &= 1 \end{cases}$

3. If possible, compute the determinant of each matrix using the LU factorization found in Exercise 1.

4. If possible, find the inverse of each matrix using the LU factorization found in Exercise 1.

5. Prove by contradiction that it is not possible to perform the LU factorization on the matrix $\begin{bmatrix} 0 & 1 \\ 1 & 0 \end{bmatrix}$.

6. Prove that the LU factorization is not uniquely determined. Hint: Introduce a diagonal matrix and its inverse.

7. Prove that an elementary matrix corresponding to the elementary operation $aR_i + R_j$ with $i < j$ is a unit upper triangular matrix.

8. Prove that a finite product of unit lower (upper) triangular matrices is unit lower (upper) triangular (Lemma 2.8.(a)).

9. Prove that the inverse of a unit lower (upper) triangular matrix is unit lower (upper) triangular (Lemma 2.8.(b)).

CHAPTER 3

Vector Spaces

I N THIS CHAPTER, the general notion of a vector space is presented. The reader will be presented with the four classic vector spaces: Tuples and Matrices, which the reader has already seen, and Polynomials and Functions, which will be introduced in the first section of this chapter. There will be an increase in abstractness in the text at this point, but this is to be expected and before long overcome. In Section 3.1, a vector space is defined with the four classic examples presented among others. In Section 3.2, a subspace is defined with many examples in the context of the four classic vector spaces. The reader is shown methods for proving or disproving that a particular set of vectors is a subspace. In Section 3.3, linear independence is introduced and concrete methods for verifying linear independence/dependence are given for the four classic vector spaces. In Section 3.4, a very important subspace is defined called the span of a set of vectors. Concrete methods are given for computing the span of a set of vectors. In Section 3.5, the notions of basis and dimension are introduced. In Section 3.6, row space, column space and null space of a matrix are defined and some loose ends are tied up from earlier sections of this chapter. Finally, in Section 3.7, some counting arguments using dimension theorems are applied to a number of examples.

3.1 DEFINITION AND EXAMPLES

The first two chapters were more or less an introduction to the notion of a vector space by way of two examples: $\mathbb{R}^n$ and M_{mn}, i.e. Tuples and Matrices. The first two chapters were also meant to hone the computational skills needed in linear algebra. These skills will be employed quite liberally in what is to follow.

Now we are ready to introduce the general notion of a vector space. Technically speaking we are introducing real vector space, i.e. where the scalars are real numbers, but we could easily allow scalars to be the complex numbers or even an arbitrary field. We start, of course, with its definition which should look quite familiar at this point in the text.

Definition 3.1 *A* **vector space** *is a set V, made up of objects called* **vectors**, *together with two operations:*

1. **Scalar Multiplication**, *in which a scalar (real number) is multiplied by a vector. We will denote scalar multiplication of a scalar a and a vector v by av.*

DOI: 10.1201/9781003217794-3

2. **Vector Addition**, *in which two vectors are added together. We will denote vector addition of vectors u and v by* $u + v$.

V together with its two operations, must also satisfy the following axioms in order to be a vector space:

0. *For all* $u, v \in V$ *and scalar* a, *we have* $u + v$, $au \in V$.

1. *For all* $u, v \in V$, $u + v = v + u$.

2. *For all* $u, v, w \in V$, $(u + v) + w = u + (v + w)$.

3. *There exists* $0 \in V$, *such that for all* $u \in V$, $u + 0 = u$.

4. *For each* $u \in V$, *there is a* $v \in V$, *such that* $u + v = 0$.

5. *For all* $u, v \in V$ *and scalar* a, $a(u + v) = au + av$.

6. *For all* $u \in V$ *and scalars* a, b, $(a + b)u = au + bu$.

7. *For all* $u \in V$ *and scalars* a, b, $(ab)u = a(bu)$.

8. *For all* $u \in V$, $1u = u$.

The 0 in Property 3 is called the **zero** vector and the v in Property 4 is called the **additive inverse** of u and is denoted symbolically as $-u$ (the use of the definite article *the* in front of 0 and v will be justified at the end of this section).

We now give the four classical examples of a vector space which will appear over and over again in the remainder of the text. For this reason the reader should be sure to have a comfortable familiarity with these four examples. The first two examples should already be quite familiar. For each example, in order for the example to be complete, it is necessary to define what the set of vectors are as well as the two operations of scalar multiplication and vector addition.

Example 1 is Tuples: Define the vectors to be

$$\mathbb{R}^n = \{[a_1, \ldots, a_n] \ : \ a_1, \ldots, a_n \in \mathbb{R}\}.$$

Define vector addition by

$$[a_1, \ldots, a_n] + [b_1, \ldots, b_n] = [a_1 + b_1, \ldots, a_n + b_n].$$

For each scalar a, define scalar multiplication by

$$a[a_1, \ldots, a_n] = [aa_1, \ldots, aa_n].$$

Theorem 1.1 verified that $\mathbb{R}^n$ satisfies the axioms of a vector space. Hence, $\mathbb{R}^n$ together with these two operations forms a vector space.

Example 2 is Matrices: Define the vectors to be

$$M_{mn} = \{[a_{ij}] \ : \ a_{ij} \in \mathbb{R}\}.$$

Define vector addition by

$$[a_{ij}] + [b_{ij}] = [a_{ij} + b_{ij}].$$

For each scalar a, define scalar multiplication by

$$a[a_{ij}] = [aa_{ij}].$$

Theorem 1.4 verified that M_{mn} satisfied the axioms of a vector space. Hence, M_{mn} together with these two operations forms a vector space.

Example 3 is Polynomials: Define the vectors to be

$$P = \{a_0 + a_1 x + \cdots + a_n x^n \mid a_0, a_1, \ldots, a_n \in \mathbb{R} \text{ and } n = 0, 1, 2, \ldots\}.$$

We call these vectors **polynomials** in unknown x. The $a_0, a_1, \ldots, a_n$ are called the **coefficients** of the polynomial. The highest power of x which has a non-zero coefficient is called the **degree** of the polynomial. For future reference, the constant polynomial 0 will be said to have degree $-\infty$ as opposed to non-zero constant polynomials which have degree 0.

We define vector addition by

$$(a_0 + a_1 x + \cdots + a_n x^n) + (b_0 + b_1 x + \cdots + b_n x^n)$$

$$= (a_0 + b_0) + (a_1 + b_1)x + \cdots + (a_n + b_n)x^n.$$

We define scalar multiplication by

$$a(a_0 + a_1 x + \cdots + a_n x^n) = (aa_0) + (aa_1)x + \cdots + (aa_n)x^n.$$

At times we may wish to represent a polynomial by functional notation $p(x)$ and then the degree of $p(x)$ will be denoted by $deg(p)$. However, the reader must keep in mind that polynomials here are defined as *formal expressions* and is meant only a an abbreviation.

Example 3.1 *If* $p(x) = 1 - x + 3x^3 - 6x^7$, *then* $deg(p) = 7$.

One comment needs to be made. Two polynomials need not have the same degree in order to add them. For the polynomial with smaller degree simply introduce higher powers of x with zeros as their coefficients so that we can add according to the definition.

Theorem 3.1 *Polynomials, P, as defined above is a vector space.*

Proof 3.1 *We give a partial proof and leave the rest as an exercise. To prove property 1, we use the commutative property of real numbers:*

$$(a_0 + a_1 x + \cdots + a_n x^n) + (b_0 + b_1 x + \cdots + b_n x^n) =$$
$$(a_0 + b_0) + (a_1 + b_1)x + \cdots + (a_n + b_n)x^n =$$
$$(b_0 + a_0) + (b_1 + a_1)x + \cdots + (b_n + a_n)x^n =$$
$$(b_0 + b_1 x + \cdots + b_n x^n) + (a_0 + a_1 x + \cdots + a_n x^n).$$

For property 3, the zero vector we are looking for is the constant polynomial 0. To prove property 6,

$$(a + b)(a_0 + a_1 x + \cdots + a_n x^n) = [(a + b)a_0] + [(a + b)a_1]x + \cdots + [(a + b)a_n]x^n =$$
$$(aa_0 + ba_0) + (aa_1 + ba_1)x + \cdots + (aa_n + ba_n)x^n =$$
$$(aa_0) + (aa_1)x + \cdots + (aa_n)x^n + (ba_0) + (ba_1)x + \cdots + (ba_n)x^n =$$
$$a(a_0 + a_1 x + \cdots + a_n x^n) + b(a_0 + a_1 x + \cdots + a_n x^n).$$

$\square$

Example 4 is Functions: Define the vectors to be

$$\mathcal{F} = \{f : \mathbb{R} \longrightarrow \mathbb{R} \mid f \text{ is a real-valued function}\}.$$

Adding functions f and g yields the function $f + g$ defined in the natural way as $(f + g)(x) = f(x) + g(x)$.

Multiplying a scalar $a \in \mathbb{R}$ by a function f yields the function af defined in the natural way as $(af)(x) = af(x)$.

Notice that the two operations should yield as an output another function. Therefore, we were required to give a definition of these outputs as functions. Recall from calculus that the **domain** of a function f, which we shall denote by $D(f)$, is the collection of all definable inputs for f.

Example 3.2 *Let $f(x) = 4x$ and $g(x) = 2 - x^2$. Then $2f - g$ is defined by the formula $(2f - g)(x) = x^2 + 8x - 2$, since $(2f - g)(x) = 2f(x) - g(x) = 2(4x) - (2 - x^2) = x^2 + 8x - 2$.*

We point out that $f = g$ iff $D(f) = D(g)$ and $f(x) = g(x)$ for all $x \in D(f)$.

Example 3.3 *The functions $f(x) = \sin 2x$ and $g(x) = 2\sin x \cos x$ are equal functions, because of the double angle identity from trigonometry.*

Example 3.4 *The functions $f(x) = x+1$ and $g(x) = \frac{x^2-1}{x-1}$ are **not** equal functions, because although they are equal whenever both are defined, the domain of f is all real numbers while the domain of g, $D(g) = \{x \in \mathbb{R} : x \neq 1\}$ and so $D(f) \neq D(g)$.*

Theorem 3.2 *Functions, $\mathcal{F}$, as defined above is a vector space.*

Proof 3.2 *Again, we verify only some of the vector space properties and leave the rest as an exercise. Keep in mind that we consistently appeal to the given properties of real numbers to verify each of the properties.*
To prove Property 1, let $f, g \in \mathcal{F}$. Then for all $x \in D(f+g)$,

$$(f+g)(x) = f(x) + g(x) = g(x) + f(x) = (g+f)(x).$$

Since $D(f+g) = D(g+f)$ and $f+g$ and $g+f$ are equal on every input, this implies that $f+g = g+f$.
To prove Property 3, the zero vector we are looking for is the function $0(x) = 0$ for all $x \in \mathbb{R}$. Indeed,

$$(f+0)(x) = f(x) + 0(x) = f(x) + 0 = f(x).$$

Hence, $f+0 = f$. To prove Property 5, for all scalars a and $f, g \in \mathcal{F}$,

$$[a(f+g)](x) = a(f+g)(x) = a(f(x) + g(x)) = af(x) + ag(x) =$$

$$(af)(x) + (ag)(x) = (af + ag)(x).$$

$\square$

Thus, we have presented the four classical examples of a vector space. Below are some other examples of vector spaces.

Example 3.5 *Let V have exactly one element, v say, and define $av = v$ and $v+v = v$. One can verify that V satisfies the properties of a vector space. V is called the **trivial** vector space. It is the vector space which contains only a zero vector.*

Example 3.6 *Define the vectors to be*

$$V = \{[a, b] \mid a, b \in \mathbb{R}\}.$$

Define addition by

$$[a_1, b_1] + [a_2, b_2] = [a_1 + a_2 + 1, b_1 + b_2 - 2].$$

Define scalar multiplication by

$$a[a_1, b_1] = [aa_1 + a - 1, ab_1 - 2a + 2].$$

This example brings up several good points. First, although V has the same vectors as $\mathbb{R}^2$ it is different due to the fact that the operations for V are defined differently from those of $\mathbb{R}^2$. Second, we shall see in the proof of the claim below that the zero vector is **not** *an object filled with zeros. Indeed, for this example*

$$[a, b] + [0, 0] = [a + 1, b - 2] \neq [a, b],$$

so $[0, 0]$ cannot be the zero vector for V.

We wish to point out, though, that this example was given simply to abolish certain assumptions about vector spaces. It will not appear again in this text.

Claim: *V with operations defined above forms a vector space.*

Proof 3.3 *We give a partial proof and leave the rest as an exercise. To verify Property 3, we need to find a $[a_2, b_2]$ such that*

$$[a_1, b_1] + [a_2, b_2] = [a_1, b_1].$$

i.e., by definition,

$$[a_1 + a_2 + 1, b_1 + b_2 - 2] = [a_1, b_1].$$

Equating yields
$$a_1 + a_2 + 1 = a_1 \qquad and \qquad b_1 + b_2 - 2 = b_1.$$

Solving yields $a_2 = -1$ and $b_2 = 2$ and so the zero vector for V is $[-1, 2]$.

To prove Property 4, we point out that for $[a_1, b_1] \in V$, the additive inverse of $[a_1, b_1]$ will **not** *be $[-a_1, -b_1]$ as is the case in $\mathbb{R}^n$. Indeed, we want to find a $[a_2, b_2]$ such that*

$$[a_1, b_1] + [a_2, b_2] = [-1, 2].$$

i.e., by definition,

$$[a_1 + a_2 + 1, b_1 + b_2 - 2] = [-1, 2].$$

Equating yields

$$a_1 + a_2 + 1 = -1 \qquad and \qquad b_1 + b_2 - 2 = 2.$$

Solving yields $a_2 = -a_1 - 2$ and $b_2 = -b_1 + 4$ and so the additive inverse of $[a_1, b_1]$ in V is $[-a_1 - 2, -b_1 + 4]$ (observe that this additive inverse should be defined in terms of a_1 and b_1).

We remark that the fact that we can solve the equations above verifies that Properties 3 and 4 hold. To verify Property 5, we will simplify both sides of the equality to find a common middle ground. Working on the left-hand side,

$$a([a_1, b_1] + [a_2, b_2]) = a([a_1 + a_2 + 1, b_1 + b_2 - 2])$$

$$= [a(a_1 + a_2 + 1) + a - 1, a(b_1 + b_2 - 2) - 2a + 2]$$

$$= [aa_1 + aa_2 + 2a - 1, ab_1 + ab_2 - 4a + 2].$$

Working on the righthand side,

$$a[a_1, b_1] + a[a_2, b_2] = [aa_1 + a - 1, ab_1 - 2a + 2] + [aa_2 + a - 1, ab_2 - 2a + 2]$$

$$= [(aa_1 + a - 1) + (aa_2 + a - 1) + 1, (ab_1 - 2a + 2) + (ab_2 - 2a + 2) - 2]$$

$$= [aa_1 + aa_2 + 2a - 1, ab_1 + ab_2 - 4a + 2].$$

Hence the righthand side has been made identical with the left-hand side, thus verifying Property 5. □

The reader should see Exercise 13 in this section to gain some insight as to how this vector space was constructed.

It's actually rare that a set of vectors together with its two operations will turn out to satisfy the properties of a vector space. To convince you of this we give below several examples of non-vector spaces to show that the operations on the vectors have to be defined carefully in order to produce a vector space satisfying all the axioms.

Example 3.7 *Define the vectors to be $V = \{[a,b] \mid a, b \in \mathbb{R}\}$. Define scalar multiplication as it was for $\mathbb{R}^2$, however define addition by*

$$[a_1, b_1] + [a_2, b_2] = [a_1 + b_2, a_2 + b_1].$$

If even one of the nine properties of a vector space should fail, then V cannot be a vector space. In this case, we will show that Property 1 fails. We suspect that this property will fail when the sum

$$[a_1, b_1] + [a_2, b_2] = [a_1 + b_2, a_2 + b_1]$$

is compared with the reverse sum

$$[a_2, b_2] + [a_1, b_1] = [a_2 + b_1, a_1 + b_2].$$

But to truly convince ourselves that this property fails it is best to exhibit a **counterexample** *to the property, i.e. exhibit two specific vectors which, in this case, do not satisfy the property. Let's keep our counterexample simple. Oftentimes a counterexample can be created using numbers –1, 0 and 1. For instance, $[1, 0] + [0, 1] = [2, 0]$ while $[0, 1] + [1, 0] = [0, 2]$ and so $[1, 0] + [0, 1] \neq [0, 1] + [1, 0]$.*

Example 3.8 *Let V be $m \times n$ matrices. Define addition as in M_{mn}, but scalar multiplication by $a[a_{ij}] = 0_{mn}$ for all a and $[a_{ij}]$.*

Observe that Properties 1 through 4 will not fail, since we defined vector addition exactly how we did in the vector space M_{mn}. Thus, we can narrow our scope and search among Properties 0 and 5–8 for a failure. In this case we show that, among others, Property 8 fails. This is apparent, since if $[a_{ij}] \neq 0_{mn}$, then $1[a_{ij}] = 0_{mn} \neq [a_{ij}]$ as the Property demands. Hence, Property 8 fails for all non-zero matrices (although it is enough that it fails for just a single one!).

Example 3.9 *Define the vectors to be polynomials and define addition of polynomials to be the same as in P, but scalar multiplication will be defined as follows:*

$$a(a_0 + a_1 x + \cdots + a_n x^n) = (a + a_0) + (a + a_1)x + \cdots + (a + a_n)x^n$$

A number of properties fail in this case, among which is Property 7. Here is a counterexample:

$$1(2 + 3x) = 3 + 4x \neq 2 + 3x$$

Example 3.10 *Define the vectors to be real-valued functions*

$$V = \{f : \mathbb{R} \longrightarrow \mathbb{R}\}.$$

Multiplying a scalar a by a function f yields the function af defined as in our classic example, i.e. $(af)(x) = af(x)$. However, adding functions f and g yields the function $f + g$ defined by $(f + g)(x) = f(x)g(x)$.

One can show that Properties 0–3, 6–8 hold (note that the zero vector is the function which takes on the value 1 for all inputs), however, Properties 4 and 5 fail. Property 4 fails for functions which have zero for even a single output. Property 5 fails, since $a(f(x) + g(x)) = af(x)g(x)$ while $af(x) + ag(x) = a^2 f(x)g(x)$. As a counterexample, choose $a = 2$ and $f(x) = x$, $g(x) = x$ for all inputs x.

We end this section with some simple consequences of the axioms of a vector space.

Theorem 3.3 *Let V be a vector space. Then the following are true:*

i. V has exactly one zero vector (denoted by 0).

ii. Every $u \in V$ has exactly one additive inverse (denoted by $-u$).

iii. For all scalars a, $a0 = 0$.

iv. For all $u \in V$, $0u = 0$.

v. For all $u \in V$, $(-1)u = -u$.

vi. If for a scalar a and vector $u \in V$ we should have $au = 0$, then either $a = 0$ or $u = 0$.

Proof 3.4 *Keep in mind that in the proof we implicitly appeal to the properties of a vector space which V is assumed to have.*

To prove i, suppose 0_1 and 0_2 are both zero vectors in V. We shall show that $0_1 = 0_2$. Since 0_1 is a zero vector, $0_1 + 0_2 = 0_2$. Since 0_2 is also a zero vector, $0_1 + 0_2 = 0_1$. Equating these two equations yields $0_1 = 0_2$ as desired.

To prove ii, suppose v_1 and v_2 are both inverses of a vector $u \in V$. By definition, $u + v_1 = 0$ and $u + v_2 = 0$. Then

$$v_1 = v_1 + 0 = v_1 + (u + v_2) = (v_1 + u) + v_2 = (u + v_1) + v_2 = 0 + v_2 = v_2.$$

Hence, v_1 and v_2 are one and the same.

To prove iii, notice that $a0 = a(0 + 0) = a0 + a0$. Now we add to both sides of the equation the inverse of the vector $a0$, i.e. $-(a0)$:

$$-(a0) + a0 = -(a0) + (a0 + a).$$

By the properties of a vector space (Properties 2 and 3), the above equation simplifies to $0 = a0$ and we are done.

The proof of iv is similar to iii. To prove v, we are required to show that $(-1)u$ is the inverse of u. We do this by direct verification using part iv.

$$(-1)u + u = (-1)u + 1u = (-1+1)u = 0u = 0.$$

To prove vi, either $a = 0$ (and we have reached one of the conclusions) or $a \neq 0$. In this latter case we need to show that $u = 0$. Since $a \neq 0$, we can multiply both sides of the equation by $1/a$ to get $(1/a)(au) = (1/a)0$. Using properties of a vector space and part iii, this equation simplifies to $u = 0$ and we are done. □

EXERCISES

1. Let V have the same vectors as $\mathbb{R}^2$ and define

$$[a_1, a_2] + [b_1, b_2] = [a_1 + b_1, 0],$$

$$a[a_1, a_2] = [aa_1, 1].$$

Determine which of the nine properties of a vector space fail for this structure.

2. Let V have the same vectors as $\mathbb{R}^2$ and define

$$[a_1, a_2] + [b_1, b_2] = \left[\sqrt[3]{a_1^3 + b_1^3}, \sqrt[3]{a_2^3 + b_2^3}\right],$$

$$a\,[a_1, a_2] = [\sqrt[3]{a}\,a_1, \sqrt[3]{a}\,a_2].$$

Decide whether or not V together with these operations forms a vector space.

3. Let $V = \{[a, b] \mid a, b \geq 0\}$ and define

$$[a_1, a_2] + [b_1, b_2] = \left[\sqrt{a_1^2 + b_1^2}, \sqrt{a_2^2 + b_2^2}\right],$$

$$a\,[a_1, a_2] = [\sqrt{a}\,a_1, \sqrt{a}\,a_2].$$

Decide whether or not V together with these operations forms a vector space.

4. Let V be the collection of $n \times n$ matrices. Define vector addition by $A + B = AB$ (matrix multiplication) and the usual scalar multiplication. Decide whether or not V together with these operations forms a vector space.

5. Let V be the collection of $n \times n$ invertible matrices. Define vector addition by $A + B = AB$ (matrix multiplication) and the usual scalar multiplication. Decide whether or not V together with these operations forms a vector space.

6. Let V be the collection of $n \times n$ matrices with determinant 1. Define vector addition by $A + B = AB$ (matrix multiplication) and the usual scalar multiplication. Decide whether or not V together with these operations forms a vector space.

7. Let V be the collection of $n \times n$ matrices. Define vector addition by $A + B = AB$ (matrix multiplication) and scalar multiplication by $aA = \frac{1}{a}A$ (the right-hand side represents the usual scalar multiplication). Determine which of the nine properties of a vector space fail in this setting.

8. Let V be the collection of polynomials. Define vector addition as follows:

$$(a_0 + a_1 x + \cdots + a_n x^n) + (b_0 + b_1 x + \cdots + b_n x^n) = (a_0 b_0) + (a_1 b_1)x + \cdots + (a_n b_n)x^n$$

Define scalar multiplication as follows:

$$c(a_0 + a_1 x + \cdots + a_n x^n) = (ca_0) + (ca_1)x + \cdots + (ca_n)x^n$$

Decide which of the axioms of a vector space **fail** and when they do provide a counterexample.

9. Let V be the collection of real-valued functions. Define vector addition by composition of functions, i.e. $(f + g)(x) = f(g(x))$ and scalar multiplication as the usual one for $\mathcal{F}$. Decide which of the axioms of a vector space fail and when they do provide a counterexample.

10. Let V be real numbers (i.e. scalars) with vector addition being scalar addition and scalar multiplication being multiplication of scalars. Show that V together with these operations forms a vector space.

11. Let V be real numbers with vector addition defined by $a + b = ab$ (multiplication of real numbers) and $a \cdot b = a^b$ (the usual exponentiation). Which properties of a vector space are satisfied under this definition?

12. Let $V_1, \ldots, V_n$ be vectors spaces with respective operations $+_1, \ldots, +_n$ and $\cdot_1, \ldots, \cdot_n$. Define $V = \{(v_1, \ldots, v_n) \mid v_1 \in V_1, \ldots, v_n \in V_n\}$. with addition defined by

$$(v_1, \ldots, v_n) + (u_1, \ldots, u_n) = (v_1 +_1 u_1, \ldots, v_n +_n u_n)$$

and scalar multiplication defined by

$$a(v_1, \ldots, v_n) = (a \cdot_1 v_1, \ldots, a \cdot_n v_n).$$

Prove that V is a vector space with these defined operations.

13. This exercise will shed some light as to how our non-classical vector space was formulated.

   a. Let $(V, +, \cdot)$ be a vector space and $f : V \longrightarrow V$ be a bijection (i.e. f is one-to-one and maps onto V). Define the following two new operations $\oplus$ and $\odot$:

$$\text{For } u, v \in V, \quad u \oplus v = f^{-1}(f(u) + f(v)).$$

$$\text{For } a \in \mathbb{R} \text{ and } v \in V, \quad a \odot v = f^{-1}(a \cdot f(v)).$$

   Show that $(V, \oplus, \odot)$ forms a vector space.

   b. Referring to part a, find the bijection f which yields our non-classical vector space in the section.

3.2 SUBSPACE

Every algebraic structure has a notion of a sub-algebraic structure and vector spaces are no exception.

Definition 3.2 *A non-empty subset U of a vector space V is a **subspace** if U together with the operations defined for V is a vector space in its own right.*

Before giving any examples we first prove a fact that will make it easier to verify whether or not a given subset of a vector space is a subspace. It states that one doesn't need to check all nine properties of a vector space to ensure that U is a subspace, but rather only one of the properties. This is certainly a time saver.

Lemma 3.1 *A non-empty subset U of a vector space V is a subspace iff U satisfies Property 0 of a vector space, namely*

a. *If $u_1, u_2 \in U$, then $u_1 + u_2 \in U$. (Closure under Addition)*

b. *If $a \in$ i.e.$\mathbb{R}$ and $u \in U$, then $au \in U$. (Closure under Scalar Multiplication)*

Proof 3.5 *One direction of the "ff" is immediate (assuming U is a subspace). Now assume that U satisfies Property 0. We need to show Properties 1 through 8 hold as well, and so we will be done.*

First note that Properties 1, 2, 5–8 are what we call inherited from the vector space V. For instance, since all the vectors in V commute (Property 1) certainly the vectors in the subset U also commute.

To prove Property 4, let $u \in U$. We need to show that u has an additive inverse also in U. In V we know $-u$ exists. But since U satisfies Property 0, $-u = (-1)u \in U$.

To prove Property 3, we know there is a 0 in V. We show this same 0 is in U as well. Take any $u \in U$ (here we use the fact that U is non-empty). From what we just proved we know there exists $-u \in U$. Now by Property 0, $0 = u + (-u) \in U$. □

We give several examples of subspaces of the classical vector spaces presented in the previous section.

Example 3.11 *Consider the following subset of* $\mathbb{R}^2$:

$$U = \{[a, b] \in \mathbb{R}^2 \mid a + b = 0\}.$$

Sometimes it is useful to put in words the property which defines U. *In this case, a vector from* $\mathbb{R}^2$ *is in* U *if the sum of its coordinates is equal to zero. Using Lemma 3.1 above we show* U *satisfies Property 0 and thus is a subspace of* $\mathbb{R}^2$. *Recall that there are two parts of Property 0 that need to be verified.*

First, take $[a_1, b_1]$ *and* $[a_2, b_2]$ *in* U *(so we know that* $a_1 + b_1 = 0$ *and* $a_2 + b_2 = 0$*). We need to show that* $[a_1, b_1] + [a_2, b_2] = [a_1 + a_2, b_1 + b_2]$ *is in* U *as well. To this end we show that the sum of its coordinates equals zero.*

$$(a_1 + a_2) + (b_1 + b_2) = (a_1 + b_1) + (a_2 + b_2) = 0 + 0 = 0.$$

Second, take a scalar c *and* $[a, b] \in U$ *(so we know that* $a + b = 0$*). We need to show that* $c[a, b] = [ca, cb]$ *is in* U *as well. Again, we show that the sum of its coordinates equals zero.*

$$ca + cb = c(a + b) = c0 = 0.$$

Geometrically speaking, U *describes the set of all vectors in* $\mathbb{R}^2$ *whose terminal points lie on the line* $x + y = 0$ *or* $y = -x$.

Example 3.12 *We show that* $n \times n$ *skew-symmetric matrices (i.e.* $A^T = -A$*) are a subspace of* M_{nn}.

First, take two skew-symmetric matrices A *and* B *(so we know that* $A^T = -A$ *and* $B^T = -B$*). We need to show that* $A + B$ *is also skew-symmetric which we do directly using properties of transpose.*

$$(A + B)^T = A^T + B^T = (-A) + (-B) = -(A + B).$$

Second, take a scalar a *and* A *skew-symmetric (so we know that* $A^T = -A$*). We need to show that* aA *is also skew-symmetric.*

$$(aA)^T = aA^T = a(-A) = -(aA).$$

Example 3.13 *Consider the following subset of* $\mathcal{F}$:

$$U = \{f \in \mathcal{F} \mid f(1) = 0\}.$$

First, we take $f, g \in U$ *(so* $f(1) = 0$ *and* $g(1) = 0$*) and show* $f + g \in U$ *(i.e. we show* $(f + g)(1) = 0$*).*

$$(f + g)(1) = f(1) + g(1) = 0 + 0 = 0.$$

Second, we take a scalar a and $f \in U$ (so $f(1) = 0$) and show $af \in U$ (i.e. we show $(af)(1) = 0$).

$$(af)(1) = af(1) = a0 = 0.$$

Definition 3.3 *For a positive integer n, define $P_n = \{p(x) \in P \mid deg(p) \leq n\}$. In other words P_n consists of polynomials of degree at most n.*

Example 3.14 *We show that P_n is a subspace of P. Here Property 0 is apparent for P_n. First, if we add two polynomials of degree $\leq n$ we certainly get another polynomial of degree $\leq n$. Similarly, multiplying a scalar by a polynomial of degree $\leq n$ yields a polynomial of degree $\leq n$.*

Example 3.15 *Here is an important subspace of $\mathbb{R}^n$. First choose a specific matrix $A \in M_{mn}$ and define $U = \{u \in \mathbb{R}^n \mid Au = 0\}$ (here we are viewing the $u \in U$ as a column vector). In other words, U is the solution set to the homogeneous linear system of equations $AX = 0$. We use properties of matrices to show that U is a subspace of $\mathbb{R}^n$.*

First, if $u, v \in U$ (so $Au = 0$ and $Av = 0$), then $A(u + v) = Au + Av = 0 + 0 = 0$ and so $u + v \in U$. Second, if $a \in \mathbb{R}$ and $u \in U$, then $A(au) = a(Au) = a0 = 0$ and so $au \in U$.

*This subspace is called the **null space** of the matrix A.*

Example 3.16 *For those familiar with Calculus, other examples of subspaces are the following:*

1. P, polynomials, are a subspace of real-valued, differentiable functions which is a subspace of continuous functions which is a subspace of $\mathcal{F}$. The collection of all continuous functions will be denoted by $\mathcal{C}(\mathbb{R})$ and the collection of all differentiable functions will be denoted by $\mathcal{D}(\mathbb{R})$.

2. The solution set of a set of homogeneous differential equations is a subspace of $\mathcal{F}$.

Now let's look at some examples which fail to be subspaces. Keep in mind that it is sufficient that one of the two parts of Property 0 fail in order to prove the given subset is not a subspace. As before, in order to verify that a property fails, at times we will provide a specific counterexample.

Example 3.17 *Define the following subset of $\mathbb{R}^2$:*

$$U = \{[a, b] \mid ab = 0\}.$$

In this case, the first part of Property 0 fails (the second part holds). The reader might be wondering "If I am given a subset of a vector space, how am I to decide whether or not it is a subspace of the given vector space?" We suggest that the best approach is to first try to show it is a subspace. If you fail and your demonstration seems to suspiciously suggest that perhaps the subset is not a subspace, then perhaps you might want to attempt to supply a counterexample. This is what we do here.

Take $[a_1, b_1]$ and $[a_2, b_2]$ in U (so we know that $a_1 b_1 = 0$ and $a_2 b_2 = 0$). In order to show that $[a_1, b_1] + [a_2, b_2] = [a_1 + a_2, b_1 + b_2]$ is in U we need to show that $(a_1 + a_2)(b_1 + b_2) = 0$. But notice that

$$(a_1 + a_2)(b_1 + b_2) = a_1 b_1 + a_1 b_2 + a_2 b_1 + a_2 b_2 = a_1 b_2 + a_2 b_1.$$

It is these two middle terms which linger that leads us to believe that it is not always the case that $(a_1 + a_2)(b_1 + b_2) = 0$ and so U is not a subspace. Now, a specific counterexample is called for to cement in the fact that it is not.

Notice that $[1, 0]$ and $[0, 1]$ are in U but the sum, $[1, 1]$, is clearly not in U. Hence, we have shown by counterexample that U is not a subspace.

Example 3.18 *Let $U = \{ f \in \mathcal{F} \mid f(0) = 1 \}$.*

If $f, g \in U$, then $f(0) = 1$ and $g(0) = 1$. But $(f + g)(0) = f(0) + g(0) = 1 + 1 = 2 \neq 1$ and so $f + g \notin U$. No specific counterexample is required here, since the first part of Property 0 fails in general.

Example 3.19 *This example of a non-subspace of $\mathcal{F}$ looks suspiciously like a previous example which was a subspace, but there is a subtle difference. Define*

$$U = \{ f \in \mathcal{F} \mid f(a) = 0 \text{ for some } a \in \mathbb{R} \}.$$

The key words here are "for some". In other words the a is not fixed. The a that puts f in U may be different for each f. Again, it is the first part of Property 0 which fails. We provide a specific counterexample. $f(x) = x$ is in U, since $f(0) = 0$ (here $a = 0$) and $g(x) = -x + 1$ is in U (here $a = 1$). But $f + g$ is not in U, since $(f + g)(x) = f(x) + g(x) = x + (-x + 1) = 1$ a constant function of value 1. Thus, regardless of what a we plug into $f + g$ we have no hope of getting 0 as an output. Hence, U is not a subspace of $\mathcal{F}$.

Example 3.20 *Consider the following subset of M_{nn}:*

$$U = \{ A \in M_{nn} \mid AA^T = I_n \}.$$

This example illustrates another way to show a subset is not a subspace. The logic is this: For a subset to be a subspace, surely it must contain the zero vector of the larger vector space (look at the proof of Lemma 3.1). Thus, if one can show that $0 \notin U$, then U cannot be a subspace of the vector space. A word of caution is in order here. If you should find that the zero vector is indeed in U, you cannot conclude that U is a subspace. Just look at the previous two examples and you'll see that the zero vector was in U even though U was not a subspace.

But in this example, the method works. The zero vector, $0_{nn} \notin U$, since $0_{nn} 0_{nn}^T = 0_{nn} 0_{nn} = 0_{nn} \neq I_n$. hence U is not a subspace of M_{nn}.

Example 3.21 *Consider the following subset of P:*

$$U = \{p(x) \in P \mid deg(p) = n\} \text{ for a fixed } n \geq 0.$$

U fails to be a subspace for the same reason as the previous example. The zero polynomial, which has degree $-\infty$ is not in U. Notice the efficiency of this method.

Now we look at some specific subspaces which are always defined in any arbitrary vector space V.

Definition 3.4

1. *The subset $\{0\}$ is called the **trivial** subspace in V.*

2. *V is called the **improper** subspace of V.*

3. *Given $U_1, U_2 \ldots, U_n$ subspaces of V define $U_1 \cap U_2 \cap \cdots \cap U_n$ to be the set theoretic **intersection** of the subspaces of $U_1, U_2, \ldots, U_n$.*

4. *Given $U_1, U_2 \ldots, U_n$ subspaces of V define*

$$U = \{u_1 + u_2 + \cdots + u_n \mid u_1, \in U_2, \in u_2 \in U_2, \in U_n\}$$

*U is called the **sum** of the subspaces $U_1, U_2, \ldots, U_n$ and in this case we write $U = U_1 + U_2 + \cdots + U_n$.*

Of course, for each of the defined subsets above one really needs to prove that each is indeed a subspace by verifying Property 0 in each case. These make for good exercises for the reader, so we wouldn't want to have the reader miss any opportunity for practice.

The next subspace will have some use in later sections.

Definition 3.5 *Let U_1, U_2 be subspaces of a vector space V and set $U = U_1 + U_2$. We write $U = U_1 \oplus U_2$ if in addition to $U = U_1 + U_2$ we have $U_1 \cap U_2 = \{0\}$. We call the type of sum of subspaces a **direct sum**.*

Lemma 3.2 *Let U, W be subspaces of a vector space V. $V = U \oplus W$ iff every $v \in V$ can be written uniquely as $v = u + w$ for some $u \in U$ and $w \in W$.*

We leave the proof as an exercise, however, we point out that the existence of the expression $v = u + w$ follows from the fact $V = U + W$ and the uniqueness follows from the fact $U \cap W = \{0\}$.

EXERCISES

1. Decide whether or not each of the following subsets of $\mathbb{R}^n$ is a subspace:

   a. $U = \{[a, b] \in \mathbb{R}^2 \mid a = b\}$.

   b. $U = \{[a, b] \in \mathbb{R}^2 \mid ca + db = 0\}$, where $c, d \in \mathbb{R}$ are fixed.

   c. $U = \{[a, b] \in \mathbb{R}^2 \mid a^2 = b\}$.

   d. $U = \{[a, b, 2a - b] \mid a, b \in \mathbb{R}\}$.

   e. $U = \{[a, b] : ab \geq 0\}$.

   f. $U = \{u \in \mathbb{R}^n \mid u \cdot v = 0\}$, where $v \in \mathbb{R}^n$ is fixed.

   g. $U = \{u \in \mathbb{R}^n \mid Au = cu\}$, where $A \in M_{nn}$ and $c \in \mathbb{R}$ are fixed.

2. Decide whether or not each of the following subsets of M_{22} is a subspace:

   a. $U = \left\{ \begin{bmatrix} a & b \\ c & d \end{bmatrix} : a + b + c + d = 0 \right\}$.

   b. $U = \left\{ \begin{bmatrix} a & b \\ c & d \end{bmatrix} : a + b + c + d = 1 \right\}$.

   c. $U = \left\{ \begin{bmatrix} a & a+b \\ a-b & b \end{bmatrix} : a, b \in \mathbb{R} \right\}$.

3. Decide whether or not each of the following subsets of M_{nn} is a subspace:

   a. $U = \{A \in M_{nn} \mid A^2 = A\}$.

   b. $U = \{A \in M_{nn} \mid |A| = 0\}$.

   c. $U = \{A \in M_{nn} \mid |A| > 0\}$.

   d. $U = \{A \in M_{nn} \mid A \text{ is invertible}\}$.

   e. $U = \{A \in M_{nn} \mid A^k = 0_{nn} \text{ for a fixed k}\}$.

   f. $U = \{A \in M_{nn} \mid A^k = 0_{nn} \text{ for a some k}\}$.

   g. $U = \{[a_{ij}] \in M_{nn} \mid a_{11}a_{22} \cdots a_{nn} = 0\}$.

   h. $U = \{A \in M_{nn} \mid AB = BA\}$, where $B \in M_{nn}$ is fixed.

   i. $U = \{A \in M_{nn} \mid AB = 0\}$, where $B \in M_{nn}$ is fixed.

   j. $U = \{A \in M_{nn} \mid A = aI \text{ for some scalar } a\}$.

   k. $U = \{A \in M_{nn} : \exists c \in \mathbb{R} \text{ s.t. } Av = cv\}$, where $v \in \mathbb{R}^n$ is fixed.

4. Decide whether or not each of the following subsets of $\mathcal{F}$ is a subspace:

   a. $U = \{f \in \mathcal{F} \mid f(x) = xg(x) \text{ for some } g \in \mathcal{F}\}$.

   b. $U = \{f \in \mathcal{F} \mid f(x^2) = f(x)\}$.

   c. $U = \{f \in \mathcal{F} \mid f(x^2) = f(x)^2\}$.

   d. $U = \{f \in \mathcal{F} \mid f'(x) = f(x)\}$.

   e. $U = \{f \in \mathcal{F} \mid f(0) = f(1)\}$.

   f. $U = \{f \in \mathcal{F} \mid f'(0) = f(1)\}$.

   g. $U = \{f \in \mathcal{F} \mid f(x) = \frac{p(x)}{q(x)} \text{ for some polynomials } p(x), q(x)\}$.

   h. $U = \{f \in \mathcal{F} \mid f \text{ is an increasing function}\}$.

   i. $U = \{f \in \mathcal{F} \mid f \text{ is one-to-one}\}$.

   j. $U = \{f \in \mathcal{F} \mid \text{For all } x, \ f(x) \leq M, \text{ for some real number } M\}$.

   k. $U = \{f \in \mathcal{F} \mid f(x) = e^x g(x) \text{ for some function } g(x)\}$.

5. Decide whether or not each of the following subsets of P is a subspace:

   a. $U = \{a + (a + b)x + bx^2 \mid a, b \in \mathbb{R}\}$.

   b. $U = \{a + (ab)x + bx^2 \mid a, b \in \mathbb{R}\}$.

   c. $U = \{a + bx + cx^2 \in P \mid a > b + c\}$.

   d. $U = \{p \in P \mid deg(p) \text{ is odd}\}$.

6. U_{nn} will represent the set of all $n \times n$ upper triangular matrices. Show it's a subspace of M_{nn}.

7. L_{nn} will represent the set of all $n \times n$ lower triangular matrices. Show it's a subspace of M_{nn}.

8. D_{nn} will represent the set of all $n \times n$ diagonal matrices. Show it's a subspace of M_{nn}.

9. For a square matrix $A \in M_{nn}$, define the **trace** of A, written $tr(A)$ to be the sum of the diagonal entries of A, i.e. if $A = [a_{ij}]$ then $tr(A) = a_{11} + a_{22} + \cdots + a_{nn}$. Show that the following subset of M_{nn} is a subspace:

$$U = \{A \in M_{nn} \ : \ tr(A) = 0\}.$$

10. Consider the vector space $M_{nn}(\mathbb{C})$ where $\mathbb{C}$ is the complex numbers. In other words matrices with complex number entries (note scalars are now complex numbers).

a. The conjugate of $a + bi \in \mathbb{C}$, written $\overline{a + bi} = a - bi$. Verify the following two properties of conjugacy for $a, b \in \mathbb{C}$:

$$\overline{a} + \overline{b} = \overline{a + b} \qquad \overline{a}\overline{b} = \overline{ab}.$$

b. Decide whether or not the following subset of $M_{nn}(\mathbb{C})$ is a subspace:

$$U = \{[a_{ij}] \mid \overline{a_{ij}} = a_{ji}\}.$$

11. Illustrate by example two subspaces U and W of a vector space V whose union, $U \cup W$, is not a subspace (hint: $V = \mathbb{R}^2$ is an easy setting).

12. Show that the only non-trivial proper subspaces of $\mathbb{R}^2$ are lines through the origin.

13. Given $U_1, U_2, \ldots, U_n$ subspaces of a vector space V, show that $U_1 \cap U_2 \cap \cdots \cap U_n$ is a subspace of V.

14. Given $U_1, U_2, \ldots, U_n$ subspaces of a vector space V, show that $U_1 + U_2 + \cdots + U_n$ is a subspace of V.

15. Let $X_1, \ldots, X_n$ be any sets and let X_i represent one of these sets. Prove that $X_1 \cap \cdots \cap X_n \subseteq X_i$ (this result extends to infinite intersections).

16. Let U and W be subspaces of a vector space V. Prove that if $U \cup W$ is a subspace, then either $U \subseteq W$ or $W \subseteq U$.

17. Let $U_1, U_2, \ldots, U_n$ be subspaces of a vector space V. In Exercise 14 above the reader verified that $U_1 + \cdots + U_n$ is a subspace of V. Suppose we have the addition property that for every i, $1 \leq i \leq n$, we have

$$U_i \cap (U_1 + \cdots + U_{i-1} + U_{i+1} + \cdots + U_n) = \{0\}.$$

Show that every element in $U_1 + U_2 + \cdots + U_n$ has a unique representation as $u_1 + u_2 + \cdots + u_n$, where $u_1 \in U_1, u_2 \in U_2, \ldots, u_n \in U_n$ (note that in this case we write $U_1 \oplus U_2 \oplus \cdots \oplus U_n$).

18. Prove Lemma 3.2.

3.3 LINEAR INDEPENDENCE

We present now a topic which is critical in Linear Algebra. It will allow us to define the notions of basis and dimension in an upcoming section. First we present some definitions.

Definition 3.6 *Let $v_1, \ldots, v_n$ be elements of a vector space V and $a_1, \ldots, a_n$ be scalars. We call the expression $a_1 v_1 + \cdots + a_n v_n$ a* **linear combination** *of the vectors $v_1, \ldots, v_n$. The scalars $a_1, \ldots, a_n$ are called the* **coefficients** *of the linear combination.*

Example 3.22 *In P_2, the vector $1 + 2x + x^2$ is a linear combination of $1 + x$, $1 - x^2$, $x + x^2$, since*

$$1 + 2x + x^2 = (-1)(1 + x) + (2)(1 - x^2) + (3)(x + x^2).$$

Definition 3.7 *Let $v_1, \ldots, v_n$ be elements of a vector space V. We say these vectors are **linearly dependent** in V, if there exists scalars $a_1, \ldots, a_n$ not all zero such that $a_1 v_1 + \cdots + a_n v_n = 0$. In other words there is a non-trivial linear combination of $v_1, \ldots, v_n$ which equals 0. If no such non-trivial linear combination exists, then we say that $v_1, \ldots, v_n$ are **linearly independent** in V. In other words $v_1, \ldots, v_n$ are linearly independent if whenever it should be the case that $a_1 v_1 + \cdots + a_n v_n = 0$, then it must be that $a_1 = 0, \ldots, a_n = 0$.*

The last restatement of linear independence gives us a method for checking linear independence: We assume that $a_1 v_1 + \cdots + a_n v_n = 0$ and show that this implies that $a_1 = 0, \ldots, a_n = 0$. Some simple results immediately follow from this definition (which we leave as exercises):

1. Any collection of vectors which includes the zero vector is linearly dependent.

2. Any single vector $v \neq 0$ on its own is linearly independent.

3. Two vectors u, v are linearly dependent iff one is a scalar multiple of the other (i.e. there exists a scalar a such that $u = av$ or $v = au$).

Example 3.23 *The functions $f(x) = \sin^2 x$, $g(x) = \cos^2 x$, $h(x) = 1$ are linearly dependent in $\mathcal{F}$ since $(1) \sin^2 x + (1) \cos^2 x + (-1)1 = 0$.*

Example 3.24 *We show that the vectors $[1, 0, 0], [1, 1, 0], [1, 1, 1]$ are linearly independent vectors in $\mathbb{R}^3$. Suppose $a_1[1, 0, 0] + a_2[1, 1, 0] + a_3[1, 1, 1] = [0, 0, 0]$. Combining the left-hand side yields $[a_1 + a_2 + a_3, a_2 + a_3, a_3] = [0, 0, 0]$. Equating components gives us*

$$\begin{cases} a_1 + a_2 + a_3 &= 0 \\ a_2 + a_3 &= 0 \\ a_3 &= 0 \end{cases}.$$

Notice that answering a question about linear independence has been reduced to solving a homogeneous linear system. If this linear system has only the trivial solution, then we have proved that the three vectors are linearly independent. Let's switch to the augmented matrix:

$$\left[\begin{array}{ccc|c} 1 & 1 & 1 & 0 \\ 0 & 1 & 1 & 0 \\ 0 & 0 & 1 & 0 \end{array} \right].$$

One can easily compute that this augmented matrix reduces to

$$\left[\begin{array}{ccc|c} 1 & 0 & 0 & 0 \\ 0 & 1 & 0 & 0 \\ 0 & 0 & 1 & 0 \end{array}\right].$$

Thus, the system has only the trivial solution and hence the three vectors are linearly independent.

In future, to check for linear independence of vectors in $\mathbb{R}^n$, the example above illustrates that one can immediately form an augmented matrix for a homogeneous system in which the columns on the left are the vectors in question. We summarize this technique in a theorem.

Theorem 3.4 *For $A \in M_{mn}(F)$, the following are equivalent:*

i. $AX = 0$ has only the trivial solution.

ii. $AX = B$ has at most one solution, for any $B \in \mathbb{R}$, i.e. $\mathbb{R}_m$.

iii. The columns of A are linearly independent.

iv. $rk(A) = n$.

Proof 3.6 *In Theorem 2.16, we proved the equivalence of i and ii. From Theorem 2.18, we have that part iv implies part i and part ii implies part iv (the contrapositive statement). We complete this proof by showing that i and iii are equivalent. Set $A = [c_1 \cdots c_n]$ represented in columns.*
Assuming part i, if $a_1 c_1 + \cdots + a_n c_n = 0$ for some scalars $a_1, \ldots, a_n$, then by Lemma 1.1, $Au = 0$ where $u = [a_1, \ldots, a_n]$. Hence, u is a solution to $AX = 0$ and by assumption we must have $u = 0$, i.e. $a_1 = \cdots = a_n = 0$.
Assuming part iii, suppose that $u = [a_1, \ldots, a_n]$ is a solution to $AX = 0$. Again, by Lemma 1.1, $a_1 c_1 + \cdots + a_n c_n = 0$. By assumption, we must have $a_1 = \cdots = a_n = 0$ and so $u = 0$. □

Perhaps the reader had noticed another (quicker) way to complete the previous example. Since the coefficient matrix for the system is square, we can show using Theorems 2.21 and 3.4, that the linear system has only the trivial solution by showing the determinant of the coefficient matrix is non-zero. For in this example, the determinant of the coefficient (upper-triangular) matrix is equal to $(1)(1)(1) = 1 \neq 0$.

Example 3.25 *We show that the vectors $1+x, 2+x-x^2, x+x^2$ are linearly dependent in P_2. Suppose $a(1+x) + b(2+x-x^2) + c(x+x^2) = 0$. Then, by collecting like terms, we have $(a + 2b) + (a + b + c)x + (-b + c)x^2 = 0$. But this implies that*

$$\begin{cases} a + 2b & = & 0 \\ a + b + c & = & 0 \\ -b + c & = & 0 \end{cases}.$$

As in the previous example, we have reduced the problem to solving a system of homogeneous equations. If we can show that the linear system has infinitely many solutions, then the original polynomials must be linearly dependent.

Since the coefficient matrix is square, we can once again look at its determinant. If the determinant is zero, by Theorem 3.7, the system has infinitely many solutions:

$$\begin{vmatrix} 1 & 2 & 0 \\ 1 & 1 & 1 \\ 0 & -1 & 1 \end{vmatrix} \underset{=}{-R_1+R_2} \begin{vmatrix} 1 & 2 & 0 \\ 0 & -1 & 1 \\ 0 & -1 & 1 \end{vmatrix} = 0 \ \textit{(two identical rows)}.$$

As in the previous example, we point out a shortcut method. One can immediately form an augmented matrix for a homogeneous system where the columns on the left are the coefficients of each polynomial:

$$\begin{matrix} 1 \\ x \\ x^2 \end{matrix} \left[\begin{array}{ccc|c} 1 & 2 & 0 & 0 \\ 1 & 1 & 1 & 0 \\ 0 & -1 & 1 & 0 \end{array} \right].$$

The notation to the left of the matrix keeps track of the coefficients of the polynomial, for the order in which they are inserted into each column must be consistent.

Example 3.26 *We show that the following matrices are linearly independent in M_{22}:*

$$\begin{bmatrix} 1 & 1 \\ 0 & 1 \end{bmatrix}, \begin{bmatrix} 1 & 1 \\ 1 & 0 \end{bmatrix}, \begin{bmatrix} 1 & 0 \\ 1 & 1 \end{bmatrix}.$$

For suppose

$$a \begin{bmatrix} 1 & 1 \\ 0 & 1 \end{bmatrix} + b \begin{bmatrix} 1 & 1 \\ 1 & 0 \end{bmatrix} + c \begin{bmatrix} 1 & 0 \\ 1 & 1 \end{bmatrix} = \begin{bmatrix} 0 & 0 \\ 0 & 0 \end{bmatrix}.$$

Then combining the righthand side we have

$$\begin{bmatrix} a+b+c & a+b \\ b+c & a+c \end{bmatrix} = \begin{bmatrix} 0 & 0 \\ 0 & 0 \end{bmatrix}.$$

Equating, yields again a homogeneous system of equations:

$$\begin{cases} a+b+c &=& 0 \\ a+b &=& 0 \\ b+c &=& 0 \\ a+c &=& 0 \end{cases}.$$

In this example, our coefficient matrix is not square so we cannot appeal to Theorem 2.21, but we can use Theorem 3.4. We form the augmented matrix:

$$
\left[\begin{array}{ccc|c}
1 & 1 & 1 & 0 \\
1 & 1 & 0 & 0 \\
0 & 1 & 1 & 0 \\
1 & 0 & 1 & 0
\end{array}\right]
\quad which\ reduces\ to \quad
\left[\begin{array}{ccc|c}
1 & 0 & 0 & 0 \\
0 & 1 & 0 & 0 \\
0 & 0 & 1 & 0 \\
0 & 0 & 0 & 0
\end{array}\right].
$$

Hence, the linear system has only the trivial solution and the matrices are therefore linearly independent.

Again there is a short cut method for checking linear independence. Notice that the above augmented matrix can be obtained by inserting the matrices in question into the columns of the augmented matrix. This is done by inserting the rows of each matrix. For instance, a matrix such as

$$
\left[\begin{array}{ccc}
1 & 2 & 3 \\
4 & 5 & 6
\end{array}\right]
\quad would\ become\ the\ column \quad
\left[\begin{array}{c}
1 \\ 2 \\ 3 \\ 4 \\ 5 \\ 6
\end{array}\right].
$$

Perhaps a more convenient way to verify independence for this example would be to compute the rank of the coefficient matrix. Notice that

$$
\left[\begin{array}{ccc}
1 & 1 & 1 \\
1 & 1 & 0 \\
0 & 1 & 1 \\
1 & 0 & 1
\end{array}\right]
\quad reduces\ to \quad
\left[\begin{array}{ccc}
1 & 0 & 0 \\
0 & 1 & 0 \\
0 & 0 & 1 \\
0 & 0 & 0
\end{array}\right].
$$

Therefore, the rank of the coefficient matrix equals the number of columns, so by Theorem 3.4, the augmented matrix has only the trivial solution and again we conclude that the original matrices are linearly independent.

Our approach is different when we verify linear independence for vectors in $\mathcal{F}$. We give two methods for testing linear independence of functions (one of which involves Calculus). We derive some results which will make this task possible.

For the first method, we need the following result:

Theorem 3.5 *Let $f_1, \ldots, f_n$ be vectors in $\mathcal{F}$. If there are n distinct scalars $b_1, \ldots, b_n$ with the property that*

$$\begin{vmatrix} f_1(b_1) & f_1(b_2) & \cdots & f_1(b_n) \\ f_2(b_1) & f_2(b_2) & \cdots & f_2(b_n) \\ \vdots & \vdots & & \vdots \\ f_n(b_1) & f_n(b_2) & \cdots & f_n(b_n) \end{vmatrix} \neq 0,$$

then $f_1, \ldots, f_n$ are linearly independent in $\mathcal{F}$.

Proof 3.7 *We prove the contrapositive statement. Assume that $f_1, \ldots, f_n$ are linearly dependent. Then there exist scalars $a_1, \ldots, a_n$ not all zero such that $a_1 f_1 + \cdots + a_n f_n = 0$. Now evaluate the functions at any $b_i \in \mathbb{R}$ ($i = 1, 2, \ldots, n$) to obtain the following set of equations:*

$$a_1 f_1(b_1) + \cdots + a_n f_n(b_1) = 0$$

$$a_1 f_1(b_2) + \cdots + a_n f_n(b_2) = 0$$

$$\vdots$$

$$a_1 f_1(b_n) + \cdots + a_n f_n(b_n) = 0.$$

This implies in $\mathbb{R}^n$ that

$$a_1[f_1(b_1), \ldots, f_1(b_n)] + \cdots + a_n[f_n(b_1), \ldots, f_n(b_n)] = [0, 0, \ldots, 0],$$

and so $[f_1(b_1), \ldots, f_1(b_n)], \ldots, [f_n(b_1), \ldots, f_n(b_n)]$ are linearly dependent in $\mathbb{R}^n$. But then

$$\begin{vmatrix} f_1(b_1) & f_1(b_2) & \cdots & f_1(b_n) \\ f_2(b_1) & f_2(b_2) & \cdots & f_2(b_n) \\ \vdots & \vdots & & \vdots \\ f_n(b_1) & f_n(b_2) & \cdots & f_n(b_n) \end{vmatrix} = 0.$$

$\square$

Example 3.27 *We use this method to show that $\cos x, \sin x$ are linearly independent in $\mathcal{F}$. Use the values $b_1 = 0$ and $b_2 = \pi/2$ to obtain the determinant*

$$\begin{vmatrix} 1 & 0 \\ 0 & 1 \end{vmatrix} = 1 \neq 0.$$

Hence, by the theorem we have shown that $\cos x, \sin x$ are linearly independent in $\mathcal{F}$.

We point out that the converse of Theorem 3.5 is **not** true. Namely, the following statement is **not** true:

If there exists scalars $b_1, \ldots, b_n \in \mathbb{R}$ such that

$$\begin{vmatrix} f_1(b_1) & f_1(b_2) & \cdots & f_1(b_n) \\ f_2(b_1) & f_2(b_2) & \cdots & f_2(b_n) \\ \vdots & \vdots & & \vdots \\ f_n(b_1) & f_n(b_2) & \cdots & f_n(b_n) \end{vmatrix} = 0,$$

then $f_1, \ldots, f_n$ are linearly dependent. Just take the simple example of x, x^2 which we know to be linearly independent by our earlier method. However, using the values $b_1 = 0$ and $b_2 = 1$ we get the determinant

$$\begin{vmatrix} 0 & 1 \\ 0 & 1 \end{vmatrix} = 0.$$

Now we give a second method for testing linear independence of certain functions in $\mathcal{F}$. First we define a special function.

Definition: Let $f_1, f_2 \ldots, f_n$ be vectors in $\mathcal{F}$ which are $n - 1$ times differentiable. The **Wronskian** of $f_1, f_2 \ldots, f_n$ is the following function in $\mathcal{F}$:

$$W[f_1, \ldots, f_n](x) = \begin{vmatrix} f_1(x) & f_2(x) & \cdots & f_n(x) \\ f_1'(x) & f_2'(x) & \cdots & f_n'(x) \\ f_1''(x) & f_2''(x) & \cdots & f_n''(x) \\ \vdots & \vdots & & \vdots \\ f_1^{(n-1)}(x) & f_2^{(n-1)}(x) & \cdots & f_n^{(n-1)}(x) \end{vmatrix}.$$

Example 3.28 *The Wronskian of e^x, e^{2x} is*

$$W[e^x, e^{2x}](x) = \begin{vmatrix} e^x & e^{2x} \\ e^x & 2e^{2x} \end{vmatrix} = 2e^{3x} - e^{3x} = e^{3x}.$$

Theorem 3.6 *Let $f_1, \ldots, f_n$ be vectors in $\mathcal{F}$. If there is a scalar a such that $W[f_1, \ldots, f_n](a) \neq 0$, then $f_1, \ldots, f_n$ are linearly independent in $\mathcal{F}$.*

Proof 3.8 *We prove the contrapositive statement. Assume that $f_1 \ldots, f_n$ are linearly dependent. Then there exist scalars $a_1, \ldots, a_n$ not all zero such that $a_1 f_1 + \cdots + a_n f_n = 0$. Taking successive derivatives yields*

$$a_1 f_1' + \cdots + a_n f_n' = 0$$

$$a_1 f_1'' + \cdots + a_n f_n'' = 0$$

$$\vdots$$

$$a_1 f_1^{(n-1)} + \cdots + a_n f_n^{(n-1)} = 0.$$

Now plug in any scalar a into each equation to get

$$a_1 f_1(a) + \cdots + a_n f_n(a) = 0$$

$$a_1 f_1'(a) + \cdots + a_n f_n'(a) = 0$$

$$\vdots$$

$$a_1 f_1^{(n-1)}(a) + \cdots + a_n f_n^{(n-1)}(a) = 0.$$

This implies in $\mathbb{R}^n$ that

$$a_1 [f_1(a), \ldots, f_1^{(n-1)}(a)] + \cdots + a_n [f_n(a), \ldots, f_n^{(n-1)}(a)] = [0, 0, \ldots, 0],$$

and so $[f_1(a), f_1'(a), \ldots, f_1^{(n-1)}(a)]$, $\ldots$, $[f_n(a), f_n'(a), \ldots, f_n^{(n-1)}(a)]$ are linearly independent in $\mathbb{R}^n$. Hence,

$$W[f_1, \ldots, f_n](a) = \begin{vmatrix} f_1(a) & \cdots & f_n(a) \\ \vdots & & \vdots \\ f_1^{(n-1)}(a) & \cdots & f_n^{(n-1)}(a) \end{vmatrix} = 0.$$

Since a was an arbitrary scalar, our proof is complete. □

Example 3.29 *The vectors e^x, e^{2x} are linearly independent in $\mathcal{F}$, since $W[e^x, e^{2x}](x) = e^{3x}$ and for $a = 0$, $W[e^x, e^{2x}](0) = e^0 = 1 \neq 0$.*

This method of verifying linear independence is much more efficient as compared to the previous method, since we are required to find only one scalar as opposed to n scalars. We point out that the converse (or any distortion) of Theorem 3.6 is not true. For instance, the following are **not** true:

1. If there exists a scalar a such that $W[f_1, \ldots, f_n](a) = 0$, then $f_1, \ldots, f_n$ are linearly dependent.

2. If for all scalars a we have $W[f_1, \ldots, f_n](a) = 0$, then $f_1, \ldots, f_n$ are linearly dependent (although under certain conditions this is a valid application—this one would encounter in a course on Ordinary Differential Equations).

We round off this section by adding yet another statement to our ever expanding theorem (the proof is left as an exercise).

Theorem 3.7 *For an $n \times n$ square matrix A the following are equivalent:*

i. A is invertible.

ii. A is equivalent to I.

iii. The linear system $AX = B$ has a unique solution for any $B \in \mathbb{R}^n$.

iv. A is a product of elementary matrices.

v. The linear system $AX = 0$ has only the trivial solution.

vi. $rk(A) = n$.

vii. $|A| \neq 0$.

viii. The columns (or rows) of A are linearly independent in $\mathbb{R}^n$.

EXERCISES

1. Consider the following vectors in $\mathbb{R}^2$:

$$w = [-5, 0], \quad v_1 = [2, -3], \quad v_2 = [3, -2].$$

   a. Set up the system of equations necessary to solve in order for w to be a linear combination of v_1 and v_2 and decide whether or not it is possible (without solving).

   b. If part a. is possible use Cramer's Rule to write w as a linear combination of v_1 and v_2.

2. Consider the following vectors in $\mathbb{R}^3$:

$$w = [1, 2, -3], \quad v_1 = [1, 1, 1], \quad v_2 = [-1, 1, 0], \quad v_3 = [1, 3, 2].$$

   a. Set up the system of equations necessary to solve in order for w to be a linear combination of v_1, v_2 and v_3 and decide whether or not it is possible (without solving).

b. If part a. is possible use Cramer's Rule to write w as a linear combination of v_1, v_2 and v_3.

3. Decide whether each collection of vectors is linearly independent or linearly dependent:

a. $[1, 0, 1, 1], [1, 1, 0, 1], [1, 3, -2, 1] \in \mathbb{R}^4$

b. $\begin{bmatrix} 1 & 0 \\ -1 & 1 \end{bmatrix}, \begin{bmatrix} 2 & 1 \\ -1 & 2 \end{bmatrix}, \begin{bmatrix} -1 & 2 \\ -1 & -1 \end{bmatrix} \in M_{22}$

c. $1 + 2x^2, \ x + x^2, \ -1 + x \in P_2$

d. $\sin x, \ \tan x \in \mathcal{F}$

4. Decide whether each collection of vectors is linearly independent or linearly dependent:

a. $[1, 2, 0, -1], [0, 1, 1, -1], [1, 1, -2, 1] \in \mathbb{R}^4$

b. $\begin{bmatrix} 1 & 2 \\ -1 & 0 \end{bmatrix}, \begin{bmatrix} 2 & -1 \\ 1 & 1 \end{bmatrix}, \begin{bmatrix} 0 & 5 \\ -3 & -1 \end{bmatrix} \in M_{22}$

c. $1 - x^2, \ -1 + x + x^2, \ 3 - x - 3x^2 \in P_2$

d. $e^x, e^{2x}, e^{3x} \in \mathcal{F}$

5. Decide whether or not X is linearly independent, for each of the following vector spaces V containing a set of vectors X

a. $V = \mathbb{R}^3$ and $X = \{[1, 1, 1], \ [-1, 1, 0], \ [1, 3, 2]\}$

b. $V = M_{22}$ and

$$X = \left\{ \begin{bmatrix} 2 & 2 \\ -1 & 3 \end{bmatrix} \begin{bmatrix} -1 & 2 \\ 1 & -2 \end{bmatrix} \begin{bmatrix} 1 & -2 \\ 1 & 0 \end{bmatrix} \right\}$$

c. $V = M_{22}$ and

$$X = \left\{ \begin{bmatrix} -1 & -1 \\ 0 & 0 \end{bmatrix} \begin{bmatrix} 0 & 0 \\ 2 & 3 \end{bmatrix} \begin{bmatrix} 4 & 0 \\ -2 & 1 \end{bmatrix} \begin{bmatrix} 3 & -1 \\ 0 & 4 \end{bmatrix} \right\}$$

d. $V = M_{22}$ and

$$X = \left\{ \begin{bmatrix} 1 & 0 \\ -1 & 0 \end{bmatrix} \begin{bmatrix} 1 & 1 \\ 0 & -1 \end{bmatrix} \begin{bmatrix} -1 & -3 \\ -2 & 3 \end{bmatrix} \right\}$$

e. $V = M_{22}$ and

$$X = \left\{ \begin{bmatrix} 2 & 3 \\ 4 & 5 \end{bmatrix} \begin{bmatrix} 1 & -1 \\ 1 & -1 \end{bmatrix} \begin{bmatrix} 5 & 5 \\ 9 & 9 \end{bmatrix} \right\}$$

f. $V = P_2$ and $X = \{1 - x + 3x^2, \ 2 + 4x^2, \ -1 - 3x + x^2\}$.

g. $V = P_2$ and $X = \{1 + x + x^2, \ 1 + x, \ 2 - x^2\}$.

6. Decide whether each of the following vectors in $\mathcal{F}$ are linearly dependent or linearly independent:

   a. $\sin x, x \sin x$

   b. $\ln x, \ln x^2$

   c. $\cos x, \cos^2 x$

   d. $e^x, \sin 2x, \cos 3x$

   e. $1, \cos 2x, \cos^2 x$

   f. $1, \cosh^2 x, \sinh^2 x$

   g. $1, \ln x, \ln 5x$

   h. $\log_2 x, \log_2 (x - 1), \log_2 \left(\frac{x^2}{x-1}\right)$

   i. $1, \tan x, \sec x$

   j. $1, \tan^2 x, \sec^2 x$

7. Prove that any collection of vectors which includes 0 is linearly dependent.

8. Prove that any vector $v \neq 0$ is linearly independent.

9. Prove that two vectors u, v are linearly dependent iff one is a scalar multiple of the other (i.e. there exists a scalar a such that $u = av$ or $v = au$).

10. Find three vectors in $\mathbb{R}^3$ which are linearly dependent, but any two of them are linearly independent.

11. Prove that if $v_1, \ldots, v_n$ are linearly independent, then so are

    a. $v_1, \ v_1 + v_2, \ v_1 + v_3, \ \ldots, \ v_1 + v_n$.

    b. $v_1, \ v_1 + v_2, \ v_1 + v_2 + v_3, \ \ldots, \ v_1 + v_2 + \cdots + v_n$.

12. Prove that if $v_1, v_2, \ldots, v_n \in V$ are linearly independent, then any subset of the vectors is linearly independent.

13. Let $v_1, v_2, \ldots, v_k \in \mathbb{R}^n$ and $A \in M_{nn}$. Prove that if $Av_1, Av_2, \ldots, Av_k$ are linearly independent, then so are $v_1, v_2, \ldots, v_k$.

14. Complete the proof of Theorem 3.7 by showing that viii implies iii.

3.4   SPAN

We present in this section a special subspace which plays an important role in the theory of vector spaces as well as introduce the second property necessary for a basis.

Definition 3.8 *Given vectors $v_1, \ldots, v_n$ in a vector space V, the* **span** *of $v_1, \ldots, v_n$, written* $\text{span}(v_1, \ldots, v_n)$, *is the set of all linear combinations of the vectors $v_1, \ldots, v_n$. In other words*

$$\text{span}(v_1, \ldots, v_n) = \{a_1 v_1 + \cdots + a_n v_n \mid a_1, \ldots, a_n \in \mathbb{R}\}.$$

It is also called the subspace **generated by** *$v_1, \ldots, v_n$ and is sometimes indicated by the notation $\langle v_1, \ldots, v_n \rangle$. The vectors $v_1, \ldots, v_n$ are called the* **generators**.

We remark that one can define span for infinite sets of vectors as well, but this text does not require such treatment. It wouldn't be fair to introduce such a non-intuitive object without giving some examples. Later in the section, we will give a method for uncovering a nice description of the span of a collection of vectors. For this reason, our examples at this point will be simple.

Example 3.30 *Let $V = \mathbb{R}^3$. The span of $\hat{\imath}$ and $\hat{\jmath}$,*

$$\text{span}(\hat{\imath}, \hat{\jmath}) = \{a\hat{\imath} + b\hat{\jmath} \mid a, b \in \mathbb{R}\} = \{[a, b, 0] \mid a, b \in \mathbb{R}\}.$$

Hence, this span describes all vectors in $\mathbb{R}^3$ which lie in the xy-plane, or we might just say that this span **is** *the xy-plane. Similarly, the span of $\hat{\imath}, \hat{\jmath}$ and $\hat{k}$ will be all of $\mathbb{R}^3$.*

Definition 3.9 *Let $A \in M_{mn}$ with rows $r_1, \ldots, r_m \in F^n$ and columns $c_1, \ldots, c_n \in \mathbb{R}^m$. Then*

1. $\text{span}(r_1, \ldots, r_m)$ is called the **row space** *of A.*

2. $\text{span}(c_1, \ldots, c_n)$ is called the **column space** *of A.*

Now we prove an essential fact that the span of a collection of vectors is a subspace of V (and more).

Lemma 3.3 *Given vectors $v_1, \ldots, v_n$ in a vector space V,*

i. $v_1, \ldots, v_n \in \text{span}(v_1, \ldots, v_n)$.

ii. $\text{span}(v_1, \ldots, v_n)$ is a subspace of V.

iii. If U is a subspace of V containing the vectors $v_1, \ldots, v_n$ then $\text{span}(v_1, \ldots, v_n) \subseteq U$.

To summarize i–iii, span $(v_1, \ldots, v_n)$ is the smallest (with respect to inclusion) subspace of V containing the vectors $v_1, \ldots, v_n$ (see Figure 3.1) .

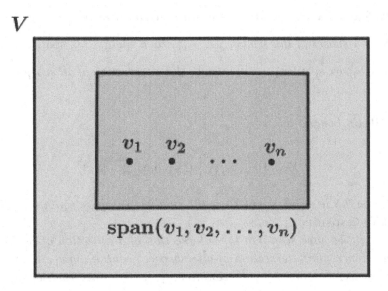

Figure 3.1 The span of a set of vectors

Proof 3.9 *For the proof of i, if v_i is one of $v_1, \ldots, v_n$, since $v_i = 0v_1 + \cdots + 1v_i + \cdots + 0v_n$ is a linear combination of $v_1, \ldots, v_n$, by definition, $v_i \in \text{span}(v_1, \ldots, v_n)$.*

To prove ii, we need to verify that $\text{span}(v_1, \ldots, v_n)$ *satisfies Property 0. First, take* $a_1v_1 + \cdots + a_nv_n$ *and* $b_1v_1 + \cdots + b_nv_n$ *in* $\text{span}(v_1, \ldots, v_n)$*. Then the sum,*

$$(a_1v_1 + \cdots + a_nv_n) + (b_1v_1 + \cdots + b_nv_n) = (a_1v_1 + b_1v_1) + \cdots + (a_nv_n + b_nv_n) =$$

$$(a_1 + b_1)v_1 + \cdots + (a_n + b_n)v_n \in \text{span}(v_1, \ldots, v_n).$$

Second, for any scalar c and $a_1v_1 + \cdots + a_nv_n$ in $\text{span}(v_1, \ldots, v_n)$*, the scalar product*

$$c(a_1v_1 + \cdots + a_nv_n) = c(a_1v_1) + \cdots + c(a_nv_n) =$$

$$(ca_1)v_1 + \cdots + (ca_n)v_n \in \text{span}(v_1, \ldots, v_n).$$

To prove iii, assume U is a subspace of V containing the vectors $v_1, \ldots, v_n$ and take any $a_1v_1 + \cdots + a_nv_n$ in $\text{span}(v_1, \ldots, v_n)$*. We need to show that it is in U. Since $v_1, \ldots, v_n$ are in U, then so are $a_1v_1, \ldots, a_nv_n$ (since U satisfies Property 0). By Property 0 again the sum, $a_1v_1 + \cdots + a_nv_n$, is also in U.* □

The following result is a direct consequence of Lemma 3.3:

Corollary 3.1 *Let $v_1, \ldots, v_n$ be vectors in a vector space V.*

i. If X is a subset of the set $\{v_1, \ldots, v_n\}$, then $\mathrm{span}(X) \subseteq \mathrm{span}(v_1, \ldots, v_n)$.

ii. The subspace $\mathrm{span}(v_1, \ldots, v_n)$ equals the intersection of all subspaces containing the vectors $v_1 \ldots, v_n$.

Proof 3.10 *By Lemma 3.3.i,*

$$X \subseteq \{v_1, \ldots, v_n\} \subseteq \mathrm{span}(v_1, \ldots, v_n).$$

By Lemma 3.3.iii, since $\mathrm{span}(X)$ is the smallest subspace containing X, it follows that $\mathrm{span}(X) \subseteq \mathrm{span}(v_1, \ldots, v_n)$.

To prove ii, we have seen that the intersection of a collection of subspaces is itself a subspace. Hence, the intersection of all subspaces containing $v_1, \ldots, v_n$ (let's call it U) is a subspace. Certainly this U contains $v_1, \ldots, v_n$ since each of its constituents in the intersection does so. Thus, by Lemma 3.3.iii, $\mathrm{span}(v_1, \ldots, v_n) \subseteq U$. For the reverse inclusion, since $\mathrm{span}(v_1, \ldots, v_n)$ is a subspace containing $v_1, \ldots, v_n$, it is certainly contained in the intersection of all such subspaces (see Exercise 15), namely U. □

Definition 3.10 *A set of vectors $v_1, \ldots, v_n \in V$ **spans** a vector space V if $\mathrm{span}(v_1, \ldots, v_n) = V$, i.e. the vectors $v_1, \ldots, v_n$ generate all the vectors in V (recall that the span of a set of vectors is a subspace of V which is not necessarily the entire vector space V).*

Example 3.31 *Consider the vectors $\hat{\imath}, \hat{\jmath}, \hat{k} \in \mathbb{R}^3$. One computes $\mathrm{span}(\hat{\imath}, \hat{\jmath}, \hat{k}) = \mathbb{R}^3$ and so $\hat{\imath}, \hat{\jmath}, \hat{k}$ span $\mathbb{R}^3$.*

We add yet another extension of Theorem 3.7.

Theorem 3.8 *For an $n \times n$ square matrix A the following are equivalent:*

i. A is invertible.

ii. A is equivalent to I.

iii. The linear system $AX = B$ has a unique solution for any $B \in \mathbb{R}^n$.

iv. A is a product of elementary matrices.

v. The linear system $AX = 0$ has only the trivial solution.

vi. $rk(A) = n$.

vii. $|A| \neq 0$.

viii. The columns (or rows) of A are linearly independent in $\mathbb{R}^n$.

ix. The span of the columns (or rows) of A equals all of $\mathbb{R}^n$.

Proof 3.11 *Set $A = [c_1 \cdots c_n]$ viewed in columns. We will show that viii is equivalent to iii. First assume iii is true and show that $\operatorname{span}(c_1, \ldots, c_n) = \mathbb{R}^n$. It's enough to show that any $B \in \mathbb{R}^n$ is an element of $\operatorname{span}(c_1, \ldots, c_n)$. Since, by assumption, $AX = B$ has a solution, there exists $X_0 = [a_1, \ldots, a_n] \in \mathbb{R}^n$ such that $AX_0 = B$. Now by Exercise 1.2, $a_1 c_1 + \cdots + a_n c_n = B$ which by definition implies that $B \in \operatorname{span}(c_1, \ldots, c_n)$.*

We leave the proof of viii implies iii as an exercise. □

Example 3.32 *A simple example which illustrates Theorem 3.8 nicely is the matrix $A = I_3$. Certainly, A is invertible, since $|A| = 1 \neq 0$. Furthermore, the rows (or columns) span $\mathbb{R}^3$, since the rows (or columns) are $\hat{\imath}, \hat{\jmath}, \hat{k}$.*

The next Lemma, besides its theoretical usefulness, gives us a method for computing the span of a collection of vectors. First, we make the following definition:

Definition 3.11 *Let $v, v_1, \ldots, v_n \in V$ a vector space. We say v is **linearly dependent** on $v_1, \ldots, v_n$ if there exist scalars $a_1, \ldots, a_n$ such that $v = a_1 v_1 + \cdots + a_n v_n$. Otherwise, v is **linearly independent** of $v_1, \ldots, v_n$.*

Example 3.33 *$1 - x - 2x^2$ is linearly dependent on $1 + x, x + x^2$ since $1 - x - 2x^2 = (1)(1 + x) + (-2)(x + x^2)$.*

Another way to phrase the definition above is v is linearly dependent on $v_1, \ldots, v_n$ iff $v \in \operatorname{span}(v_1, \ldots, v_n)$. The next lemma links the two notions of linear dependence we have seen. The proof is left as an exercise.

Lemma 3.4 *Let $v_1, \ldots, v_n \in V$ a vector space. Then $v_1, \ldots, v_n$ are linearly dependent iff there exists a $v_i \in \{v_1, \ldots, v_n\}$ such that v_i is linearly dependent on $v_1, \ldots, v_{i-1}, v_{i+1}, \ldots, v_n$.*

Now we prove the lemma we had mentioned earlier.

Lemma 3.5 *Let $v_1, \ldots, v_n \in V$ a vector space. If v_1 is linearly dependent on $v_2, \ldots, v_n$, then $\operatorname{span}(v_1, \ldots, v_n) = \operatorname{span}(v_2, \ldots, v_n)$.*

Proof 3.12 *The fact that $\operatorname{span}(v_2, \ldots, v_n) \subseteq \operatorname{span}(v_1, \ldots, v_n)$ follows from Corollary 3.1.i.*

For the reverse inclusion, take any $a_1 v_1 + \cdots + a_n v_n \in \operatorname{span}(v_1, \ldots, v_n)$. Since v_1 is linearly dependent on $v_2, \ldots, v_n$ there exist scalars $b_2, \ldots, b_n$ such that $v_1 = b_2 v_2 + \cdots + b_n v_n$. Hence,

$$a_1 v_1 + a_2 v_2 + \cdots + a_n v_n = a_1 (b_2 v_2 + \cdots + b_n v_n) + a_2 v_2 + \cdots + a_n v_n =$$

$$(a_1 b_2 + a_2)v_2 + \cdots + (a_1 b_n + a_n)v_n \in \text{span}(v_2, \ldots, v_n).$$

□

Lemma 3.5 says that we may drop a linearly dependent vector from a list of vectors without reducing the size of the span. Another way to express this is that linearly dependent vectors make no real contribution to the size of the span. By repeating this process, assuming there is at least one non-zero vector, we can reduce the set of vectors to a linearly independent set of vectors which has the same span as the original set of vectors. We call this reduced set of linearly independent vectors the **necessary generators** (later we will see that they form what we will call a **basis** for the span).

The procedure which follows illustrates an algorithm for giving a nice description of the span of a collection of vectors in $\mathbb{R}^n$, P or M_{mn}. The idea is that first we insert the vectors into the rows of a matrix (just as we had inserted them as columns to check linear independence), then we put this matrix in reduced row-echelon form. The process of row reducing will eliminate any linearly dependent vectors and the resulting non-zero rows will be the necessary generators of the span of the original vectors. See Section 3.6, Corollary 3.3 for a more formal explanation for why this process works.

Example 3.34 *We describe the span of* $[1, 0, 1], [1, 1, 0], [3, 1, 2] \in \mathbb{R}^3$. *First we form the matrix*

$$\begin{bmatrix} 1 & 0 & 1 \\ 1 & 1 & 0 \\ 3 & 1 & 2 \end{bmatrix} \text{ which reduces to } \begin{bmatrix} 1 & 0 & 1 \\ 0 & 1 & -1 \\ 0 & 0 & 0 \end{bmatrix}.$$

Hence, the necessary generators are $[1, 0, 1], [0, 1, -1]$ *and*

$$\text{span}([1, 0, 1], [1, 1, 0], [3, 1, 2]) = \text{span}([1, 0, 1], [0, 1, -1])$$

$$= \{a[1, 0, 1] + b[0, 1, -1] \mid a, b \in \mathbb{R}\} = \{[a, b, a - b] \mid a, b \in \mathbb{R}\}.$$

So the span consists of vectors in $\mathbb{R}^3$ *in which the third coordinate is the difference of the first and second coordinates. Notice that now one can easily decide whether or not a vector is in a span. For instance,* $[1, 2, -1] \in \text{span}([1, 0, 1], [1, 1, 0], [3, 1, 2])$ *since here* $a = 1, b = 2$ *and* $a - b$ *is indeed -1. While* $[1, 2, 3] \notin \text{span}([1, 0, 1], [1, 1, 0], [3, 1, 2])$ *since* $a - b = 3 \neq -1$.

Example 3.35 *We describe the span of* $1 + x + x^2, 1 - x^3, x + x^2 + x^3 \in P_3$. *We form the matrix*

$$
\begin{bmatrix} 1 & 1 & 1 & 0 \\ 1 & 0 & 0 & -1 \\ 0 & 1 & 1 & 1 \end{bmatrix} \quad \text{which reduces to} \quad \begin{bmatrix} 1 & 0 & 0 & -1 \\ 0 & 1 & 1 & 1 \\ 0 & 0 & 0 & 0 \end{bmatrix}.
$$

Hence, the necessary generators are $1 - x^3, x + x^2 + x^3$ *and*

$$
\text{span}(1 + x + x^2, 1 - x^3, x + x^2 + x^3) = \text{span}(1 - x^3, x + x^2 + x^3)
$$

$$
= \{a(1 - x^3) + b(x + x^2 + x^3) \mid a, b \in \mathbb{R}\} = \{a + bx + bx^2 + (b - a)x^3 \mid a, b \in \mathbb{R}\}.
$$

Again, $1 - 2x - 2x^2 - 3x^3$ *is in the span while vectors like* $1 - 2x - 2x^2 + x^3$ *and* $1 - 2x - x^2 - 3x^3$ *are not.*

Example 3.36 *We describe the span of*

$$
\begin{bmatrix} 1 & 1 \\ 0 & 0 \end{bmatrix}, \begin{bmatrix} 1 & 0 \\ 1 & 0 \end{bmatrix}, \begin{bmatrix} 0 & -1 \\ 1 & 0 \end{bmatrix} \in M_{22}.
$$

Form the matrix

$$
\begin{bmatrix} 1 & 1 & 0 & 0 \\ 1 & 0 & 1 & 0 \\ 0 & -1 & 1 & 0 \end{bmatrix} \quad \text{which reduces to} \quad \begin{bmatrix} 1 & 0 & 1 & 0 \\ 0 & 1 & -1 & 0 \\ 0 & 0 & 0 & 0 \end{bmatrix}.
$$

Hence, the necessary generators are

$$
\begin{bmatrix} 1 & 0 \\ 1 & 0 \end{bmatrix}, \begin{bmatrix} 0 & 1 \\ -1 & 0 \end{bmatrix}
$$

and the span equals

$$
\left\{ a \begin{bmatrix} 1 & 0 \\ 1 & 0 \end{bmatrix} + b \begin{bmatrix} 0 & 1 \\ -1 & 0 \end{bmatrix} : a, b \in \mathbb{R} \right\}
$$

$$
= \left\{ \begin{bmatrix} a & b \\ a - b & 0 \end{bmatrix} : a, b \in \mathbb{R} \right\}.
$$

Example 3.37 *We show that* $[1, 0, 0]$, $[1, 1, 0]$, $[1, 1, 1]$ *span* $\mathbb{R}^3$. *Putting the vectors in rows,*

$$\begin{bmatrix} 1 & 0 & 0 \\ 1 & 1 & 0 \\ 1 & 1 & 1 \end{bmatrix} \text{ reduces to } \begin{bmatrix} 1 & 0 & 0 \\ 0 & 1 & 0 \\ 0 & 0 & 1 \end{bmatrix}.$$

Hence,

$$\text{span}([1, 0, 0], [1, 1, 0], [1, 1, 1]) = \text{span}([1, 0, 0], [0, 1, 0], [0, 0, 1])$$

$$= \{a[1, 0, 0] + b[0, 1, 0] + c[0, 0, 1] \mid a, b, c \in \mathbb{R}\} = \{[a, b, c] \mid a, b, c \in \mathbb{R}\} = \mathbb{R}^3.$$

Example 3.38 *We show that* $[1, 0, 0]$, $[1, 1, 0]$, $[0, -2, 0]$ *do not span* $\mathbb{R}^3$.

$$\begin{bmatrix} 1 & 0 & 0 \\ 1 & 1 & 0 \\ 0 & -2 & 0 \end{bmatrix} \text{ reduces to } \begin{bmatrix} 1 & 0 & 0 \\ 0 & 1 & 0 \\ 0 & 0 & 0 \end{bmatrix}.$$

Hence,

$$\text{span}([1, 0, 0], [1, 1, 0], [0, -2, 0]) = \text{span}([1, 0, 0], [0, 1, 0])$$

$$= \{a[1, 0, 0] + b[0, 1, 0] \mid a, b \in \mathbb{R} \} = \{ [a, b, 0] \mid a, b \in \mathbb{R} \} \neq \mathbb{R}^3. \text{ (it is smaller)}$$

Sometimes, we wish to select our necessary generators from the original set of vectors. To do this we put the vectors in columns and row reduce until we find the pivots of the matrix (as we did when we computed the rank of a matrix). Then the columns in the original matrix corresponding to these pivots are necessary generators for the original set of vectors. The reason why this works (without going into detail) is that essentially we are finding the generators necessary for expressing as a linear combination any arbitrary vector in the span. See Section 3.6, Corollary 3.3 for a more formal justification for why this process works. This algorithm generally does not give a nice description of the span, but can serve other purposes. For instance, this algorithm comes in handy when we wish to *extend* a set of linearly independent vectors to a basis for a vector space (see Section 3.5).

One example should suffice for this procedure. Let's rework the previous example in $\mathbb{R}^3$.

Example 3.39 *Select from* $[1,0,1], [1,1,0], [3,1,2] \in \mathbb{R}^3$ *necessary generators which form its span. Form the matrix*

$$\begin{bmatrix} 1 & 1 & 3 \\ 0 & 1 & 1 \\ 1 & 0 & 2 \end{bmatrix} \quad which \ reduces \ to \quad \begin{bmatrix} 1 & 1 & 3 \\ 0 & 1 & 1 \\ 0 & 0 & 0 \end{bmatrix}.$$

At this point we know the pivots and we therefore choose $[1,0,1], [1,1,0]$ *as the necessary generators of the span.*

EXERCISES

1. For each of the following vector spaces V containing a set of vectors X and a vector v (all listed below),

   i. Find necessary generators for span(X).

   ii. Give a nice description of span(X).

   iii. Use part ii to decide whether or not $v \in$ span(X).

   a. $V = \mathbb{R}^3$, $X = \{[1,1,1], [-1,1,0], [1,3,2]\}$, $v = [2,4,3]$

   b. $V = \mathbb{R}^4$, $X = \{[1,1,0,1], [2,1,0,0], [1,-1,0,-3]\}$, $v = [3,1,0,-1]$

   c. $V = M_{22}$,
   $$X = \left\{ \begin{bmatrix} 2 & 2 \\ -1 & 3 \end{bmatrix}, \begin{bmatrix} -1 & 2 \\ 1 & -2 \end{bmatrix}, \begin{bmatrix} 1 & -2 \\ 1 & 0 \end{bmatrix} \right\}, \ v = \begin{bmatrix} 1 & -2 \\ 2 & 3 \end{bmatrix}$$

   d. $V = M_{22}$,
   $$X = \left\{ \begin{bmatrix} 1 & 2 \\ 0 & -1 \end{bmatrix}, \begin{bmatrix} 0 & 1 \\ -1 & 2 \end{bmatrix}, \begin{bmatrix} 1 & 5 \\ -3 & 5 \end{bmatrix} \right\}, \ v = \begin{bmatrix} 1 & -2 \\ 4 & -1 \end{bmatrix}$$

   e. $V = M_{22}$,
   $$X = \left\{ \begin{bmatrix} -1 & -1 \\ 0 & 0 \end{bmatrix}, \begin{bmatrix} 0 & 0 \\ 2 & 3 \end{bmatrix}, \begin{bmatrix} 4 & 0 \\ -2 & 1 \end{bmatrix}, \begin{bmatrix} 3 & -1 \\ 0 & 4 \end{bmatrix} \right\}, \ v = \begin{bmatrix} 2 & 3 \\ -1 & 1 \end{bmatrix}$$

   f. $V = M_{22}$,
   $$X = \left\{ \begin{bmatrix} 1 & 0 \\ -1 & 0 \end{bmatrix}, \begin{bmatrix} 1 & 1 \\ 0 & -1 \end{bmatrix}, \begin{bmatrix} -1 & -3 \\ -2 & 3 \end{bmatrix} \right\}, \ v = \begin{bmatrix} 2 & 3 \\ 1 & 2 \end{bmatrix}$$

   g. $V = M_{22}$,
   $$X = \left\{ \begin{bmatrix} 2 & 3 \\ 4 & 5 \end{bmatrix}, \begin{bmatrix} 1 & -1 \\ 1 & -1 \end{bmatrix}, \begin{bmatrix} 5 & 5 \\ 9 & 9 \end{bmatrix} \right\}, \ v = \begin{bmatrix} 2 & 3 \\ 4 & -5 \end{bmatrix}$$

,

h. $V = M_{22}$,

$$X = \left\{ \begin{bmatrix} 2 & -1 \\ 1 & 3 \end{bmatrix}, \begin{bmatrix} 1 & 0 \\ 1 & 1 \end{bmatrix}, \begin{bmatrix} 3 & -1 \\ 2 & 4 \end{bmatrix} \right\}, \ v = \begin{bmatrix} 1 & 2 \\ 3 & -1 \end{bmatrix}$$

i. $V = P_2$, $X = \{1 - x + 3x^2, \ 2 + 4x^2, \ -1 - 3x + x^2\}$, $v = -1 + 3x - 5x^2$

j. $V = P_2$, $X = \{1 + x + x^2, \ 1 + x, \ 2 - x^2\}$, $v = \frac{1}{2} + 3x^2$

k. $V = P_3$, $X = \{1+x-x^2, \ 2+x-2x^2+x^3, \ 1+2x-x^2-x^3\}$, $v = 1+2x+x^2-x^3$

2. Describe geometrically the span of a single vector in $\mathcal{R}^3$.

3. Suppose that $u \in \text{span}(v_1, v_2)$, $v_1 \in \text{span}(w_1, w_2)$ and $v_2 \in \text{span}(w_3, w_4)$ for some $u, v_1, v_2, w_1, w_2, w_3, w_4 \in V$ a vector space. Show that $u \in \text{span}(w_1, w_2, w_3, w_4)$.

4. Let $v, v_1, v_2, v_3, w_1, w_2 \in V$ a vector space. Show that if $v \in \text{span}(v_1, v_2, v_3)$ and $v_1, v_2, v_3 \in \text{span}(w_1, w_2)$, then $v \in \text{span}(w_1, w_2)$.

5. Let $v, v_1, \ldots, v_k \in V$, a vector space. Suppose that $w \in \text{span}(v, v_1, \ldots, v_k)$ but $w \notin \text{span}(v_1, \ldots, v_k)$. Prove that $\text{span}(w, v_1, \ldots, v_k) = \text{span}(v, v_1, \ldots, v_k)$ (sometimes called the *exchange principle*).

6. Let $v, v_1, \ldots, v_k, w_1, \ldots, w_m \in V$, a vector space. Suppose that $\text{span}(v, v_1, \ldots, v_k) = \text{span}(v, w_1, \ldots, w_m)$, $\text{span}(v_1, \ldots, v_k) \subseteq \text{span}(w_1, \ldots, w_m)$ and $v \notin \text{span}(v_1, \ldots, v_k)$. Prove that $\text{span}(v_1, \ldots, v_k) = \text{span}(w_1, \ldots, w_m)$.

7. Prove that if $v_1, \ldots, v_n$ are linearly dependent and $v_1, \ldots, v_{n-1}$ are linearly independent, then $v_n \in \text{span}(v_1, \ldots, v_{n-1})$.

3.5 BASIS AND DIMENSION

This section introduces what we call a **counting principle** for comparing the relative sizes of vector spaces, called **dimension**. It will conform to our intuition of dimension for $\mathbb{R}^2$ (2-dimensional) and $\mathbb{R}^3$ (3-dimensional), but in addition it will assign dimension to many other vector spaces. Our focus will be on investigating vector spaces of finite dimension.

Definition 3.12 *A set of vectors $v_1, \ldots, v_n \in V$ a vector space is a* **basis** *for V if*

1. *The vectors $v_1, \ldots, v_n$ span V, and*

2. *The vectors $v_1, \ldots, v_n$ are linearly independent.*

Example 3.40 *Take the earlier example. We have already verified that $[1, 0, 0]$, $[1, 1, 0]$, $[1, 1, 1]$ span $\mathbb{R}^3$. Now we show they are linearly independent (and thus a basis for $\mathbb{R}^3$). Again, we use the technique from the Section 3.3. Putting the vectors in columns in a determinant,*

$$\begin{vmatrix} 1 & 1 & 1 \\ 0 & 1 & 1 \\ 0 & 0 & 1 \end{vmatrix} = (1)(1)(1) = 1 \neq 0.$$

Example 3.41 *The necessary generators of a particular span are a basis for that span. We shall formally prove this fact in Section 3.6. However, we illustrate this fact with one of the earlier examples. Although* $[1,0,0]$, $[1,1,0]$, $[0,-2,0]$ *do not span* $\mathbb{R}^3$*, from our calculations above, we know that* $[1,0,0]$, $[0,1,0]$ *form a basis for* span($[1,0,0], [1,1,0], [0,-2,0]$)*, since* $[1,0,0]$, $[0,1,0]$ *are the necessary generators of* span($[1,0,0], [1,1,0], [0,-2,0]$) *and one can check that they are linearly independent.*

We will not give too many examples at this point, because within this section we will shortcut the method of determining basis even further.

Definition 3.13 *The following bases for their respective vector spaces are called* **standard** *bases:*

1. *The standard basis for* $\mathbb{R}^n$ *is the collection of vectors*

$$e_1 = [1,0,\ldots,0], \ e_2 = [0,1,0,\ldots,0], \ \ldots, \ e_n = [0,\ldots,0,1].$$

   Note that in $\mathbb{R}^2$ *the notation for the standard basis is* $\hat{\imath}, \hat{\jmath}$ *and in* $\mathbb{R}^3$ *the notation is* $\hat{\imath}, \hat{\jmath}, \hat{k}$*.*

2. *The standard basis for* P_n *is the collection of vectors* $1, x, x^2, \ldots, x^n$*.*

3. *The standard basis for* P *is the infinite collection of vectors* $1, x, x^2, \ldots$*.*

4. *The standard basis for* M_{mn} *is the collection of vectors*

$$\{E_{ij} \mid 1 \leq i \leq m, \ 1 \leq j \leq n\}$$

   where each E_{ij} *is a matrix filled with zeros except that there is a 1 in the* ijth *entry.*

Example 3.42 *The standard basis for* M_{23} *is*

$$E_{11} = \begin{bmatrix} 1 & 0 & 0 \\ 0 & 0 & 0 \end{bmatrix}, E_{12} = \begin{bmatrix} 0 & 1 & 0 \\ 0 & 0 & 0 \end{bmatrix}, E_{13} = \begin{bmatrix} 0 & 0 & 1 \\ 0 & 0 & 0 \end{bmatrix},$$

$$E_{21} = \begin{bmatrix} 0 & 0 & 0 \\ 1 & 0 & 0 \end{bmatrix}, E_{22} = \begin{bmatrix} 0 & 0 & 0 \\ 0 & 1 & 0 \end{bmatrix}, E_{23} = \begin{bmatrix} 0 & 0 & 0 \\ 0 & 0 & 1 \end{bmatrix}.$$

The result we now prove is an essential fact about bases. Among other reasons we will need it later in this section to define *coordinates*, and coordinates play a major role in linear algebra as we shall see in later sections of this text.

Theorem 3.9 *Let $v_1, \ldots, v_n$ be vectors in a vector space V. Then $v_1, \ldots, v_n$ form a basis for V iff for each $v \in V$ there exist* **unique** *scalars $a_1, \ldots, a_n$ such that*

$$v = a_1 v_1 + \cdots + a_n v_n.$$

Proof 3.13 *Take $v \in V$. By definition of basis (in particular the fact that the basis spans V), we know there exist scalars $a_1, \ldots, a_n$ such that*

$$v = a_1 v_1 + \cdots + a_n v_n.$$

We need to show that these scalars are unique. To show this, assume there were potentially other scalars $b_1, \ldots, b_n$ such that

$$v = b_1 v_1 + \cdots + b_n v_n.$$

Since both linear combinations equal v, we can equate them:

$$a_1 v_1 + \cdots + a_n v_n = b_1 v_1 + \cdots + b_n v_n.$$

Now collect like terms:

$$a_1 v_1 + \cdots + a_n v_n - (b_1 v_1 + \cdots + b_n v_n) = 0.$$

$$(a_1 - b_1) v_1 + \cdots + (a_1 - b_1) v_n = 0.$$

We now have a linear combination of $v_1, \ldots, v_n$ equaling zero. Since $v_1, \ldots, v_n$ are linearly independent this implies

$$a_1 - b_1 = 0, \ \ldots \ , a_n - b_n = 0,$$

and so

$$a_1 = b_1, \ \ldots \ , a_n = b_n.$$

Hence, we have proved the uniqueness of the representation. We leave the proof of the reverse implication as an exercise. □

Our goal now is to prove an important fact about vector spaces and their bases. Consider the following example for $V = \mathbb{R}^3$. Both $\hat{\imath}, \hat{\jmath}, \hat{k}$ and $[1, 0, 0], [1, 1, 0], [1, 1, 1]$ are bases for V, and one can find many more bases for $\mathbb{R}^3$. But one fact remains constant about these bases for $\mathbb{R}^3$: Although a given vector space may have many bases, the number of vectors in each basis is always the same; we say the number of elements in the basis for a given vector space is an *invariant*. This invariant will be what we will call the dimension of the vector space. To prove the invariance of dimension we need the following lemma:

Lemma 3.6 *Given a vector space V, suppose $v_1, \ldots, v_n \in V$ span V and $w_1, \ldots, w_k \in V$ are linearly independent. Then $k \leq n$.*

Proof 3.14 *We prove this statement by contradiction. Suppose $k > n$. Since $v_1, \ldots, v_n$ span V, for $1 \leq i \leq k$ there exist scalars $a_{1i}, \ldots, a_{ni}$ such that $w_i = a_{1i}v_1 + \cdots + a_{ni}v_n$. Now consider the following homogeneous linear system of equations:*

$$\begin{cases} a_{11}x_1 + \cdots + a_{1k}x_k &= 0 \\ \vdots \\ a_{n1}x_1 + \cdots + a_{nk}x_k &= 0 \end{cases}$$

Since $k > n$, by Corollary 2.4, the linear system above has a non-trivial solution. In other words there exist scalars $b_1, \ldots, b_k$ not all zero such that

$$a_{11}b_1 + \cdots + a_{1k}b_k = 0$$
$$\vdots$$
$$a_{n1}b_1 + \cdots + a_{nk}b_k = 0$$

Notice then that

$$b_1 w_1 + \cdots + b_k w_k = b_1(a_{11}v_1 + \cdots + a_{n1}v_n) + \cdots + b_k(a_{1k}v_1 + \cdots + a_{nk}v_n)$$

$$= (a_{11}b_1 + \cdots + a_{1k}b_k)v_1 + \cdots + (a_{n1}b_1 + \cdots + a_{nk}b_k)v_n = 0v_1 + \cdots + 0v_n = 0.$$

Since $b_1, \ldots, b_k$ are not all zero, this implies that $w_1, \ldots, w_k$ are linearly dependent, contradicting our assumption. Hence, it must be the case that $k \leq n$. □

Corollary 3.2 *Let V be a vector space. If $v_1, \ldots, v_n \in V$ is a basis for V and $w_1, \ldots, w_k$ is a basis for V, then $n = k$.*

Proof 3.15 *By definition of basis, $v_1, \ldots, v_n$ span V and $w_1, \ldots, w_k$ are linearly independent. Thus, by Lemma 3.6, $k \leq n$. Now reverse the roles: $w_1, \ldots, w_k$ span V and $v_1, \ldots, v_n$ are linearly independent. Again, by Lemma 3.6, $n \leq k$. The above two inequalities yield the desired equality.* □

It is precisely this corollary that allows us to make the following definition:

Definition 3.14 *The* **dimension** *of a vector space V, written $\dim(V)$ (or just $\dim V$) is the number of elements in* <u>*any*</u> *basis for V.*

A few remarks are in order here.

1. We emphasize that the corollary ensures us that dimension is *well-defined* (in the case when a basis has a finite number of elements). In other words for a given vector space V, although we might find different bases for V, our bases will always have the same number of vectors and thus we will always be in agreement on the dimension of V.

2. If $V = \{0\}$, we define $\dim V = 0$—essentially because V has no basis, and also to make the dimension theorems in later sections (for instance, Section 3.7) work out nicely.

3. If V has a basis which is infinite, for the purposes of this text we assign it the dimension ∞ (although, infinite dimensional vector spaces can be given a much more in depth treatment).

4. The reader should also keep in mind that we have in no way shown that every vector space is guaranteed to have a basis, however this can be proved in a more advanced mathematical setting.

Example 3.43 *Let's use the standard bases for $\mathbb{R}^n, P_n, P$ and M_{mn} to compute their dimensions:* $\dim \mathbb{R}^n = n$, $\dim P_n = n + 1$, $\dim P = \infty$, $\dim M_{mn} = mn$.

Example 3.44 *Referring to an earlier example in this section,*

$$\dim(\text{span}([1, 0, 0], [1, 1, 0], [0, -2, 0])) = 2,$$

since $[1, 0, 0], [0, 1, 0]$ form a basis for $\text{span}([1, 0, 0], [1, 1, 0], [0, -2, 0])$.

We illustrate in the example below a simple but practical observation about bases: If you know the dimension of a vector space, then for a collection of vectors to form a basis for that vector space, it is necessary that the number of vectors in that collection equal the dimension of the vector space.

Example 3.45 *Notice that $1, 1 + x, 1 + x^2, x + x^2$ has no hope of being a basis for P_2 since there are too many vectors. Indeed, $\dim P_2 = 3$.*

The vectors $[1, 2, 3, 4], [4, 3, 2, 1]$ have no hope of being a basis for $\mathbb{R}^4$ since there are too few vectors. Indeed, $\dim \mathbb{R}^4 = 4$.

In order to further streamline the verification of basis, we prove two results below and then give their practical implications.

Lemma 3.7 *Let $v_1, \ldots, v_n$ be linearly independent vectors in a vector space V. If for $v \in V$ it is the case that $v \notin \text{span}(v_1, \ldots, v_n)$, then $v, v_1, \ldots, v_n$ are linearly independent.*

Proof 3.16 *Given the assumptions of the lemma, suppose to the contrary that $v, v_1, \ldots, v_n$ were linearly dependent. This implies that there exist scalars $a, a_1, \ldots, a_n$ not all zero such that $av + a_1 v_1 + \cdots + a_n v_n = 0$. Note that a cannot be zero, otherwise this would imply that $v_1, \ldots, v_n$ were linearly dependent. Hence, $1/a$ exists and we can solve for v as follows:*

$$av = -a_1 v_1 - \cdots - a_n v_n$$

$$v = (-a_1/a)v_1 + \cdots + (-a_n/a)v_n.$$

This last equation implies that $v \in \text{span}(v_1, \ldots, v_n)$ contrary to our assumptions. Hence, it must be that case that $v, v_1, \ldots, v_n$ are linearly independent. □

Theorem 3.10 *Let V be a vector space with $\dim V = n$ and $v_1, \ldots, v_n \in V$. Then*

1. *If $v_1, \ldots, v_n$ are linearly independent, then they form a basis for V.*

2. *If $v_1, \ldots, v_n$ span V, then they form a basis for V.*

Proof 3.17 *Let $w_1, \ldots, w_n$ be a basis for V so that $\dim V = n$. To prove i, suppose to the contrary that $v_1, \ldots, v_n$ is not a basis for V, i.e. $v_1, \ldots, v_n$ do not span V. This implies that $\text{span}(v_1, \ldots, v_n) \neq V$ and so there is a $v \in V$ which is not in $\text{span}(v_1, \ldots, v_n)$. Then by Lemma 3.7, $v, v_1, \ldots, v_n$ are linearly independent. Since $w_1, \ldots, w_n$ span V, by Lemma 3.6, $n + 1 \leq n$ which is an obvious contradiction. Hence it must be the case that $v_1, \ldots, v_n$ form a basis for V.*

To prove ii, suppose to the contrary that $v_1, \ldots, v_n$ is not a basis for V, i.e. $v_1, \ldots, v_n$ are linearly dependent. By Lemma 3.4, there exists a v_i linearly dependent on $v_1, \ldots, v_{i-1}, v_{i+1}, \ldots, v_n$. In other words $v_i \in \text{span}(v_1, \ldots, v_{i-1}, v_{i+1}, \ldots, v_n)$. By Lemma 3.5,

$$\text{span}(v_1, \ldots, v_{i-1}, v_{i+1}, \ldots, v_n) = \text{span}(v_1, \ldots, v_n) = V.$$

Hence, the $n - 1$ vectors $v_1, \ldots, v_{i-1}, v_{i+1}, \ldots, v_n$ span V. Since $w_1, \ldots, w_n$ are linearly independent, by Lemma 3.6, $n \leq n - 1$ an obvious contradiction. Thus, it must be the case that $v_1, \ldots, v_n$ is a basis for V. □

Now we investigate some of the consequences of these two results above by way of examples.

Example 3.46 *We show that* $[1,0,0]$, $[1,2,0]$, $[1,2,3]$ *form a basis for* $\mathbb{R}^3$. *These vectors are linearly independent since*

$$\begin{vmatrix} 1 & 1 & 1 \\ 0 & 2 & 2 \\ 0 & 0 & 3 \end{vmatrix} = (1)(2)(3) = 6 \neq 0.$$

Since $\dim\mathbb{R}^3 = 3$, *by Lemma 3.10.i,* $[1,0,0], [1,2,0], [1,2,3]$ *form a basis for* $\mathbb{R}^3$.

Example 3.47 *We show that* $1, 1+x, 1+x^2$ *form a basis for* P_2.
We show these vectors span P_2, *by direct computation.*

$$\begin{bmatrix} 1 & 0 & 0 \\ 1 & 1 & 0 \\ 1 & 0 & 1 \end{bmatrix} \text{ reduces to } \begin{bmatrix} 1 & 0 & 0 \\ 0 & 1 & 0 \\ 0 & 0 & 1 \end{bmatrix}.$$

Hence, $\text{span}(1, 1+x, 1+x^2) = \text{span}(1, x, x^2) = P_2$. *Since* $\dim P_2 = 3$, *by Lemma 3.10.ii,* $1, 1+x, 1+x^2$ *form a basis for* P_2.

Example 3.48 *This example illustrates a quick way to extract a basis for a vector space (or subspace) when we have a nice description of it. Consider the following subspace of* M_{22}:

$$U = \left\{ \begin{bmatrix} a & b \\ c & 2a-b+c \end{bmatrix} : a,b,c \in \mathbb{R} \right\}.$$

Notice that

$$\begin{bmatrix} a & b \\ c & 2a-b+c \end{bmatrix} = \begin{bmatrix} a & 0 \\ 0 & 2a \end{bmatrix} + \begin{bmatrix} 0 & b \\ 0 & -b \end{bmatrix} + \begin{bmatrix} 0 & 0 \\ c & c \end{bmatrix} =$$

$$a\begin{bmatrix} 1 & 0 \\ 0 & 2 \end{bmatrix} + b\begin{bmatrix} 0 & 1 \\ 0 & -1 \end{bmatrix} + c\begin{bmatrix} 0 & 0 \\ 1 & 1 \end{bmatrix}.$$

The basis for U *is then*

$$\begin{bmatrix} 1 & 0 \\ 0 & 2 \end{bmatrix}, \begin{bmatrix} 0 & 1 \\ 0 & -1 \end{bmatrix}, \begin{bmatrix} 0 & 0 \\ 1 & 1 \end{bmatrix}.$$

Indeed, these matrices span U *(the work above shows that every matrix in* U *is a linear combination of the three matrices), and it is easy to show that the three matrices are linearly independent.*

Here is an easier way to obtain the three matrices. Consider the arbitrary matrix in U *as defined above. Set* $a = 1, b = 0, c = 0$ *to obtain the first matrix; set* $a = 0, b = 1, c = 0$ *to obtain the second; say* $a = 0, b = 0, c = 1$ *to obtain the third.*

In sum, if we know the dimension of a vector space and we have a set of vectors whose number equals that dimension, then we need only verify one of the two requirements for being a basis: either linear independence or span. As a rule of thumb, linear independence is usually the easier of the two to verify.

The next example illustrates a method for taking a set of linearly independent vectors and adding additional vectors in order to obtain a basis. We call this process **extending to a basis**. The process is algorithmic and proceeds as follows:

1. Start with a collection of linearly independent vectors $v_1, \ldots, v_k \in V$ with $\dim V = n$.

2. If $k = n$ then $v_1, \ldots, v_k$ form a basis for V (Lemma 3.10) and stop the process. Otherwise, proceed to the next step.

3. Compute $\text{span}(v_1, \ldots, v_k)$.

4. Select any $v_{k+1} \notin \text{span}(v_1, \ldots, v_k)$. Note that by Lemma 3.7, $v_1, \ldots, v_k, v_{k+1}$ are linearly independent. Go back to the second step (replacing k by $k + 1$).

Example 3.49 *Let's apply the above process to linearly independent vectors* $[1, 0, 1], [1, 1, 0] \in \mathbb{R}^3$ *(where $k = 2$ and $n = 3$). We proceed to step 3.*

$$\begin{bmatrix} 1 & 0 & 1 \\ 1 & 1 & 0 \end{bmatrix} \text{ which reduces to } \begin{bmatrix} 1 & 0 & 1 \\ 0 & 1 & -1 \end{bmatrix}.$$

Hence,

$$\text{span}([1, 0, 1], [1, 1, 0]) = \text{span}([1, 0, 1], [0, 1, -1]) =$$

$$\{a[1, 0, 1] + b[0, 1, -1] \mid a, b \in \mathbb{R}\} = \{[a, b, a - b] \mid a, b \in \mathbb{R}\}.$$

Certainly, $[1, 0, 0] \notin \text{span}([1, 0, 1], [1, 1, 0])$. Therefore, $[1, 0, 1], [1, 1, 0], [1, 0, 0]$ are linearly independent and in fact form a basis for $\mathbb{R}^3$ (since $k + 1 = 3 = n$).

There is a much simpler way to extend a set of linearly independent vectors to a basis for a given vector space, especially if you are using computer software to do your computation. The problem with the previous algorithm from a computer's point of view is the fourth step where one has to make a choice. Although this is easy for a human to do, for a computer this is not easy, i.e. how do we write code so that we choose systematically a vector outside of a span—even when we have a nice description of the span? We now present a more deterministic way of extending to a basis.

1. Drop your original set of linearly independent vectors into columns of a matrix and put alongside them (in columns) the standard basis for the given vector space.

2. Row reduce the matrix into row-echelon form (as was done for computing rank) and identify the pivots.

3. The columns in the original matrix corresponding to these pivots will form a basis for the given vector space.

At the end of Section 3.4 we discussed why this algorithm will yield a basis.

Example 3.50 *Let's take the same set of vectors in* $\mathbb{R}^3$, *namely* $[1, 0, 1], [1, 1, 0]$. *As described above, we form the matrix*

$$
\begin{bmatrix}
1 & 1 & 1 & 0 & 0 \\
0 & 1 & 0 & 1 & 0 \\
1 & 0 & 0 & 0 & 1
\end{bmatrix}
\quad \textit{which reduces to} \quad
\begin{bmatrix}
1 & 0 & 0 & 0 & 1 \\
0 & 1 & 0 & 1 & 0 \\
0 & 0 & 1 & -1 & -1
\end{bmatrix}.
$$

Hence, the first three columns are the pivot columns and looking at the original matrix we take for our basis $[1, 0, 1], [1, 1, 0], [1, 0, 0]$.

Example 3.51 *Let's repeat the algorithm in a different setting. In the vector space* M_{22} *we extend the following set of vectors to a basis for* M_{22}:

$$
\begin{bmatrix} 2 & 1 \\ 0 & 0 \end{bmatrix}, \begin{bmatrix} -1 & -2 \\ 2 & -1 \end{bmatrix}.
$$

As described above we put these two matrices into the columns of a matrix and put the four standard basis matrices alongside them (in columns).

$$
\begin{bmatrix}
2 & -1 & 1 & 0 & 0 & 0 \\
1 & -2 & 0 & 1 & 0 & 0 \\
0 & 2 & 0 & 0 & 1 & 0 \\
0 & -1 & 0 & 0 & 0 & 1
\end{bmatrix}
\quad \textit{which reduces to} \quad
\begin{bmatrix}
1 & 0 & 0 & 1 & 0 & -2 \\
0 & -1 & 0 & 0 & 0 & 1 \\
0 & 0 & 1 & -2 & 0 & 3 \\
0 & 0 & 0 & 0 & 1 & 2
\end{bmatrix}.
$$

Hence, the pivots columns are the first, second, third and fifth. Looking at the original matrix we choose our basis for M_{22} *to be*

$$
\begin{bmatrix} 2 & 1 \\ 0 & 0 \end{bmatrix}, \begin{bmatrix} -1 & -2 \\ 2 & -1 \end{bmatrix} \begin{bmatrix} 1 & 0 \\ 0 & 0 \end{bmatrix}, \begin{bmatrix} 0 & 0 \\ 1 & 0 \end{bmatrix}.
$$

We add yet another extension of Theorem 3.8. The addition of this new statement is certainly self-evident in view of statements viii. and ix. and Theorem 3.10.

Theorem 3.11 *For an $n \times n$ square matrix A the following are equivalent:*

i. A is invertible.

ii. A is equivalent to I.

iii. The linear system $AX = B$ has a unique solution for any $B \in \mathbb{R}^n$.

iv. A is a product of elementary matrices.

v. The linear system $AX = 0$ has only the trivial solution.

vi. $rk(A) = n$.

vii. $|A| \neq 0$.

viii. The columns (or rows) of A are linearly independent in $\mathbb{R}^n$.

ix. The columns (or rows) of A span $\mathbb{R}^n$.

x. The columns (or rows) of A form a basis for $\mathbb{R}^n$.

The last topic in this section is the notion of *coordinates*. The definition of coordinates relies on the result of Theorem 3.9.

Definition 3.15 *Let $v_1, \ldots, v_n$ be a basis for a vector space V. Then $(v_1, \ldots, v_n)$ is called an* **ordered** *basis.*

Ordered bases function as ordered tuples do. For instance, the following ordered bases are not equal: $(v_1, v_2) \neq (v_2, v_1)$ even though the sets, $\{v_1, v_2\}$ and $\{v_2, v_1\}$ are equal. In other words the order in which basis vectors are listed has become important. It will be useful as well to adopt the notation $B = (v_1, \ldots, v_n)$.

Definition 3.16 *Let $B = (v_1, \ldots, v_n)$ be an ordered basis for a vector space V and take $v \in V$. By Theorem 3.9, there exist* **unique** *scalars $a_1, \ldots, a_n$ such that $v = a_1 v_1 + \cdots + a_n v_n$. These scalars $a_1, \ldots, a_n$ are called the* **coordinates of v with respect to the basis** *B.*

We normally represent these coordinates as an n-tuple in $\mathbb{R}^n$, i.e. as $[a_1, \ldots, a_n]$. We also adopt the notation $[v]_B = [a_1, \ldots, a_n]$ for the coordinates of v with respect to the ordered basis B.

Note that the coordinates of a vector v will change when the ordered basis B changes (even if simply the order of the basis vectors should change). In other words, the coordinates we find for v are completely dependent upon the ordered basis in question.

Example 3.52 *Consider the following ordered basis for $\mathbb{R}^3$:*

$$B = ([1, 0, 0], [1, 1, 0], [1, 1, 1]).$$

We compute $[v]_B$ where $v = [1, 2, -3]$. In other words, we want to find scalars a, b, c such that

$$[1, 2, -3] = a[1, 0, 0] + b[1, 1, 0] + c[1, 1, 1].$$

Combining the righthand side yields

$$[1, 2, -3] = [a + b + c, b + c, c].$$

Equating coordinates gives

$$\begin{cases} a + b + c &=& 1 \\ b + c &=& 2 \\ c &=& -3 \end{cases}.$$

Thus, finding coordinates reduces to solving a system of linear equations (as was stressed to the reader, many computations in this text reduce to a linear system of equations). We switch to the augmented matrix.

$$\begin{bmatrix} 1 & 1 & 1 & 1 \\ 0 & 1 & 1 & 2 \\ 0 & 0 & 1 & -3 \end{bmatrix} \text{ which reduces to } \begin{bmatrix} 1 & 0 & 0 & -1 \\ 0 & 1 & 0 & 5 \\ 0 & 0 & 1 & -3 \end{bmatrix}.$$

Therefore, $a = -1, b = 5, c = -3$ and so $[1, 2, -3]_B = [-1, 5, -3]$. One can leave out the extra square brackets and simply write $[1, 2, -3]_B = [-1, 5, -3]$.

As usual, we can shortcut some of this work. Notice in the example above, we could have immediately formed an augmented matrix where the columns on the left of the bar are the elements of the ordered basis (which must be listed in their proper order) and the column on the right of the bar is the vector for which we wish to find the coordinates.

It will be useful to introduce some additional notation. The ordered basis ST will represent the **standard** ordered basis. For instance, the standard ordered basis for $\mathbb{R}^3$ is $ST = (\hat{i}, \hat{j}, \hat{k})$. Notice that in the example above, $[1, 2, -3]_{ST} = [1, 2, -3]$. In other words, there is no difference between a vector in $\mathbb{R}^n$ and its coordinates with respect to the standard ordered basis for $\mathbb{R}^n$. Refer to the previous section for the other standard bases.

Example 3.53 *Consider the following ordered basis for P_2: $B = (1, 1 + x, 1 + x^2)$. We find the coordinates of $v = 1 - x + 2x^2$ with respect to B. As in the previous example, we can immediately form the augmented matrix*

$$
\begin{matrix} 1 \\ x \\ x^2 \end{matrix}
\left[\begin{array}{ccc|c} 1 & 1 & 1 & 1 \\ 0 & 1 & 0 & -1 \\ 0 & 0 & 1 & 2 \end{array} \right]
\quad \textit{which reduces to} \quad
\left[\begin{array}{ccc|c} 1 & 0 & 0 & 0 \\ 0 & 1 & 0 & -1 \\ 0 & 0 & 1 & 2 \end{array} \right].
$$

Hence, $[1 - x + 2x^2]_B = [0, -1, 2]$.

Example 3.54 *Consider the following ordered basis for M_{22}:*

$$
B = \left(\begin{bmatrix} 1 & 0 \\ 0 & 0 \end{bmatrix}, \begin{bmatrix} 1 & 1 \\ 0 & 0 \end{bmatrix}, \begin{bmatrix} 1 & 1 \\ 1 & 0 \end{bmatrix}, \begin{bmatrix} 1 & 1 \\ 1 & 1 \end{bmatrix} \right).
$$

We find the coordinates of $\begin{bmatrix} 1 & -1 \\ 2 & 3 \end{bmatrix}$ with respect to the basis B. Again, the shortcut allows us to unwind the matrices into columns:

$$
\left[\begin{array}{cccc|c} 1 & 1 & 1 & 1 & 1 \\ 0 & 1 & 1 & 1 & -1 \\ 0 & 0 & 1 & 1 & 2 \\ 0 & 0 & 0 & 1 & 3 \end{array} \right]
\quad \textit{which reduces to} \quad
\left[\begin{array}{cccc|c} 1 & 0 & 0 & 0 & 2 \\ 0 & 1 & 0 & 0 & -3 \\ 0 & 0 & 1 & 0 & -1 \\ 0 & 0 & 0 & 1 & 3 \end{array} \right].
$$

Hence, $\begin{bmatrix} 1 & -1 \\ 2 & 3 \end{bmatrix}_B = [2, -3, -1, 3].$

An important matrix which will appear in later sections of this text is now defined. It will play an integral role in what we shall define as *matrix similarity*.

Definition 3.17 *Let V be a vector space with $B = (v_1, \ldots, v_n)$ and $B' = (v_1', \ldots, v_n')$, two ordered bases for V. The **change of basis matrix from B' to B** is a matrix $P \in M_{nn}$ defined by*

$$
P = [\ [v_1']_B \ [v_2']_B \ \cdots \ [v_n']_B \ ] \quad \textit{(represented in columns)}.
$$

This matrix gets its name from the fact that given any vector $v \in V$ if we multiply the coordinates of v with respect to B' on the left by P, the result returns the coordinates of v with respect to B. We summarize this result in the following proposition (proof withheld until a later section):

Proposition 3.1 *Let V be a vector space with $B = (v_1, \ldots, v_n)$ and $B' = (v_1', \ldots, v_n')$, two ordered bases for V. Then*

1. *For any $v \in V$ we have $P[v]_{B'} = [v]_B$.*

2. *P^{-1} is the change of basis matrix from B to B'.*

Note that given ordered bases B and B', the change of basis matrix from B' to B is unique since coordinates of vectors are uniquely determined. For the time being let us simply illustrate the Proposition with a couple of examples.

Example 3.55 *Let $V = \mathbb{R}^2$ with ordered bases $B = ([1,0], [1,1])$ and $B' = ([1,1], [0,1])$. Since*

$$[1,1] = (0)[1,0] + (1)[1,1] \quad and \quad [0,1] = (-1)[1,0] + (1)[1,1]$$

it follows that

$$P = \begin{bmatrix} 0 & -1 \\ 1 & 1 \end{bmatrix}.$$

The general algorithm for finding P is as follows: Set up an augmented matrix with B on the left of the bar and B' on the right, i.e. $[B|B']$ (in columns). Then row reduce to reduced row-echelon form to get $[I|P]$. To the right of the bar is P.

Example 3.56 *Let's redo Example 3.55 with the algorithm described above. We set up the augmented matrix*

$$\left[\begin{array}{cc|cc} 1 & 1 & 1 & 0 \\ 0 & 1 & 1 & 1 \end{array} \right] \quad which \ reduces \ to \quad \left[\begin{array}{cc|cc} 1 & 0 & 0 & -1 \\ 0 & 1 & 1 & 1 \end{array} \right]$$

Hence, once again,

$$P = \begin{bmatrix} 0 & -1 \\ 1 & 1 \end{bmatrix}.$$

Now take some vector, say $v = [2, -1] \in V$. Since

$$v = (2)[1,1] + (-3)[0,1] \quad we \ have \quad [v]_{B'} = [2, -3].$$

Furthermore,

$$v = (3)[1,0] + (-1)[1,1] \quad we \ have \quad [v]_B = [3, -1].$$

As the Proposition guarantees, these last coordinates, $[v]_B$ can be obtained via the product $P[v]_{B'}$ and indeed as the reader can check this is the case. We leave it to the reader to check that the change of basis matrix from B to B' coincides with P^{-1}.

Example 3.57 *Let* $V = P_2$, $B = (1 + x, 1 + x^2, x + x^2)$ *and* $B' = (x^2, -1 + x, 1 + x + x^2)$. *First we compute the change of basis matrix by row reducing*

$$\left[\begin{array}{ccc|ccc} 1 & 1 & 0 & 0 & -1 & 1 \\ 1 & 0 & 1 & 0 & 1 & 1 \\ 0 & 1 & 1 & 1 & 0 & 1 \end{array}\right] \quad \text{which reduces to} \quad \left[\begin{array}{ccc|ccc} 1 & 0 & 0 & -1/2 & 0 & 1/2 \\ 0 & 0 & 1 & 1/2 & -1 & 1/2 \\ 0 & 0 & 1 & 1/2 & 1 & 1/2 \end{array}\right].$$

Therefore,

$$P = \left[\begin{array}{ccc} -1/2 & 0 & 1/2 \\ 1/2 & -1 & 1/2 \\ 1/2 & 1 & 1/2 \end{array}\right].$$

Now take $v = 1 + 2x - x^2$. *First we find* $[v]_{B'}$ *by row reducing*

$$\left[\begin{array}{ccc|c} 0 & -1 & 1 & 1 \\ 0 & 1 & 1 & 2 \\ 1 & 0 & 1 & -1 \end{array}\right] \quad \text{which reduces to} \quad \left[\begin{array}{ccc|c} 1 & 0 & 0 & -5/2 \\ 0 & 0 & 1 & 1/2 \\ 0 & 0 & 1 & 3/2 \end{array}\right].$$

Therefore $[1 + 2x - x^2]_{B'} = [-5/2, 1/2, 3/2]$. *Next we find* $[v]_B$ *by row reducing*

$$\left[\begin{array}{ccc|c} 1 & 1 & 0 & 1 \\ 1 & 0 & 1 & 2 \\ 0 & 1 & 1 & -1 \end{array}\right] \quad \text{which reduces to} \quad \left[\begin{array}{ccc|c} 1 & 0 & 0 & 2 \\ 0 & 0 & 1 & -1 \\ 0 & 0 & 1 & 0 \end{array}\right].$$

Therefore $[1 + 2x - x^2]_{B'} = [2, -1, 0]$. *Now one can verify the Proposition by multiplying*

$$\left[\begin{array}{ccc} -1/2 & 0 & 1/2 \\ 1/2 & -1 & 1/2 \\ 1/2 & 1 & 1/2 \end{array}\right]\left[\begin{array}{c} -5/2 \\ 1/2 \\ 3/2 \end{array}\right] = \left[\begin{array}{c} 2 \\ -1 \\ 0 \end{array}\right].$$

EXERCISES

1. For each set of vectors X in a vector space V answer the following questions:

   i. Is X a linearly independent set of vectors?

   ii. Does X span V?

   iii. Do the vectors X form a basis for V?

   a. $X = \{[1, 1, 1], [1, 1, 0], [1, 0, 0]\}$ and $V = \mathbb{R}^3$.

   b. $X = \{[1, 0, 0], [1, 2, 0], [1, 2, 3]\}$ and $V = \mathbb{R}^3$.

c. $X = \{1, 1 + x, 1 + x^2, x + x^2\}$ and $V = P_2$.

d. $X = \{1, 1 + x^3, x + x^2\}$ and $V = P_3$.

e.

$$X = \left\{ \begin{bmatrix} 1 & 0 \\ 0 & 0 \end{bmatrix}, \begin{bmatrix} 1 & 1 \\ 0 & 0 \end{bmatrix}, \begin{bmatrix} 1 & 1 \\ 0 & 1 \end{bmatrix} \right\} \qquad \text{and} \qquad V = M_{22}.$$

f.

$$X = \left\{ \begin{bmatrix} 1 & 0 \\ 0 & 0 \end{bmatrix}, \begin{bmatrix} 1 & 2 \\ 0 & 0 \end{bmatrix}, \begin{bmatrix} 0 & 1 \\ 0 & 2 \end{bmatrix}, \begin{bmatrix} 1 & 2 \\ 0 & 3 \end{bmatrix} \right\} \qquad \text{and} \qquad V = U_{22}.$$

2. Referring to Exercise 1 in Section 3.4, answer the following additional questions:

    v. Decide whether or not X spans V.

    vi. Decide whether or not X is a basis for V.

3. Consider the following subspace of M_{22}:

$$U = \left\{ \begin{bmatrix} a & b \\ a - b + 2c & c \end{bmatrix} : a, b, c \in \mathbb{R} \right\}.$$

    a. Find a basis for U.

    b. Verify that the vectors in part a. are indeed a basis for U.

    c. What is the dimension of U?

4. Repeat the previous exercise for the following subspace of P_3:

$$U = \{a + bx + cx^2 + dx^3 \mid d = -2a + 3c\}.$$

5. Consider the following system of linear equations:

$$\begin{cases} 2x_1 + x_2 + 3x_3 + x_4 &= 0 \\ x_1 + x_2 + x_3 + x_4 &= 0 \end{cases}.$$

    a. Find the solution set for the linear system.

    b. Find a basis for the solution set in part a.

    c. Find the dimension of the solution set.

6. Consider the vector space L_{22} and the set of vectors X and the vector v:

$$X = \left\{ \begin{bmatrix} 1 & 0 \\ 0 & 2 \end{bmatrix}, \begin{bmatrix} 1 & 0 \\ 1 & 3 \end{bmatrix}, \begin{bmatrix} 1 & 0 \\ -2 & 0 \end{bmatrix} \right\} \qquad v = \begin{bmatrix} -2 & 0 \\ 3 & -1 \end{bmatrix}.$$

a. Find a basis for span(X) and give a nice description of $span(X)$.

b. Decide whether or not $v \in span(X)$.

c. Decide whether or not X spans L_{22}.

d. What is the dimension of $span(X)$?

7. Using all the shortcuts introduced in this section, decide whether or not each of the following set of vectors form a basis for the given vector space V.

a. $V = \mathbb{R}^3$ and vectors $[1, 0, 0]$, $[1, 2, 0]$, $[1, 2, 3]$, $[0, 2, 3]$.

b. $V = \mathbb{R}^3$ and vectors $[1, 0, -1]$, $[0, 1, 2]$, $[1, 1, -1]$.

c. $V = \mathbb{R}^3$ and vectors $[1, 0, 0]$, $[1, 2, 0]$.

d. $V = \mathbb{R}^3$ and vectors $[1, 0, 0, 0]$, $[1, 2, 0, 0]$, $[0, 2, 3, 0]$, $[0, -1, 2, 1]$.

e. $V = M_{22}$ and vectors

$$\begin{bmatrix} 1 & 1 \\ 0 & 0 \end{bmatrix}, \begin{bmatrix} 0 & 1 \\ 1 & 0 \end{bmatrix}, \begin{bmatrix} 0 & 0 \\ 1 & 1 \end{bmatrix}, \begin{bmatrix} 0 & 1 \\ 0 & 1 \end{bmatrix}, \begin{bmatrix} 1 & 0 \\ 0 & 1 \end{bmatrix}.$$

f. $V = M_{22}$ and vectors

$$\begin{bmatrix} 1 & 1 \\ 0 & 0 \end{bmatrix}, \begin{bmatrix} 0 & 1 \\ 1 & 0 \end{bmatrix}, \begin{bmatrix} 0 & 0 \\ 1 & 1 \end{bmatrix}, \begin{bmatrix} 1 & 0 \\ 0 & 1 \end{bmatrix}.$$

g. $V = M_{22}$ and vectors

$$\begin{bmatrix} 1 & 1 \\ 0 & 0 \end{bmatrix}, \begin{bmatrix} 0 & 1 \\ 1 & 0 \end{bmatrix}, \begin{bmatrix} 0 & 0 \\ 1 & 1 \end{bmatrix}, \begin{bmatrix} 1 & 0 \\ 0 & 1 \end{bmatrix}, \begin{bmatrix} 0 & 1 \\ 0 & 1 \end{bmatrix}.$$

h. $V = U_{22}$ and vectors

$$\begin{bmatrix} 1 & 1 \\ 0 & 0 \end{bmatrix}, \begin{bmatrix} 1 & 0 \\ 0 & 1 \end{bmatrix}, \begin{bmatrix} 0 & 1 \\ 0 & 1 \end{bmatrix}.$$

i. $V = P_2$ and vectors $1 + x$, $x + x^2$.

j. $V = P_3$ and vectors $1 + x, 1 + x^2, 1 + x^3$.

k. $V = P_3$ and vectors $1 + x, 1 + x^2, 1 + x^3, 1 + x + x^2 + x^3$.

8. For each set of linearly independent vectors X in a vector space V, extend X to a basis for V as done in the section.

a. $X = \{[1, 0, -2], [0, -1, 1]\}$ for $V = \mathbb{R}^3$.

b. $X = \{[0, 0, 1, 1], [1, 1, 0, 0]\}$ for $V = \mathbb{R}^4$.

c.

$$X = \left\{ \begin{bmatrix} 1 & -2 \\ -1 & 0 \end{bmatrix}, \begin{bmatrix} 2 & 0 \\ 1 & -1 \end{bmatrix} \right\} \qquad \text{for} \qquad V = M_{22}.$$

d. $X = \{1 - x^2\}$ for $V = P_2$.

e. $X = \{1 + x, x^2 + x^3\}$ for $V = P_3$.

9. For each vector space V with $v \in V$ and B an ordered basis, compute $[v]_B$.

a. $V = \mathbb{R}^3$, $\quad v = [2, -3, 4]$, $\quad B = ([1, 2, -3], [1, 0, -2], [2, -1, 4])$

b. $V = P_2$, $\quad v = 1 - x + 3x^2$, $\quad B = (1 - 2x, x + 3x^2, 1 + x - x^2)$

c. $V = M_{22}$,

$$v = \begin{bmatrix} 1 & -2 \\ 3 & -1 \end{bmatrix},$$

$$B = \left(\begin{bmatrix} 2 & 0 \\ 0 & 0 \end{bmatrix}, \begin{bmatrix} 1 & -1 \\ 2 & 0 \end{bmatrix}, \begin{bmatrix} -1 & 1 \\ 0 & 0 \end{bmatrix}, \begin{bmatrix} 1 & 1 \\ 1 & 1 \end{bmatrix} \right).$$

d. $V = \mathbb{R}^3$, $\quad v = [1, -2, -1]$, $\quad B = ([1, 0, 0], [1, 1, 0], [0, 1, 1])$

e. $V = D_{22}$,

$$v = \begin{bmatrix} 2 & 0 \\ 0 & -1 \end{bmatrix},$$

$$B = \left(\begin{bmatrix} 1 & 0 \\ 0 & 1 \end{bmatrix}, \begin{bmatrix} 0 & 0 \\ 0 & -1 \end{bmatrix} \right).$$

10. For each vector space V with ordered bases B and B' compute P the change of basis matrix from B' to B. Then for the given vector v verify Proposition 3.1.1

a. $V = \mathbb{R}^2$, $\quad v = [2, -3]$, $\quad B = ([2, -3], [-1, 4])$ and $B' = ([1, 2], [0, -2])$.

b. $V = P_1$, $\quad v = 1 - x$, $\quad B = (1 - 2x, 1 + x)$ and $B' = (1 - x, 1 + 2x)$.

c. $V = D_{22}(F)$,

$$v = \begin{bmatrix} 2 & 0 \\ 0 & -1 \end{bmatrix},$$

$$B = \left(\begin{bmatrix} 1 & 0 \\ 0 & 1 \end{bmatrix}, \begin{bmatrix} 0 & 0 \\ 0 & -1 \end{bmatrix} \right),$$

$$B' = \left(\begin{bmatrix} 0 & 0 \\ 0 & -2 \end{bmatrix}, \begin{bmatrix} 1 & 0 \\ 0 & -1 \end{bmatrix} \right).$$

11. Let $V = P_1$ be a vector space with ordered bases $B = (1 + x, 1 - x)$ and $B' = (1, 1 + x)$. Let $v = 1 - 3x$.

a. Compute $[v]_B$ and $[v]_{B'}$

b. Compute P, the change of basis matrix from B' to B.

c. Verify the result that $P[v]_{B'} = [v]_B$.

12. Let $V = P_1$ be a vector space with ordered bases $B = (1 + x, 1 - x)$ and $B' = (1, 1 + x)$. Let $v = 1 - 3x$.

a. Compute $[v]_{B'}$

b. Compute P, the change of basis matrix from B' to B.

c. Use parts a and b to compute $[v]_B$.

13. Consider the following vector space with ordered bases: $V = M_{22}$,

$$B = \left(\begin{bmatrix} 1 & 1 \\ 0 & 0 \end{bmatrix}, \begin{bmatrix} 0 & 1 \\ 0 & 1 \end{bmatrix}, \begin{bmatrix} 0 & 0 \\ 1 & 1 \end{bmatrix}, \begin{bmatrix} 1 & 0 \\ 1 & 1 \end{bmatrix} \right)$$

$$B' = \left(\begin{bmatrix} 1 & 1 \\ 1 & 0 \end{bmatrix}, \begin{bmatrix} 1 & 1 \\ 0 & 1 \end{bmatrix}, \begin{bmatrix} 1 & 0 \\ 1 & 1 \end{bmatrix}, \begin{bmatrix} 0 & 1 \\ 1 & 1 \end{bmatrix} \right)$$

a. Verify that B is indeed a basis.

b. Compute the change of basis matrix P from B' to B.

c. Compute the coordinates $[v]_B$ and $[v]_{B'}$ for the vector $v = \begin{bmatrix} 1 & 1 \\ 1 & 1 \end{bmatrix}$.

d. Verify that $P[v]_{B'} = [v]_B$.

14. Let $A \in M_{22}$ and suppose $I, A, \ldots, A^4$ are all distinct. Prove that there exists scalars $a_0, a_1, \ldots, a_4$ not all zero such that $a_0 I + a_1 A + \cdots + a_4 A^4 = 0_{22}$.

15. Prove if u, v, w is a basis for a vector space V, then so is $u + v$, $v + w$, $u + w$.

16. Let V be a vector space with basis $v_1, \ldots, v_n$. Select non-zero scalars $a_1, \ldots, a_n$.

a. Prove that $a_1 v_1, \ldots, a_n v_n$ is a basis for V.

b. Prove that $a_1 v_1, a_1 v_1 + a_2 v_2, \ldots, a_1 v_1 + \cdots + a_n v_n$ is a basis for V.

17. Let V be a vector space with $v_1, \ldots, v_n \in V$. Prove that if $V = span(v_1, \ldots, v_n)$ and no proper subset of $\{v_1, \ldots, v_n\}$ spans V, then $v_1, \ldots, v_n$ is a basis for V. We say that a basis is a **minimal spanning set of vectors**

18. Let V be a vector space with $v_1, \ldots, v_n \in V$. Prove that if $v_1, \ldots, v_n$ are linearly independent and for all $v \in V$, we have that $v_1, \ldots, v_n, v$ are linearly dependent, then $v_1, \ldots, v_n$ form a basis for V. We say that a basis is a **maximal linearly independent set of vectors**.

19. Complete the proof of Theorem 3.9.

20. Prove that if U is a subspace of $\mathbb{R}^n$, then there exists $A \in M_{mn}$ such that $U = \{Ax \ : x \in \mathbb{R}^m\}$ (i.e. U equals the column space of A). Hint: Let A have columns which form a basis for U.

3.6   SUBSPACES ASSOCIATED WITH A MATRIX

In this section we address some results that were earlier discussed but never formally proved. This section also gives a justification for the process of finding a basis for a span of a set of vectors. We now remind the reader of some definitions as well as introduce some new notation and terminology.

Definition 3.18 *For $A \in M_{mn}$ with rows $r_1, \ldots, r_m \in \mathbb{R}^n$ and columns $c_1, \ldots, c_n \in \mathbb{R}^m$,*

1. *the* **row space** *of A, which we will denote by*

$$\mathrm{rowsp}(A) = \mathrm{span}(r_1, \ldots, r_m) \ a \ subspace \ of \ \mathbb{R}^n.$$

2. *the* **column space** *of A, which we will denote by*

$$\mathrm{colsp(A)} = \mathrm{span}(c_1, \ldots, c_n) \ a \ subspace \ of \ \mathbb{R}^m.$$

3. *the* **null space** *of A, which we will denote by*

$$\mathrm{nullsp(A)} = \{u \in \mathbb{R}^n \mid Au = 0\} \ a \ subspace \ of \ \mathbb{R}^n.$$

4. *the dimension of $\mathrm{rowsp}(A)$ will be called the* **rank** *of A (for good reason) and will be denoted by $r(A)$.*

5. *the dimension of $\mathrm{nullsp}(A)$ will be called the* **nullity** *of A and will be denoted by $n(A)$.*

We leave the proof of the following lemma as an exercise:

Lemma 3.8 *If $A \in M_{mn}$ and $B \in \mathbb{R}^m$, then*

i. $\mathrm{rowsp}(A^T) = \mathrm{colsp}(A)$ and $\mathrm{colsp}(A^T) = \mathrm{rowsp}(A)$.

ii. $\mathrm{colsp}(A) = \{Av \mid v \in \mathbb{R}^n\}$.

iii. The linear system $AX = B$ is consistent iff $B \in \mathrm{colsp}(A)$.

The goal of the next series of lemmas is to give a simple method for finding a basis for $\mathrm{rowsp}(A)$ and $\mathrm{colsp}(A)$.

Lemma 3.9 *Let $A, B \in M_{mn}$. If A is equivalent to B, then $\mathrm{rowsp}(A) = \mathrm{rowsp}(B)$.*

Proof 3.18 *We show that if $A \xrightarrow{op} B$ for some elementary row operation op, then* $\mathrm{rowsp}(A) = \mathrm{rowsp}(B)$. *The result then follows by induction on k, where*

$$A \xrightarrow{op_1} A_1 \xrightarrow{op_2} \cdots \xrightarrow{op_k} B,$$

and $op_1, op_2, \ldots, op_k$ are any elementary row operations.

We will prove the statement for $op = aR_i$ $(a \neq 0)$ and leave the verification for the other two elementary operations as an exercise. If $r_1, \ldots, r_i, \ldots, r_m$ are the rows of A, then the rows of B are $r_1, \ldots, ar_i, \ldots, r_m$. Hence, we need to show that

$$\mathrm{span}(r_1, \ldots, r_i, \ldots, r_m) = \mathrm{span}(r_1, \ldots, ar_i, \ldots, r_m).$$

First, take any $u \in \mathrm{span}(r_1, \ldots, r_i, \ldots, r_m)$. Then for some scalars $a_1, \ldots, a_m$ we have

$$u = a_1 r_1 + \cdots + a_i r_i + \cdots + a_m r_m = a_1 r_1 + \cdots + (a_i a^{-1}) ar_i + \cdots + a_m r_m.$$

Thus, $u \in \mathrm{span}(r_1, \ldots, ar_i, \ldots, r_m)$. Second, take any $u \in \mathrm{span}(r_1, \ldots, ar_i, \ldots, r_m)$. Then for some scalars $a_1, \ldots, a_m$ we have

$$u = a_1 r_1 + \cdots + a_i (ar_i) + \cdots + a_m r_m = a_1 r_1 + \cdots + (a_i a) r_i + \cdots + a_m r_m.$$

Thus, $u \in \mathrm{span}(r_1, \ldots, r_i, \ldots, r_m)$. $\square$

Note, however, that the following statement is **not** true, namely, if A is equivalent to B, then $\mathrm{colsp}(A) = \mathrm{colsp}(B)$. To see this, just look at the example of $A = E_{11}$ and $A \xrightarrow{R_1 + R_2} B$. The column space of A is the x-axis, while the column space of B is the line $y = x$. There is, however, a relationship between the column spaces of equivalent matrices.

Lemma 3.10 *Let $A, B \in M_{mn}$ and assume that A is equivalent to B. Let $c_1, \ldots, c_k$ be any subset of the columns of A and $d_1, \ldots, d_k$ the corresponding columns of B. Then*

    i. $c_1, \ldots, c_k$ are linearly independent iff $d_1, \ldots, d_k$ are linearly independent.

    ii. $c_1, \ldots, c_k$ form a basis for $\mathrm{colsp}(A)$ iff $d_1, \ldots, d_k$ form a basis for $\mathrm{colsp}(B)$.

Proof 3.19 *To prove part i, set $A_1 = [c_1 \cdots c_k]$ (represented in columns) and $B_1 = [d_1 \cdots d_k]$ (represented in columns). Note that since A and B are equivalent, then so are A_1 and B_1. If $c_1, \ldots, c_k$ are linearly independent, by Theorem 3.4, $A_1 X =$*

0 *has only the trivial solution. Since they are equivalent, $B_1X = 0$ has only the trivial solution (see Theorem 2.1). Then, by Theorem 3.4 again, $d_1, \ldots, d_k$ are linearly independent. The reverse implication is symmetric.*

To prove part ii, assuming that $c_1, \ldots, c_k$ form a basis for $\mathrm{colsp}(A)$, suppose that $d_1, \ldots, d_k$ do not form a basis for $\mathrm{colsp}(B)$. By part i, we know that $d_1, \ldots, d_k$ are linearly independent, so it must be the case that $d_1, \ldots, d_k$ do not span $\mathrm{colsp}(B)$. Then there must be a column of B, say $d \notin \mathrm{span}(d_1, \ldots, d_k)$. Then the columns $d_1, \ldots, d_k, d$ are linearly independent. Let c be the column in A corresponding to column d in B. By part i, $c_1, \ldots, c_k, c$ are linearly independent, but this contradicts the fact that $c_1, \ldots, c_k$ form a basis for $\mathrm{colsp}(A)$. The reverse implication is symmetric. □

The next result is left as an exercise for the reader.

Lemma 3.11 *Let $R \in M_{mn}$ be a matrix in reduced row-echelon form. Then*

i. The rows of R containing the pivots form a basis for $\mathrm{rowsp}(R)$.

ii. The columns of R containing the pivots form a basis for $\mathrm{colsp}(R)$.

Indeed, by the very nature of the reduced row-echelon form and the position of zeros and ones in this matrix, one is convinced that the above statement is true. Some immediate consequences of these lemmas are the following (we leave it to the reader to justify them in the exercises):

Corollary 3.3 *Let $A, R \in M_{mn}$ where R is the reduced row-echelon form of A. Then*

i. A basis for $\mathrm{rowsp}(A)$ can be obtained by taking the rows of R containing the pivots.

ii. A basis for $\mathrm{colsp}(A)$ can be obtained by taking the columns of A corresponding to the columns of R containing the pivots.

Corollary 3.4 *For any $A \in M_{mn}$,*

i. $r(A) = rk(A) = \dim(\mathrm{colsp}(A))$, and

ii. $rk(A) = rk(A^T)$.

Corollary 3.3 validates our discussion in Section 3.5 in which we gave two methods for finding a basis for the span of a set of vectors in $\mathbb{R}^n$, P_n and M_{mn}.

Example 3.58 *We illustrate these results with an example.*

$$\text{If } A = \begin{bmatrix} 1 & -2 & 0 & 2 & 1 \\ 2 & 1 & 1 & 0 & -1 \\ 1 & 3 & 1 & -2 & -2 \end{bmatrix}, \text{ then}$$

$$R = \begin{bmatrix} 1 & 0 & 2/5 & 2/5 & -(1/5) \\ 0 & 1 & 1/5 & -(4/5) & -(3/5) \\ 0 & 0 & 0 & 0 & 0 \end{bmatrix}.$$

Therefore, a basis for $\mathrm{rowsp}(A)$ *is*

$$[1, 0, 2/5, 2/5, -(1/5)], \quad [0, 1, 1/5, -(4/5), -(3/5)]$$

and the row space of A,

$$\mathrm{rowsp}(A) = \left\{ \left[a, b, \frac{2}{5}a + \frac{1}{5}b, \frac{2}{5}a - \frac{4}{5}b, -\frac{1}{5}a - \frac{3}{5}b \right] : a, b \in \mathbb{R} \right\}.$$

For computing $\mathrm{colsp}(A)$ *we consider*

$$A^T = \begin{bmatrix} 1 & 2 & 1 \\ -2 & 1 & 3 \\ 0 & 1 & 1 \\ 2 & 0 & -2 \\ 1 & -1 & -2 \end{bmatrix} \quad \textit{row reduces to} \quad \begin{bmatrix} 1 & 0 & -1 \\ 0 & 1 & 1 \\ 0 & 0 & 0 \\ 0 & 0 & 0 \\ 0 & 0 & 0 \end{bmatrix}.$$

Therefore, a basis for $\mathrm{colsp}(A)$ *is* $[1, 0, -1], [0, 1, 1]$ *and*

$$\mathrm{colsp}(A) = \{[a, b, -a + b] \mid a, b \in \mathbb{R}\}.$$

Notice that $r(A) = rk(A) = \dim(\mathrm{colsp}(A)) = 2.$

The next theorem is an important result which we will revisit in Chapter 4 and reprove using material we have yet to develop.

Theorem 3.12 *For any* $A \in M_{mn}$, $r(A) + n(A) = n.$

Proof 3.20 *Let R be the reduced row-echelon form of A and set k equal to the number of pivots of R. Note that $k = rk(A) = r(A)$. We have seen from examples that the number of non-pivot variables (or parameters, or independent variables) associated with* $\mathrm{nullsp}(A)$ *is precisely $n - k$ (the pivot variables correspond to the dependent variables) and this number of independent variables corresponds to the dimension of* $\mathrm{nullsp}(A)$. *In other words, $n - k = n(A)$. Hence, $r(A) + n(A) = k + (n - k) = n.$* □

Example 3.59 *Let's apply Theorem 3.12 in a specific setting. Consider the homogeneous system*

$$\begin{cases} x - y + z & = & 0 \\ 2x + y - z & = & 0 \end{cases}.$$

In this example $n = 3$ and the $rk(A) = 2$ since

$$\begin{bmatrix} 1 & -1 & 1 \\ 2 & 1 & -1 \end{bmatrix} \quad reduces \ to \quad \begin{bmatrix} 1 & -1 & 1 \\ 0 & 3 & -3 \end{bmatrix}.$$

Hence, the solution space of the system has dimension equal to $3 - 2 = 1$. Indeed, one can compute as we did in Chapter 1 that the solution space has the form $\{ [0, a, a] \mid a \in \mathbb{R} \}$, and so a basis for it would be $[0, 1, 1]$, and so has dimension 1 as predicted.

EXERCISES

1. For each of the following matrices A find a basis for and dimension of $\mathrm{rowsp}(A)$ and $\mathrm{colsp}(A)$, then use Theorem 3.12 to find the dimension of $\mathrm{nullsp}(A)$. Afterwards verify the dimension you found for the null space, by computing it explicitly and exhibiting its basis.

   a.

   $$A = \begin{bmatrix} 1 & -1 & -1 \\ 2 & -2 & 1 \\ 3 & -3 & 0 \end{bmatrix}$$

   b.

   $$A = \begin{bmatrix} 1 & 1 & 1 & 1 \\ -2 & -1 & 1 & 3 \\ 3 & 4 & 6 & 8 \end{bmatrix}$$

2. Consider the following homogeneous linear system

   $$\begin{cases} 2x_1 - x_2 + 3x_3 + x_4 & = & 0 \\ x_1 - x_2 + x_3 - x_4 & = & 0 \\ x_1 + x_2 + x_3 + x_4 & = & 0 \end{cases}.$$

   a. Find the rank of the coefficient matrix of the linear system.

   b. Use Theorem 3.12 to find the dimension of the solution space.

   c. Compute the solution space of the system, find its basis and dimension and thus confirm your answer in part b.

3. Repeat the previous exercise with the following homogeneous linear system:

   $$\begin{cases} 2x_1 + 3x_2 + 4x_3 + 5x_4 & = & 0 \\ x_1 - x_2 + x_3 - x_4 & = & 0 \\ 5x_1 + 5x_2 + 9x_3 + 9x_4 & = & 0 \end{cases}.$$

4. Repeat the previous exercise with the following homogeneous linear system:

$$\begin{cases} x_1 + x_2 + x_3 + x_4 = 0 \\ -2x_1 - x_2 + x_3 + 3x_4 = 0 \\ 3x_1 + 4x_2 + 6x_3 + 8x_4 = 0 \end{cases}.$$

5. Repeat the previous exercise with the following homogeneous linear system:

$$\begin{cases} x_1 + x_2 + 2x_3 - x_4 = 0 \\ -x_1 - x_2 - x_3 + x_4 = 0 \\ 2x_1 + 2x_2 + 4x_3 - 2x_4 = 0 \end{cases}.$$

6. Prove Lemma 3.8.

7. Complete the proof of Lemma 3.9.

8. Prove Lemma 3.11.

9. Prove Corollary 3.3.

10. Prove Corolary 3.4.

3.7 APPLICATION: DIMENSION THEOREMS

In this section, we illustrate a technique called a *counting argument*. These counting arguments will be a consequence of the two dimension theorems proved in this section. The first dimension theorem is a result we would expect to be true if dimension is indeed measuring *size* in some sense. First, we need a lemma.

Lemma 3.12 *For $v_1, \ldots, v_n \in V$ a vector space, the following are equivalent:*

i. The vectors $v_1, \ldots, v_n$ represent a maximal number of linearly independent vectors in V.

ii. $v_1, \ldots, v_n$ form a basis for V.

Proof 3.21 *To show that i. implies ii, We need to show that $v_1, \ldots, v_n$ span V. Suppose to the contrary that they don't span V. This means there exists a $v \in V$ not in the span of $v_1, \ldots, v_n$. By Lemma 3.7, $v_1, \ldots, v_n, v$ are linearly independent. But this contradicts the fact that $v_1, \ldots, v_n$ is a largest number of linearly independent vectors in V. Hence, it must be that case that $v_1, \ldots, v_n$ form a basis for V. The reverse implication is left as an exercise.* □

Theorem 3.13 (Subspace Dimension Theorem) *Let V be a vector space with finite dimension and U be a subspace of V. Then*

i. $\dim U \leq \dim V$.

ii. $\dim U = \dim V$ iff $U = V$.

Proof 3.22 *Let $v_1, \ldots, v_n$ be a basis for V (so $\dim V = n$). To prove i, if $U = \{0\}$ then certainly $\dim U = 0 \leq \dim V$, so we can assume that $U \neq \{0\}$. Consider all collections $u_1, \ldots, u_k \in U$ which are linearly independent. Such collections exist since, for instance, a single $0 \neq u \in U$ is linearly independent. By Lemma 3.6, for every such collection we have $k \leq n$ (i.e. there is a bound on how large k can be). Hence, there is a collection $u_1, \ldots, u_k$ linearly independent with k largest. By Lemma 3.12, this largest collection forms a basis for U. Thus, $\dim U = k \leq n = \dim V$.*

To prove ii, one direction is trivial: If $U = V$ then certainly $\dim U = \dim V$. Now lets assume that $\dim U = \dim V$. Choose $u_1, \ldots, u_n$ a basis for U. Since $u_1, \ldots, u_n$ are linearly independent in V, by Lemma 3.10, $u_1, \ldots, u_n$ is a basis for V, and so they span V. Thus,

$$V = span(u_1, \ldots, u_n) = U.$$

□

Example 3.60 *We will see important applications of the above dimension theorem as the material develops. Here is trivial example to illustrate its application. We have seen that $[1, 0, 0], [1, 1, 0], [1, 1, 1]$ are linearly independent in $\mathbb{R}^3$. Define the subspace $U = span([1, 0, 0], [1, 1, 0], [1, 1, 1])$. Certainly, $[1, 0, 0], [1, 1, 0], [1, 1, 1]$ form a basis for U. Hence, $\dim U = 3 = \dim \mathbb{R}^3$. By Theorem 3.13.ii, it follows that $U = \mathbb{R}^3$.*

We give some intuition for the second dimension theorem. Consider two sets X and Y. Define $|X|$ to be the number of elements in the set X. A formula which counts the number of elements in the union of X and Y is

$$|X \cup Y| = |X| + |Y| - |X \cap Y|,$$

since $|X| + |Y|$ counts $|X \cap Y|$ twice. An analogous result holds for subspaces and their dimensions. Let U and W be subspaces of a vector space V. Recall from previous exercises that $U \cup W$ is not necessarily a subspace, but that $U + W$ and $U \cap W$ are subspaces. Thus, our theorem will be phrased slightly differently from the set theoretic result we stated above.

Theorem 3.14 (Sum Dimension Theorem) *Let U and W be subspaces of a finite dimensional vector space V. Then*

$$\dim(U + W) = \dim U + \dim W - \dim(U \cap W).$$

Proof 3.23 *If U or W equals $\{0\}$ the result follows easily (check). Now let's consider the case when $U \cap W = \{0\}$. Select any bases $u_1, \ldots, u_k$ for U and $w_1, \ldots, w_\ell$ for W. We will show that $u_1, \ldots, u_k, w_1, \ldots, w_\ell$ forms a basis for $U + W$. Having shown this, we can arrive at the desired conclusion that*

$$\dim(U + W) = k + \ell = k + \ell - 0 = \dim U + \dim W - \dim(U \cap W).$$

First, we show $u_1, \ldots, u_k, w_1, \ldots, w_\ell$ span $U + W$. Take any $u + w \in U + W$ where $u \in U$ and $w \in W$. Since $u_1, \ldots, u_k$ is a basis for U, there exist scalars $a_1, \ldots, a_k$ such that $u = a_1 u_1 + \cdots + a_k u_k$. Similarly, there exist scalars $b_1, \ldots, b_\ell$ such that $w = b_1 w_1 + \cdots + b_\ell w_\ell$. But then

$$u + w = a_1 u_1 + \cdots + a_k u_k + b_1 w_1 + \cdots + b_\ell w_\ell.$$

Now we show $u_1, \ldots, u_k, w_1, \ldots, w_\ell$ are linearly independent. Suppose that $a_1 u_1 + \cdots + a_k u_k + b_1 w_1 + \cdots + b_\ell w_\ell = 0$. Rewrite this equation as

$$a_1 u_1 + \cdots + a_k u_k = (-b_1) w_1 + \cdots + (-b_\ell) w_\ell.$$

Set

$$v = a_1 u_1 + \cdots a_k u_k = (-b_1) w_1 + \cdots + (-b_\ell) w_\ell.$$

Notice that this equation implies that $v \in U$ and $v \in W$, hence $v \in U \cap W$. Since $U \cap W = \{0\}$, this implies $v = 0$ which yields the equations

$$a_1 u_1 + \cdots a_k u_k = 0 \quad and \quad (-b_1) w_1 + \cdots + (-b_\ell) w_\ell = 0.$$

Since $u_1, \ldots, u_k$ and $w_1, \ldots, w_\ell$ are each linearly independent, we have $a_1 = \cdots = a_k = 0$ and $b_1 = \cdots = b_\ell = 0$. Hence, $u_1, \ldots, u_k, w_1, \ldots, w_\ell$ are linearly independent and so a basis for $U + W$.

Finally, we consider the case when $U \cap W \neq \{0\}$. This case will be similar to the previous case with an additional twist. Choose a basis $v_1, \ldots, v_m$ for $U \cap W$. By Lemma 3.7, we can extend these vectors to $v_1, \ldots, v_m, u_{m+1}, \ldots, u_k$ a basis for U. For the same reason we can also extend them to $v_1, \ldots, v_m, w_{m+1}, \ldots, w_\ell$ a basis for W. We will show that $v_1, \ldots, v_m, u_{m+1}, \ldots, u_k, w_{m+1}, \ldots, w_\ell$ form a basis for $U + W$. Having shown this, we can arrive at the desired conclusion that

$$\dim(U + W) = k + l - m = \dim U + \dim W - \dim(U \cap W).$$

First, we show $v_1, \ldots, v_m, u_{m+1}, \ldots, u_k, w_{m+1}, \ldots, w_\ell$ span $U + W$. Take any $u + w \in U + W$ where $u \in U$ and $w \in W$. Since $v_1, \ldots, v_m, u_{m+1}, \ldots, u_k$ is a basis for U, there exist scalars $a_1, \ldots, a_k$ such that $u = a_1 v_1 + \cdots + a_m v_m + a_{m+1} u_{m+1} + \cdots + a_k u_k$. Similarly, there exist scalars $b_1, \ldots, b_\ell$ such that $w = b_1 v_1 + \cdots b_m v_m + b_{m+1} w_{m+1} + \cdots + b_\ell w_\ell$. But then

$$u + w = (a_1 + b_1) v_1 + \cdots + (a_m + b_m) v_m + a_{m+1} u_{m+1} + \cdots + a_k u_k + b_{m+1} w_{m+1} + \cdots + b_\ell w_\ell.$$

Now we show $v_1, \ldots, v_m, u_{m+1}, \ldots, u_k, w_{m+1}, \ldots, w_\ell$ are linearly independent. Suppose that

$$a_1 v_1 + \cdots + a_m v_m + a_{m+1} u_{m+1} + \cdots + a_k u_k + b_{m+1} w_{m+1} + \cdots + b_\ell w_\ell = 0.$$

Let's refer to the equation above as (*). Rewrite equation (*) as

$$a_1 v_1 + \cdots + a_m v_m + a_{m+1} u_{m+1} + \cdots + a_k u_k = (-b_{m+1}) w_{m+1} + \cdots + (-b_\ell) w_\ell.$$

Set

$$v = a_1 v_1 + \cdots + a_m v_m + a_{m+1} u_{m+1} + \cdots + a_k u_k = (-b_{m+1}) w_{m+1} + \cdots + (-b_\ell) w_\ell.$$

Notice that this equation implies that $v \in U$ and $v \in W$, hence $v \in U \cap W$. Since $v_1, \ldots, v_m$ is a basis for $U \cap W$, there exist scalars $c_1, \ldots, c_m$ such that $v = c_1 v_1 + \cdots + c_m v_m$. In particular,

$$c_1 v_1 + \cdots + c_m v_m = (-b_{m+1}) w_{m+1} + \cdots + (-b_\ell) w_\ell,$$

which implies that

$$c_1 v_1 + \cdots + c_m v_m + b_{m+1} w_{m+1} + \cdots + b_\ell w_\ell = 0.$$

Since $v_1, \ldots, v_m, w_{m+1}, \ldots, w_\ell$ are linearly independent, this implies that $b_{m+1} = \cdots = b_\ell = 0$ (as well as $c_1 = \cdots = c_m = 0$). Returning to our original equation (*), it becomes

$$a_1 v_1 + \cdots + a_m v_m + a_{m+1} u_{m+1} + \cdots + a_k u_k = 0.$$

Since $v_1, \ldots, v_m, v_{m+1}, u_{m+1}, \ldots, u_k$ are linearly independent, this implies that $a_1 = \cdots = a_k = 0$.

Hence, $v_1, \ldots, v_m, u_{m+1}, \ldots, u_k, w_{m+1}, \ldots, w_\ell$ are linearly independent and so a basis for $U + W$. □

Example 3.61 *Let $V = P_3$ and consider the following subspaces:*

$$U = \{a + bx - ax^2 + cx^3 \mid a, b, c \in \mathbb{R}\}$$

$$W = \{a - ax + bx^2 + cx^3 \mid a, b, c \in \mathbb{R}\}.$$

We wish to find $\dim(U + W)$. We do this in an indirect manner using Theorem 3.14. A basis for U is $1 - x^2, x, x^3$ and a basis for W is $1 - x, x^2, x^3$. Hence, $\dim U = 3 = \dim W$. We can find a description of $U \cap W$ via the descriptions of U and W. Notice that U requires that the coefficient of x^2 be the opposite sign of the constant coefficient, and W requires that the coefficient of x be the opposite sign of the constant coefficient. Polynomials in $U \cap W$ must meet both requirements, and so

$$U \cap W = \{a - ax - ax^2 + cx^3\}.$$

Hence, a basis for $U \cap W$ is $1 - x - x^2, x^3$, and so $\dim(U \cap W) = 2$. Now, by Theorem 3.14, $\dim(U + W) = 3 + 3 - 2 = 4$. Observe further that since $\dim(U + W) = 4 = \dim P_3$, this implies, by Theorem 3.13.ii, that $U + W = P_3$. The reader has just witnessed the first "counting argument" in this text. We hope the reader sees the beauty and the power of this type of argument.

We add a couple additional remarks about this example. First, we can illustrate explicitly how $P_3 = U + W$. Indeed, for any $a + bx + cx^2 + dx^3 \in P_3$,

$$a + bx + cx^2 + dx^3 = (b - a)x + dx^3 \ + \ a - ax + cx^2,$$

where $(b - a)x + dx^3 \in U$ and $a - ax + cx^2 \in W$. Second, since $U \cap W \neq \{0\}$ there is no unique representation of polynomials in P_3 as a sum $U + W$. For a simple example of this consider the polynomial x^3. One simple representation of x^3 in $U + W$ could be $x^3 + 0$. However, here is another representation

$$x^3 = (1 - x - x^2 + x^3) + (-1 + x + x^2)$$

Example 3.62 Let $V = M_{22}$ with subspaces U and W, where U is the collection of symmetric matrices and W is the collection of skew-symmetric matrices. Notice that

$$U = \left\{ \begin{bmatrix} a & b \\ b & c \end{bmatrix} : a, b, c \in \mathbb{R} \right\} \text{ and } W = \left\{ \begin{bmatrix} 0 & a \\ -a & 0 \end{bmatrix} : a, b, c \in \mathbb{R} \right\}.$$

A basis for U is

$$\begin{bmatrix} 1 & 0 \\ 0 & 0 \end{bmatrix}, \begin{bmatrix} 0 & 1 \\ 1 & 0 \end{bmatrix}, \begin{bmatrix} 0 & 0 \\ 0 & 1 \end{bmatrix},$$

and a basis for W is

$$\begin{bmatrix} 0 & 1 \\ -1 & 0 \end{bmatrix}.$$

Hence, $\dim U = 3$ and $\dim W = 1$. Now $U \cap W = \{0\}$, so $\dim(U \cap W) = 0$. By Theorem 3.14, $\dim(U + W) = 3 + 1 - 0 = 4$ and since $\dim(U + W) = \dim V$, by Theorem 3.13.ii, $U + W = V$.

This says that every two-by-two matrix can be written uniquely as a sum of a symmetric plus a skew-symmetric matrix. Recall, we write $V = U \oplus W$, a direct sum of U and W. Indeed, we knew this from an earlier exercise in which we expressed any matrix $A = \frac{1}{2}(A + A^T) + \frac{1}{2}(A - A^T)$.

We state a result here which we leave as an easy exercise.

Corollary 3.5 *Let U, W be subspaces of a vector space V. If $V = U \oplus W$, then $\dim V = \dim U + \dim W$.*

Theorem 3.15 (Product Dimension Theorem) *Let $V_1, V_2, \ldots, V_k$ be finite dimensional vector spaces. Then*

$$\dim(V_1 \times V_2 \times \ldots \times V_k) = \dim V_1 + \dim V_2 \cdots + \dim V_k.$$

Proof 3.24 *Let $v_1^{(i)}, \ldots, v_{n_i}^{(i)}$ be a basis for V_i for $i = 1, 2, \ldots, k$. Then it is easy to show (although perhaps cumbersome to write out) that the following vectors form a basis for $V_1 \times \ldots \times V_k$:*

$$(v_1^{(1)}, 0, \ldots, 0), \ldots, (v_{n_1}^{(1)}, 0, \ldots, 0), \ldots, (0, \ldots, 0, v_1^{(k)}), \ldots, (0, \ldots, 0, v_{n_k}^{(k)}).$$

The result now follows by simply counting the vectors in this basis. □

EXERCISES

1. Let $V = M_{22}$ with subspaces

$$U = \left\{ \begin{bmatrix} a & b \\ c & d \end{bmatrix} : a + d = 0 \right\} \quad \text{and} \quad W = \left\{ \begin{bmatrix} a & b \\ c & d \end{bmatrix} : b + c = 0 \right\}$$

   a. Find bases and dimensions for V, U and W.

   b. Give a description of $U \cap W$, then find a basis for and the dimension of it.

   c. Find the dimension of $U + W$ and decide whether or not $V = U + W$.

2. Let $V = M_{22}$, U be the subspace of matrices with trace equaling zero, and W the subspace of symmetric matrices.

   a. Find bases and dimensions for V, U and W.

   b. Give a description of $U \cap W$, then find a basis for and the dimension of it.

   c. Find the dimension of $U + W$ and decide whether or not $V = U + W$.

   d. If $V = M_{nn}$, and U and W are the same, repeat parts a.–c.

3. Let $V = M_{22}$, U be the subspace of upper-triangular matrices, and W be the subspace of skew-symmetric matrices. Repeat the previous exercise.

4. Let $V = M_{22}$, U be the subspace of matrices in which the sum of all its entries is 0, and W be the subspace of matrices whose trace is 0. Repeat the previous exercise.

5. Let $V = M_{22}$, U be the subspace of upper triangular matrices and W the subspace of lower triangular matrices. Repeat the previous exercise.

6. Let $V = P_2$, U be the subspace of polynomials with 0 as a root, and W be the subspace of polynomials with 1 as a root.

   a. Find bases and dimensions for V, U and W.

   b. Give a description of $U \cap W$, then find a basis for and the dimension of it.

   c. Find the dimension of $U + W$ and decide whether or not $V = U + W$.

   d. If $V = P_n$ $(n \geq 2)$, and U and W are the same, repeat parts a.- c.

7. Consider the vector space $V = P_2$ and subspaces

$$U = \{a + bx + (a+b)x^2 \ : \ a, b \in \mathbb{R}\} \ \text{ and } \ W = \{a + (a+b)x + bx^2 \ : \ a, b \in \mathbb{R}\}$$

   a. Find bases and dimensions for V, U and W.

   b. Give a description of $U \cap W$, then find a basis for and the dimension of it.

   c. Find the dimension of $U + W$ and decide whether or not $V = U + W$.

8. Let $V = P_3$ with subspaces

$$U = \{a + bx - ax^2 + (a+b)x^3 \ : \ a, b \in \mathbb{R}\} \ \text{ and } \ W = \{a + bx + ax^2 + cx^3 \ : \ a, b \in \mathbb{R}\}$$

   a. Find bases and dimensions for V, U and W.

   b. Give a description of $U \cap W$, then find a basis for and the dimension of it.

   c. Find the dimension of $U + W$ and decide whether or not $V = U + W$.

9. Let $V = P_2$ with subspaces

$$U = \{p(x) \in P_2 : p(0) = p(1)\} \quad \text{and} \quad W = \{p(x) \in P_2 : p'(0) = p'(1)\}\}.$$

   a. Find bases and dimensions for V, U and W.

   b. Give a description of $U \cap W$, then find a basis for and the dimension of it.

   c. Find the dimension of $U + W$ and decide whether or not $V = U + W$.

10. Let $V = P_2$, U be the subspace of polynomials whose second derivative is 0, and $W = \{p \in V \mid p(0) = 2p(1)\}$.

   a. Find bases and dimensions for V, U and W.

   b. Give a description of $U \cap W$, then find a basis for and the dimension of it.

   c. Find the dimension of $U + W$ and decide whether or not $V = U + W$.

11. Let $V = P_3$, U be the subspace of polynomials which have both 0 and 1 as roots, and W be the subspace of polynomials which have 0 as at least a double root.

   a. Find bases and dimensions for V, U and W.

   b. Give a description of $U \cap W$, then find a basis for and the dimension of it.

   c. Find the dimension of $U + W$ and decide whether or not $V = U + W$.

12. Consider the vector space $V = \mathbb{R}^3$ and the vectors $u = [3, -2, 1]$ and $w = [1, 1, -1]$. Consider also the subspaces $U = \{v \in \mathbb{R}^3 \mid u \cdot v = 0\}$ and $W = \{v \in \mathbb{R}^3 \mid w \cdot v = 0\}$.

   a. Find bases and dimensions for V, U and W.

   b. Give a description of $U \cap W$, then find a basis for and the dimension of it.

   c. Find the dimension of $U + W$ and decide whether or not $V = U + W$.

13. Prove that $n \times n$ square matrices are a direct sum of symmetric matrices and skew-symmetric matrices.

14. Let V be a vector space with basis $v_1, \ldots, v_n$. Fix a k, $1 \leq k \leq n$ and set $U = span(v_1, \ldots, v_k)$ and $W = span(v_{k+1}, \ldots, v_n)$.

   a. Show directly that $V = U + W$.

   b. Show directly that $U \cap W = \{0\}$.

   c. Use Theorem 3.14 and part a. to show that $U \cap W = \{0\}$.

15. Prove that for $v_1, \ldots, v_n \in V$ a vector space, the following are equivalent:

    i. The vectors $v_1, \ldots, v_n$ represent a minimal number of vectors which span V.

    ii. $v_1, \ldots, v_n$ form a basis for V.

16. Prove that if V is a vector space of finite dimension and U is a subspace of V, then U has finite dimension as well.

17. Prove by induction that for any subspaces $U_1, U_2, \ldots, U_k$ of a vector space V we have

$$dim(U_1 + U_2 + \cdots + U_k) \geq dimU_1 + dimU_2 + \cdots + dimU_k.$$

18. Complete the proof of Lemma 3.12.

19. Prove Lemma 3.5.

CHAPTER 4

Linear Transformations

I N THIS CHAPTER, the general notion of a linear transformation is presented. In Section 4.1, the definition of a linear transformation is introduced and many examples are given in the context of the four classic vector spaces. Methods are given to prove or disprove that a particular map is a linear transformation. In Section 4.2, two special subspaces are defined related to a linear transformation: Kernel and image. These subspaces are computed and their dimensions are determined. In Section 4.3, an important connection is made between linear transformations and matrices. In Section 4.4, the inverse linear transformation is discussed and matrix representation is used to compute it. The notion of isomorphism is discussed how two seemingly different vector spaces are essentially the same. In Section 4.5, different matrix representation of a particular linear transformation are shown to be related by similarity. In Section 4.6, the reader will learn how to compute eigenvalues, eigenvectors and eigenspaces as well as diagonalize a linear transformation or matrix (when possible). The final sections of this chapter are for advanced learners. Section 4.7 gives an axiomatic treatment of the determinant. Section 4.8 introduces quotient vector spaces. Section 4.9 introduces the dual vector space and the transpose of a linear transformation.

4.1 DEFINITION AND EXAMPLES

Every algebraic structure has its corresponding functions (maps, morphisms). For vector spaces these functions are called **linear transformations**. The inputs are the vectors of one vector space and the outputs are the vectors from another (or perhaps the same) vector space. Just as subspaces are special subsets of a vector space, linear transformations are special functions from one vector space to another; they have the property that they *respect* (or *preserve*) the operations of the vector space. We give the formal definition.

Definition 4.1 *Let V and W be two vector spaces. A function $T : V \longrightarrow W$ is a* **linear transformation** *if*

1. *For all $v_1, v_2 \in V$, we have $T(v_1 + v_2) = T(v_1) + T(v_2)$.*

2. *For all scalars a and $v \in V$, we have $T(av) = aT(v)$.*

DOI: 10.1201/9781003217794-4

We remark that the $+$'s in part a. refer to two different vector additions: $T(v_1+v_2)$ refers to addition in V and $T(v_1) + T(v_2)$ refers to addition in W. There really is no reason for confusion since the context of the addition tells you which addition it must be. A similar remark can be made for part b. of the definition regarding scalar multiplication.

We now list a few examples of linear transformations to make the reader more accustomed with the definition and also to illustrate the method by which one verifies that a function is indeed a linear transformation.

Example 4.1 *Consider the function* $T : \mathbb{R}^2 \longrightarrow P_1$ *defined by* $T([a,b]) = b + ax$. *For instance, according to its definition,* $T([-1,3]) = 3 - x$. *In order to simplify our notation, we write just* $T[a,b]$ *instead of* $T([a,b])$. *We show that* T *is a linear transformation by verifying parts a. and b. of the definition. To prove part a, notice that*

$$T([a_1, b_1] + [a_2, b_2]) = T[a_1 + a_2, b_1 + b_2] = (b_1 + b_2) + (a_1 + a_2)x$$

$$= (b_1 + a_1 x) + (b_2 + a_2 x) = T[a_1, b_1] + T[a_2, b_2].$$

To prove part b, notice that

$$T(a[a_1, b_1]) = T[aa_1, ab_1] = (ab_1) + (aa_1)x$$

$$= a(b_1 + a_1 x) = aT[a_1, b_1].$$

Example 4.2 *Consider the function* $T : \mathbb{R}^2 \longrightarrow \mathbb{R}^3$ *by* $T[a,b] = [a, a + b, a - b]$. *Again, we show that* T *is a linear transformation. To prove part a,*

$$T([a_1, b_1] + [a_2, b_2]) = T[a_1 + a_2, b_1 + b_2]$$

$$= [(a_1 + a_2), (a_1 + a_2) + (b_1 + b_2), (a_1 + a_2) - (b_1 + b_2)]$$

$$= [a_1, a_1 + b_1, a_1 - b_1] + [a_2, a_2 + b_2, a_2 - b_2] = T[a_1, b_1] + T[a_2, b_2].$$

To prove part b,

$$T(a[a_1, b_1]) = T[aa_1, ab_1] = [(aa_1), (aa_1) + (ab_1), (aa_1) - (ab_1)]$$

$$= [aa_1, a(a_1 + b_1), a(a_1 - b_1)] = a[a_1, a_1 + b_1, a_1 - b_1] = aT[a_1, b_1].$$

The previous example brings up an important observation about linear transformations which will fully come to light in Section 4.3. Notice in the previous example that

$$T[a, b] = [a, a + b, a - b]$$

$$= \begin{bmatrix} a \\ a + b \\ a - b \end{bmatrix} = \begin{bmatrix} 1 & 0 \\ 1 & 1 \\ 1 & -1 \end{bmatrix} \begin{bmatrix} a \\ b \end{bmatrix}.$$

In other words, we can represent the action of the linear transformation as multiplication on the left by an appropriate matrix. Later we will show that this is true in general: If $T : \mathbb{R}^n \longrightarrow \mathbb{R}^m$ is a linear transformation, there exists a matrix $A \in M_{mn}$ such that for all $v \in \mathbb{R}^n$ we have $T(v) = Av$ (in fact, there is yet a more general statement that can be made about finite dimensional vector spaces which will be presented in Section 4.3). The converse of this statement is true, as we shall see in the next example.

Example 4.3 *Choose any $A \in M_{mn}$ and define the map*

$$T : \mathbb{R}^n \longrightarrow \mathbb{R}^m \text{ by the formula } T(v) = Av.$$

We show that T is a linear transformation using properties of matrices. First, for $u, v \in \mathbb{R}^n$ we have

$$T(u + v) = A(u + v) = Au + Av = T(u) + T(v).$$

Second, for scalar a and $v \in \mathbb{R}^n$ we have

$$T(av) = A(av) = aAv = aT(v).$$

Example 4.4 *Define the map $T : M_{mn} \longrightarrow M_{nm}$ by $T(A) = A^T$. We show that T is a linear transformation using properties of the transpose. First, for $A, B \in M_{mn}$ we have*

$$T(A + B) = (A + B)^T = A^T + B^T = T(A) + T(B).$$

Second, for scalar a and $A \in M_{nn}$ we have

$$T(aA) = (aA)^T = aA^T = aT(A).$$

Example 4.5 *The verification that the following are linear transformations is left to the reader.*

- *Let $\mathcal{D}$ represent the collection of real-valued differentiable functions. Then the map $T : \mathcal{D} \longrightarrow \mathcal{F}$ by $T(f) = f'$ (the derivative) is a linear transformation.*

- Let $\mathcal{C}$ represent the collection of real-valued continuous functions. Then the map $T : \mathcal{C} \longrightarrow \mathcal{F}$ by $T(f)$ equals the primary antiderivative (with constant $= 0$) of f is a linear transformation.

Special Examples: The verification that the following are linear transformations is left as exercises.

1. Let V be any vector space and define $1_V : V \longrightarrow V$ by $1_V(v) = v$ for all $v \in V$. This linear transformation is called the **identity** map.

2. Let V and W be any vector spaces with identities 0_V and 0_W, respectively. Define the map $0 : V \longrightarrow W$ by $0(v) = 0_W$ for all $v \in V$. This linear transformation is called the **zero** map.

3. Choose any scalar a and define $T : \mathcal{F} \longrightarrow \mathbb{R}$ by $T(f) = f(a)$. This linear transformation is called the **evaluation** transformation.

We now present maps that are **not** linear transformations. To verify that a given map is **not** a linear transformation, it is sufficient to show that one of parts a or b of the definition of linear transformation fails, and to show failure we often have to exhibit a specific counterexample. We suggest that, without the hindsight, one should try to prove a given map **is** a linear transformation, and if a property looks suspiciously false then perhaps one should consider seeking a counterexample.

Example 4.6 *Consider the map $T : \mathbb{R}^2 \longrightarrow \mathbb{R}^2$ by $T[a, b] = [a^2, b]$. We will show that part b fails. Compare the left-hand side*

$$T(a[a_1, b_1]) = T[aa_1, ab_1] = [(aa_1)^2, ab_1] = [a^2 a_1^2, ab_1]$$

with the right-hand side

$$aT[a_1, b_1] = a[a_1^2, b_1] = [aa_1^2, ab_1].$$

Notice that the two computations above seem to give different results. In fact, it is quite clear that if we choose $a \neq 0, 1$, we will not have equality. However, to make it crystal clear choose scalar 2 and vector $[1, 0]$. Then $2T[1, 0] = 2[1, 0] = [2, 0]$ while $T(2[1, 0]) = T[2, 0] = [4, 0]$.

Example 4.7 *Consider the map $T : D_{22} \longrightarrow P_1$ by*

$$T\begin{bmatrix} a & 0 \\ 0 & b \end{bmatrix} = 1 + (ab)x.$$

We show that part a fails. Compare the left-hand side of property a

$$T(\begin{bmatrix} a_1 & 0 \\ 0 & b_1 \end{bmatrix} + \begin{bmatrix} a_2 & 0 \\ 0 & b_2 \end{bmatrix}) = T\begin{bmatrix} a_1 + a_2 & 0 \\ 0 & b_1 + b_2 \end{bmatrix}$$

$$= 1 + [(a_1 + a_2)(b_1 + b_2)]x = 1 + (a_1 b_1 + a_1 b_2 + a_2 b_1 + a_2 b_2)x$$

with the right-hand side of property a

$$T \begin{bmatrix} a_1 & 0 \\ 0 & b_1 \end{bmatrix} + T \begin{bmatrix} a_2 & 0 \\ 0 & b_2 \end{bmatrix} = 1 + (a_1 b_1)x + 1 + (a_2 b_2)x$$

$$= 2 + (a_1 b_1 + a_2 b_2)x.$$

There is no need for a specific counter-example here, since the failure of this property is true in general. Indeed, in the first computation the constant coefficient is always 1, while in the second it is always 2.

Example 4.8 Consider the map $T : M_{nn} \longrightarrow \mathbb{R}$ by $T(A) = |A|$. We show property a. fails. Set $n = 2$ and observe that $T(E_{11} + E_{22}) = T(I_2) = 1$ while $T(E_{11}) + T(E_{22}) = 0 + 0 = 0$.

We end this section with some basic properties of linear transformations. First, let's establish a convenient notation. Let V and W be vector spaces. Then

$$L(V, W) = \{T : V \longrightarrow W \mid T \text{ is a linear transformation}\}.$$

Theorem 4.1 Let V and W be vector spaces with $v_1, v_2, \ldots, v_n \in V$, $S, T \in L(V, W)$, and scalars $a, a_1, \ldots, a_n$. Then

  i. $T(0_V) = 0_W$.

  ii. $T(-v) = -T(v)$.

  iii. $T(a_1 v_1 + \cdots + a_n v_n) = a_1 T(v_1) + \cdots + a_n T(v_n)$.

  iv. $S + T \in L(V, W)$ where $S + T$ is the map defined by $(S + T)(v) = S(v) + T(v)$.

  v. $aT \in L(V, W)$ where $(aT)(v) = aT(v)$.

Proof 4.1 We prove some of the items and leave the rest as exercises. To prove i, notice

$$T(0_V) = T(00_V) = 0T(0_V) = 0_W.$$

The proof of iii is by induction. To prove v, we need to verify for aT the two parts of the definition for linear transformation. For part a,

$$(aT)(v_1 + v_2) = aT(v_1 + v_2) = a(T(v_1) + T(v_2))$$

$$= aT(v_1) + aT(v_2) = (aT)(v_1) + (aT)(v_2).$$

For part b,

$$(aT)(bv) = aT(bv) = a(bT(v)) = (ab)T(v)$$

$$(ba)T(v) = b(aT(v)) = b(aT)(v).$$

$\square$

Corollary 4.1 *Let V and W be vector spaces. Define addition and scalar multiplication as in Theorem 4.1.iv,v. Then $L(V, W)$ is also a vector space.*

Proof 4.2 *We need to verify the nine axioms of a vector space. Property 0 follows from Theorem 4.1.iv,v. The rest of the axioms are proved just as we did for $\mathcal{F}$ (we need to point out, though, that in an exercise to follow one has to prove that $0(v) = 0_W$ is a linear transformation - this is necessary for the verification of Property 3).* $\square$

We hope the reader sees the progression of complexity which culminates in Corollary 4.1: We begin with vector spaces V and W; from this we define a linear transformation $T : V \longrightarrow W$, which in turn produces a vector space $L(V, W)$ of all linear transformations from V into W.

A **linear operator** is simply a linear transformation $T \in L(V, W)$ with $V = W$. Since $V = W$ we use a short-hand notation, $T \in L(V)$. It is in this setting that, perhaps, these functions acquired the name **transformation**. For, as we shall see in the example below, linear operators in a sense transform the shape of the vector space.

Example 4.9 *Define $T \in L(\mathbb{R}^2)$ by $T[a, b] = [b, a]$ (one can easily check that this is indeed a linear transformation). Geometrically, T is reflecting every point in $\mathbb{R}^2$ across the line $y = x$. Another way to express the formula for T is*

$$T[a, b] = \begin{bmatrix} 0 & 1 \\ 1 & 0 \end{bmatrix} \begin{bmatrix} a \\ b \end{bmatrix}.$$

The collection of reflections (and rotations) is an important set of linear operators. See Section 5.3 of Chapter 5 for a more in depth discussion of this topic.

EXERCISES

1. For each of the following maps $T : \mathbb{R}^n \longrightarrow \mathbb{R}^m$, decide whether or not it is a linear transformation and if so express the map in the form $T(v) = Av$ for some matrix $A \in M_{mn}$

   a. $T : \mathbb{R}^2 \longrightarrow \mathbb{R}^3$ by $T[a, b] = [a + b, 2a, b - a]$.

   b. $T : \mathbb{R}^3 \longrightarrow \mathbb{R}^2$ by $T[a, b, c] = [a + b, c^2]$.

c. $T : \mathbb{R}^2 \longrightarrow \mathbb{R}^2$ by $T[a, b] = [2b - a, 3a + 5b]$.

2. Decide whether or not each of the following maps is a linear transformation:

a. $T : M_{nn} \longrightarrow M_{nn}$ by $T(A) = A - A^T$.

b. $T : M_{nn} \longrightarrow M_{nn}$ by $T(A) = A^2 + A$.

c. $T : \mathcal{F} \to P_1$ by $T(f) = f(0) + f(1)x$.

d. $T : \mathbb{R}^2 \to P_2$ by $T[a, b] = a + bx + (a + b)x^2$.

e. $T : \mathbb{R}^n \longrightarrow \mathbb{R}$ by $T(v) = u \cdot v$ for a fixed $u \in \mathbb{R}^n$.

f. $T : P_1 \longrightarrow M_{nn}$ by $T(a + bx) = aA + bB$ for fixed $A, B \in M_{nn}$.

g. $T : \mathcal{F} \longrightarrow \mathbb{R}$ by $T(f) = f(0)/f(1)$.

h. $T : \mathbb{R}^n \to D_{nn}$ by $T(v) = |v|I_n$ (where $|v|$ is the magnitude of v).

i. $T : \mathbb{R}^2 \to \mathcal{F}$ by $T[a, b] = ae^{bx}$.

j. $T : \mathcal{F} \to \mathbb{R}$ by $T(f) = f(0)f(1)$.

k. $T : M_{nn} \to M_{nn}$ by $T(A) = AB - BA$ for a fixed $B \in M_{nn}$.

l. $T : U_{22} \longrightarrow P_1$ by
$$T \begin{bmatrix} a & b \\ 0 & c \end{bmatrix} = a + (bc)x.$$

m. $T : \mathcal{F} \to D_{nn}$ by
$$T(f) = \begin{bmatrix} f(0) & 0 \\ 0 & f(0) \cdot f(1) \end{bmatrix}.$$

n. $T : \mathcal{D} \longrightarrow D_{22}$ by
$$T(f) = \begin{bmatrix} f(0) & 0 \\ 0 & f'(0) \end{bmatrix}.$$

o. $T : D_{22} \longrightarrow \mathbb{R}$ by
$$T \begin{bmatrix} a & 0 \\ 0 & b \end{bmatrix} = a - 2b.$$

p. $T : D_{22} \to P_1$ by
$$T \begin{bmatrix} a & 0 \\ 0 & b \end{bmatrix} = a + (a + b)x.$$

q. $T : P_2 \longrightarrow D_{22}$ by
$$T(a + bx + cx^2) = \begin{bmatrix} a & 0 \\ 0 & bc \end{bmatrix}.$$

3. Let $\mathcal{D}$ represent the collection of real-valued differentiable functions. Show the map $T : \mathcal{D} \longrightarrow \mathcal{F}$ by $T(f) = f'$ (the derivative) is a linear transformation.

4. Let $\mathcal{C}$ represent the collection of real-valued continuous functions. Show the map $T : \mathcal{C} \longrightarrow \mathcal{F}$ by $T(f)$ equals the primary antiderivative (with constant $= 0$) of f is a linear transformation.

5. Let $T : M_{nn} \to \mathbb{R}$ by $T(A) = tr(A)$, the trace of A. Prove that T is a linear transformation.

6. Let V be any vector space and define $1_V : V \longrightarrow V$ by $1_V(v) = v$ for all $v \in V$. Show this defines a linear transformation.

7. Let V and W be any vector spaces with identities 0_V and 0_W, respectively. Define the map $0 : V \longrightarrow W$ by $0(v) = 0_W$ for all $v \in V$. Show this defines a linear transformation.

8. Choose any scalar a and define $T : \mathcal{F} \longrightarrow \mathbb{R}$ by $T(f) = f(a)$. Show this defines a linear transformation.

9. Let U, V, W be vector spaces with $S \in L(U, V)$ and $T \in L(V, W)$. Prove that the composition $T \circ S \in L(U, W)$.

10. Prove if $T : W \longrightarrow W$ is a linear transformation and $T(v_1), T(v_2), \ldots, T(v_n)$ are linearly independent, then $v_1, v_2, \ldots, v_n$ are linearly independent.

11. Fix an angle θ and define $T \in L(\mathbb{R}^2)$ by

$$T[a, b] = \begin{bmatrix} \cos\theta & -\sin\theta \\ \sin\theta & \cos\theta \end{bmatrix} \begin{bmatrix} a \\ b \end{bmatrix}.$$

Describe geometrically how T transforms the points in $\mathbb{R}^2$ (the reader may want to try specific values of θ first). Justify your answer with a proof.

12. Prove parts ii,iii and iv of Theorem 4.1.

4.2   KERNEL AND IMAGE

In this section we introduce two important subspaces which play an important role in Linear Algebra. But first we remind the reader of some definitions concerning functions.

Definition 4.2 *Let $f : X \longrightarrow Y$ be a function from a set X to a set Y.*

1. *f is **one-to-one** (or 1-1) if whenever $f(x_1) = f(x_2)$ we have $x_1 = x_2$, for any $x_1, x_2 \in X$. In other words, every output of f originates from exactly one input in X.*

2. *f maps **onto** Y if for all $y \in Y$ there exists an $x \in X$ such that $f(x) = y$. In other words, every element of Y originates from an input of X via f.*

We will not take the time to illustrate these definitions, since in a moment we will have an easier way to check these properties in the context of linear transformations.

Definition 4.3 *Let $T \in L(V,W)$ with $0_V, 0_W$ the zero vectors in the vector spaces V, W, respectively.*

1. *The **kernel** of T, which we shall designate by $\ker(T)$, is the set of all vectors in V which are sent to 0_W by T, i.e.*

$$\ker(T) = \{v \in V \mid T(v) = 0_W\}.$$

2. *The **range** (or **image**) of T, which we shall designate by $T(V)$, is the set of all outputs of T, i.e.*

$$T(V) = \{T(v) \mid v \in V\}.$$

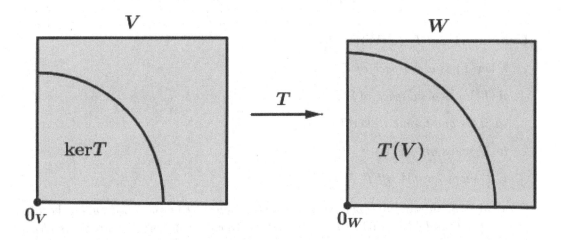

Figure 4.1 Kernel and Range of a linear transformation.

Other notation you may run across for the image of T is $Im(T)$ or $Image(T)$. We remark that $\ker(T)$ is never empty since, by Theorem 4.1.i, it is always the case that $0_V \in \ker(T)$. We will at times drop the subscripts on the zero vectors, since the context makes them understood.

Example 4.10 *Consider the linear transformation $T \in L(M_{22}, P_2)$ defined by*

$$T \begin{bmatrix} a & b \\ c & d \end{bmatrix} = (a+b) + cx + dx^2.$$

We compute $\ker(T)$ as follows:

$$\begin{bmatrix} a & b \\ c & d \end{bmatrix} \in \ker(T) \quad \textit{iff} \quad T \begin{bmatrix} a & b \\ c & d \end{bmatrix} = 0 \quad \textit{iff}$$

$$(a + b) + cx + dx^2 = 0 \quad iff \quad \begin{aligned} a + b &= 0 \\ c &= 0 \\ d &= 0 \end{aligned}.$$

Notice that to find the kernel of T we are once again reduced to solving a system of linear equations. There's no need to solve this using the augmented matrix here, since it's clear that the solution is

$$\begin{aligned} a &= -b \\ c &= 0 \\ d &= 0 \end{aligned}.$$

Hence, we see that

$$\ker(T) = \left\{ \begin{bmatrix} -b & b \\ 0 & 0 \end{bmatrix} \right\}.$$

Lemma 4.1 Let $T \in L(V, W)$. Then

    i. $\ker(T)$ is a subspace of V.

    ii. $T(V)$ is a subspace of W.

    iii. T is one-to-one iff $\ker(T) = \{0_V\}$.

    iv. T maps onto $T(V)$.

    v. T maps onto W iff $T(V) = W$.

Proof 4.3 Some parts of this proof are left as exercises. To prove i, take $u, v \in \ker(T)$ (so $T(u) = 0$ and $T(v) = 0$). First, we need to show $u + v \in \ker(T)$; but this is clear, since by the property of linear transformation, $T(u + v) = T(u) + T(v) = 0 + 0 = 0$. Second, we need to show that for any scalar a, we have $au \in \ker(T)$, but this follows since $T(au) = aT(u) = a0 = 0$.

To prove iii, first assume that $\ker(T) = \{0_V\}$ and we show that T is one-to-one. To do this, suppose that $T(u) = T(v)$ for $u, v \in V$ (and show $u = v$). This implies that $T(u) - T(v) = 0_W$. But then $T(u - v) = 0_W$ and so $u - v \in \ker(T) = \{0_V\}$. Hence, $u - v = 0_V$ and so $u = v$. For the reverse implication, assume that T is one-to-one and suppose $v \in \ker(T)$. Thus, $T(v) = 0_W$. Recall that $T(0_V) = 0_W$ as well, and so $T(v) = T(0_V)$. Since T is one-to-one, we have $v = 0_V$, and so $\ker(T) = \{0_V\}$. $\square$

Example 4.11 In the previous example, $T \in L(M_{22}, P_2(F))$ is **not** one-to-one, since by Lemma 4.1.iii,

$$\ker(T) \neq \left\{ \begin{bmatrix} 0 & 0 \\ 0 & 0 \end{bmatrix} \right\}.$$

Below are some useful results which we will leave as exercises.

Lemma 4.2 *Let $T \in L(V, W)$. Then*

i. T is one-to-one iff whenever $v_1, \ldots, v_k \in V$ are linearly independent, then $T(v_1), \ldots, T(v_k)$ are linearly independent in W.

ii. T maps onto W iff whenever $v_1, \ldots, v_k \in V$ span V, then $T(v_1), \ldots, T(v_k)$ span W.

iii. T is one-to-one and onto W iff whenever $v_1, \ldots, v_k \in V$ form a basis for V, then $T(v_1), \ldots, T(v_k)$ form a basis for W.

Now we have arrived at another dimension theorem which will lead to another counting principle. Among other things, with this theorem we will be able to verify whether or not a linear transformation maps onto W. First, we present a bit of terminology.

Definition 4.4 *Let $T \in L(V, W)$.*

1. *The dimension of the kernel of T is called the **nullity** of T and is written $n(T)$.*

2. *The dimension of the range of T is called the **rank** of T and is written $rk(T)$.*

Later in this section we will give the reason behind this terminology.

Theorem 4.2 (Linear Transformation Dimension Theorem) *Let $T \in L(V, W)$ and assume that $dim(V) < \infty$. Then*

$$\dim(V) = \dim(\ker(T)) + \dim(T(V)).$$

In other words the dimension of V equals the nullity plus rank of T.

Proof 4.4 *Let $\dim(V) = n$ which is given to be finite. First, let's assume that $\ker(T) = \{0\}$. In this case, $dim(\ker(T)) = 0$ and by Lemma 4.1, T is 1–1 and maps onto $T(V)$. Choose a basis, $v_1, \ldots, v_n$, for V. By Lemma 4.2.iii, $T(v_1), \ldots, T(v_n)$ forms a basis for $T(V)$. Hence, $\dim(T(V)) = n$ and so*

$$\dim(V) = n = 0 + n = \dim(\ker(T)) + \dim(T(V)).$$

Now let's assume that $\ker(T) \neq \{0\}$. Since $\ker(T)$ is a subspace of V and $\dim(V)$ is finite, this implies that $\ker(T)$ is finite as well. Let $v_1, \ldots, v_m$ be a basis for $\ker(T)$ (note that $m \leq n$ by Theorem 3.13). As we saw in the previous chapter, we can extend to a basis for V, say

$$v_1, \ldots, v_m, v_{m+1}, \ldots, v_n.$$

We will show that $T(v_{m+1}), \ldots, T(v_n)$ form a basis for $T(V)$ and so we will have

$$\dim(V) = n = m + (n - m) = \dim(\ker(T)) + \dim(T(V)).$$

First we show that $T(v_{m+1}), \ldots, T(v_n)$ are linearly independent. Suppose that $a_{m+1}T(v_{m+1}) + \cdots + a_n T(v_n) = 0$. By Theorem 4.1.iii, we have $T(a_{m+1}v_{m+1} + \cdots + a_n v_n) = 0$. This means that $a_{m+1}v_{m+1} + \cdots + a_n v_n \in \ker(T)$. Since $v_1, \ldots, v_m$ is a basis for $\ker(T)$, there exist $a_1, \ldots, a_m \in F$ such that

$$a_{m+1}v_{m+1} + \cdots + a_n v_n = a_1 v_1 + \cdots + a_m v_m,$$

and so

$$(-a_1)v_1 + \cdots + (-a_m)v_m + a_{m+1}v_{m+1} + \cdots + a_n v_n = 0.$$

Since $v_1, \ldots, v_n$ are linearly independent, $a_1 = \cdots = a_n = 0$. In particular, $a_{m+1} = \cdots = a_n = 0$.

To show that $T(v_{m+1}), \ldots, T(v_n)$ span $T(V)$ take any $T(v) \in T(V)$ where $v \in V$. Since $v_1, \ldots, v_n$ span V, there exist $a_1, \ldots, a_n \in \mathbb{R}$ such that $v = a_1 v_1 + \cdots + a_n v_n$. Now apply T to both sides of the equation:

$$T(v) = T(a_1 v_1 + \cdots + a_n v_n).$$

By properties of a linear transformation, this equation becomes

$$T(v) = a_1 T(v_1) + \cdots + a_n T(v_n).$$

Let's call this last equation $*$. Since $v_1, \ldots, v_m$ are a basis for $\ker(T)$, in particular, they are vectors in $\ker(T)$ and so $T(v_1) = \cdots = T(v_m) = 0$. Then equation $*$ loses the first m terms and becomes

$$T(v) = a_{m+1}T(v_{m+1}) + \cdots + a_n T(v_n).$$

Therefore, $T(v)$ is in the span of $T(v_{m+1}), \ldots, T(v_n)$ and the proof is complete. □

Example 4.12 *Let's return to the earlier example where $T \in L(M_{22}(F), P_2(F))$ defined by*

$$T \begin{bmatrix} a & b \\ c & d \end{bmatrix} = (a + b) + cx + dx^2.$$

We found that

$$\ker(T) = \left\{ \begin{bmatrix} -b & b \\ 0 & 0 \end{bmatrix} \right\}.$$

Hence a basis for $\ker(T)$ contains a single vector (set $b = 1$), namely $\begin{bmatrix} -1 & 1 \\ 0 & 0 \end{bmatrix}$.

Therefore, $\dim(\ker(T)) = 1$. *Now we know that* $\dim(V) = \dim(M_{22}(F)) = 4$. *Using Theorem 4.2, we have*

$$\dim(T(M_{22}(F))) = \dim(M_{22}(F)) - \dim(\ker(T)) = 4 - 1 = 3.$$

Notice that $\dim(T(M_{22})) = 3 = \dim(W) = \dim(P_2)$. *By Theorem 3.13.ii, we have that* $T(M_{22}) = P_2$. *Then by Lemma 4.1.v, we know that* T *maps onto* P_2. *We hope the reader sees both the power and beauty in this counting argument.*

Example 4.13 *Let's look at a complete example from start to finish which illustrates the material in this section. Consider* $T \in L(U_{22}, P_1)$ *defined by*

$$T \begin{bmatrix} a & b \\ 0 & c \end{bmatrix} = (2a - b + c) + (b - c - 2a)x.$$

First we compute $\ker(T)$.

$$\begin{bmatrix} a & b \\ 0 & c \end{bmatrix} \in \ker(T) \;\; \text{iff} \;\; T \begin{bmatrix} a & b \\ 0 & c \end{bmatrix} = 0 \;\; \text{iff} \;\; (2a - b + c) + (b - c - 2a)x = 0$$

$$\text{iff} \;\; 2a - b + c = 0 \;\; \text{and} \;\; b - c - 2a = 0 \;\; \text{iff} \;\; c = b - 2a$$

Hence,

$$\ker(T) = \left\{ \begin{bmatrix} a & b \\ 0 & b - 2a \end{bmatrix} : a, b \in F \right\}.$$

By Lemma 4.1.iii, T *is not one-to-one, since*

$$\ker(T) \neq \left\{ \begin{bmatrix} 0 & 0 \\ 0 & 0 \end{bmatrix} \right\}.$$

A basis for $\ker(T)$ *is*

$$\begin{bmatrix} 1 & 0 \\ 0 & -2 \end{bmatrix}, \begin{bmatrix} 0 & 1 \\ 0 & 1 \end{bmatrix},$$

and so $\dim(\ker(T)) = 2$. *Now, by Linear Transformation Dimension Theorem,*

$$\dim(T(U_{22})) = \dim U_{22} - \dim(\ker(T)) = 3 - 2 = 1.$$

Since $\dim P_1 = 2 \neq 1 = \dim(T(U_{22}))$, *by Theorem 3.13.ii,* $P_1 \neq T(U_{22})$, *and so by Lemma 4.1.v,* T *does not map onto* P_1.

The next Corollary is a consequence of Theorem 4.2. We have already seen a proof of this fact in Section 3.6.

Corollary 4.2 *Let $A \in M_{mn}$ and consider the homogeneous linear system of equations $AX = 0$. Set $U = \text{nullsp}(A)$, i.e. recall $\text{nullsp}(A) = \{v \in \mathbb{R}^n \mid Av = 0\}$. Then*

$$\dim U = n - rk(A)$$

equivalently, $n = r(A) + n(A)$.

Proof 4.5 *Consider the linear transformation $T \in L(\mathbb{R}^n, \mathbb{R}^m)$ defined by $T(v) = Av$. First notice that $\ker(T) = U$. Indeed, $v \in \ker(T)$ iff $T(v) = 0$ iff $Av = 0$ iff $v \in U$.*

Express $A = [c_1 \cdots c_n]$ in columns. We now show that $T(\mathbb{R}^n) = \text{colsp}(A)$. Indeed, $b \in T(\mathbb{R}^n)$ iff there exists a $v \in \mathbb{R}^n$ such that $T(v) = b$ iff there exists a $v \in \mathbb{R}^n$ such that $b = Av \in \text{colsp}(A)$. Now $\dim[\text{span}(c_1, \ldots, c_n)] = rk(A)$, by Corollary 3.4. Finally, by Theorem 4.2,

$$\dim(\mathbb{R}^n) = \dim(\ker(T)) + \dim(T(V))$$

and by the work above this translates to $n = \dim U + rk(A)$. □

Example 4.14 *We give an application of Corollary 4.2. Suppose you want to quickly decide whether or not a homogeneous linear system has non-trivial solutions. One can simply compute the rank of the coefficient matrix for the system. If the rank is less than the number of columns in the coefficient matrix, then by Corollary 4.2, the system must have non-trivial solutions. Indeed, we in fact knew this already due to Theorem 2.18.ii*

EXERCISES

1. Consider $T \in L(P_2, D_{22})$ defined by

$$T(a + bx + cx^2) = \begin{bmatrix} a+b+c & 0 \\ 0 & a-2c \end{bmatrix}.$$

a. Find $\ker(T)$.

b. Use part a. to decide whether or not T is one-to-one.

c. Find a basis for $\ker(T)$.

d. Find $\dim(\ker(T))$.

e. Use Theorem 4.2 to find $\dim(T(P_2))$.

f. Use part e. to decide whether or not T maps onto D_{22}.

2. Consider $T \in L(U_{22}, P_1)$ defined by

$$T \begin{bmatrix} a & b \\ 0 & c \end{bmatrix} = (a + b) + (b - c)x.$$

a. Verify that T is a linear transformation.

b. Calculate $\ker(T)$ and decide whether or not T is one-to-one.

c. Find a basis for and dimension of $\ker(T)$.

d. Use Theorem 4.2 to find $\dim(T(U_{22}))$ and decide whether or not T maps onto P_1.

3. Consider $T \in L(U_{22}, P_1)$ defined by

$$T \begin{bmatrix} a & b \\ 0 & c \end{bmatrix} = (2a - b) + (c - a)x.$$

a. Find $\ker(T)$.

b. Use part a. to decide whether or not T is one-to-one.

c. Find a basis for and dimension of $\ker(T)$.

d. Find $\dim(T(U_{22}))$ and decide whether or not T maps onto P_1.

4. Consider $T \in L(P_1, \mathbb{R}^3)$ defined by

$$T(a + bx) = [a + b, a - b, b - a].$$

a. Find $\ker(T)$.

b. Use part a. to decide whether or not T is one-to-one.

c. Find a basis for and dimension of $\ker(T)$.

d. Find $\dim(T(P_1))$ and decide whether or not T maps onto $\mathbb{R}^3$.

5. Let V be the set of all symmetric 2×2 matrices. Consider the following linear transformation:

$$T : V \longrightarrow P_1 \text{ by } T \begin{bmatrix} a & b \\ b & c \end{bmatrix} = (a + b) + (a - c)x.$$

a. Verify that T is a linear transformation.

b. Calculate $\ker(T)$ and decide whether or not T is one-to-one.

c. Find a basis for and dimension of $\ker(T)$.

d. Use Theorem 4.2 to find $\dim(T(V))$ and decide whether or not T maps onto P_1.

6. For each of the following linear transformations $T \in L(V, W)$,

   i. Find $\ker(T)$ and decide whether or not T is one-to-one.

   ii. Find a basis for and dimension of $\ker(T)$.

   iii. Find $\dim(T(V))$ and decide whether or not T maps onto W.

   a. $T \in L(M_{22}, P_1)$ defined by

   $$T \begin{bmatrix} a & b \\ c & d \end{bmatrix} = (a + b) + (c + d)x.$$

   b. $T \in L(U_{22}, P_1)$ defined by

   $$T \begin{bmatrix} a & b \\ 0 & c \end{bmatrix} = (2a - b + c) + (b - c - 2a)x.$$

   c. $T \in L(V, P_2)$ defined by

   $$T \begin{bmatrix} a & b \\ c & a \end{bmatrix} = (a + b - 2c) + (2a + 2b - 4c)x + (2c - a - b)x^2, \text{ where}$$

   $$V = \left\{ \begin{bmatrix} a & b \\ c & a \end{bmatrix} \mid a, b, c \in \mathbb{R} \right\}.$$

   d. $T \in L(P_2, D_{22})$ defined by

   $$T(a + bx + cx^2) = \begin{bmatrix} a - b + c & 0 \\ 0 & b - a - c \end{bmatrix}.$$

   e. $T \in L(P_3, D_{22})$ defined by

   $$T(a + bx + cx^2 + dx^3) = \begin{bmatrix} a + b + c & 0 \\ 0 & b - c - d \end{bmatrix}.$$

   f. $T \in L(P_3, U_{22})$ defined by

   $$T(a + bx + cx^2 + dx^3) = \begin{bmatrix} a + b & b + c + d \\ 0 & a + 2b + c + d \end{bmatrix}.$$

   g. $T \in L(P_2, D_{22})$ defined by

   $$T(a + bx + cx^2) = \begin{bmatrix} a + b - c & 0 \\ 0 & 2a - b + c \end{bmatrix}.$$

h. $T \in L(U_{22}, \mathbb{R}^2)$ defined by

$$T\begin{bmatrix} a & b \\ 0 & c \end{bmatrix} = [a+b, 2a-c].$$

i. $T \in L(\mathbb{R}^3, P_2)$ defined by

$$T[a, b, c] = (a + 2b - c) + (a - b + 2c)x + (4a - b + 5c)x^2.$$

. Prove if $T : W \longrightarrow W$ is a linear transformation with $\dim(V) = \dim(W)$, then T is one-to-one iff T maps onto W.

. Prove that T is one-to-one iff whenever $v_1, \ldots, v_k \in V$ are linearly independent, then $T(v_1), \ldots, T(v_k)$ are linearly independent in W.

. Prove that (assuming $\dim(V) < \infty$), T maps onto W iff whenever $v_1, \ldots, v_k \in V$ span V, then $T(v_1), \ldots, T(v_k)$ span W.

. Let $T \in L(V, W)$ and U a subspace of W. Define the **inverse image** of T in U,

$$T^{-1}(U) = \{ v \in V \mid T(v) \in U \}.$$

a. Show that $T^{-1}(U)$ is a subspace of V.

b. What can be said about $T^{-1}(\{w\})$, any $w \in W$, if T is one-to-one?

. Let U be a subspace of a finite dimensional vector space V and W any other vector space with dimension at least 1. Construct a $T \in L(V, W)$ such that $\ker(T) = U$.

. Let $T \in L(U, V)$ and $S \in L(V, W)$. Show that $\ker(T) \subseteq \ker(S \circ T)$ and use this to show that $\dim((S \circ T)(U)) \leq \dim(T(U))$.

. Let $S, T \in L(V)$. Show that $(S \circ T)(V) \subseteq S(V)$ and use this to show that $\dim(\ker(S)) \leq \dim(\ker(S \circ T))$.

. Let $A \in M_{mn}$ and $B \in M_{nr}$.

a. Prove that $rk(AB) \leq rk(A)$.
   (hint: Consider the linear transformations defined by multiplication on the left by a matrix)

b. Prove that $rk(AB) \leq rk(B)$.

c. Conclude that $rk(AB) \leq min\{ rk(A), rk(B) \}$.

. Let $T \in L(V)$ for V a finite dimensional vector space with $T^2 = T$ (i.e. $T(T(v)) = T(v)$ for all $v \in V$).

   a. Prove that $\ker(T) \cap T(V) = \{0\}$.

   b. Prove that $V = \ker(T) + T(V)$.

16. Let $T \in L(V,W)$ where V and W are vector spaces and $\dim(V) < \infty$. Assume also that for every subspace U of V of dimension 1, we have $\dim(T(U)) = 1$.

   a. Show that if for $v \in V$, $U = \text{span}(v)$, then $T(U) = \text{span}(T(v))$.

   b. Using part a. prove that $T = a1_V$ for some scalar $a \in \mathbb{R}$.

17. Let $T \in L(V)$ where V is a vector space. Prove that if $rk(T) = 1$, then $T^2 = aT$ for some scalar $a \in \mathbb{R}$, i.e. $T(T(v)) = aT(v)$ for all $v \in V$.

18. Let $S, T \in L(V)$ for some vector space V. Prove that if $ST = TS$ then S maps $\ker(T)$ into $\ker(T)$ and S maps $T(V)$ into $T(V)$.

19. Prove parts ii,iv and v of Lemma 4.1.

20. Prove Lemma 4.2.

4.3 MATRIX REPRESENTATION

Just as in Chapter 2 where we made the connection between matrices and linear systems of equations, we now link the concepts of matrices and linear transformations. We hinted at such a connection in Example 4.2 of Section 4.1 that linear transformations can be represented as multiplication on the left by a matrix. We first prove the foundational result that makes this possible.

Lemma 4.3 *Let $(v_1, \ldots, v_n)$ be an ordered basis for a vector space V and let W be any other vector space.*

   i. If $S, T \in L(V,W)$ and $S(v_i) = T(v_i)$ for $i = 1, \ldots, n$, then $S = T$.

   ii. Take any $w_1, \ldots, w_n \in W$ (possibly non-distinct). There exists a unique $T \in L(V,W)$ such that $T(v_i) = w_i$ for $i = 1, \ldots, n$.

Proof 4.6 *For part i, take any $v \in V$. Since $v_1, \ldots, v_n$ is a basis for V, there exist $a_1, \ldots, a_n \in F$ such that $v = a_1 v_1 + \cdots + a_n v_n$. Since $S, T \in L(V,W)$,*

$$S(v) = S(a_1 v_1 + \cdots + a_n v_n) = a_1 S(v_1) + \cdots + a_n S(v_n)$$
$$= a_1 T(v_1) + \cdots + a_n T(v_n) = T(a_1 v_1 + \cdots + a_n v_n) = T(v).$$

Hence, by definition of equality of functions, we have $S = T$. To prove part ii, define a function $T : V \longrightarrow W$ as follows: For $v \in V$, express $v = a_1 v_1 + \cdots + a_n v_n$ and then define $T(v) = a_1 w_1 + \cdots + a_n w_n$. Notice that T is well-defined, since the $a_1, \ldots, a_n$ are uniquely determined. We need to show that $T \in L(V,W)$. First, we take $u, v \in V$ and show that $T(u + v) = T(u) + T(v)$. Express $u = a_1 v_1 + \cdots + a_n v_n$ and $v = b_1 v_1 + \cdots + b_n v_n$. Notice that $u + v = (a_1 + b_1)v_1 + \cdots + (a_n + b_n)v_n$. Hence,

$$T(u + v) = (a_1 + b_1)w_1 + \cdots + (a_n + b_n)w_n$$
$$= (a_1v_1 + \cdots + a_nv_n) + (b_1v_1 + \cdots + b_nv_n) = T(u) + T(v).$$

*Second, we need to show that $T(au) = aT(u)$. Notice that $au = (aa_1)v_1 + \cdots + $
$n)v_n$. Hence,*

$$T(au) = (aa_1)w_1 + \cdots + (aa_n)w_n$$
$$= a(a_1v_1 + \cdots + a_nv_n) = aT(u).$$

*To show the uniqueness of T, suppose there was an $S \in L(V, W)$ such that $S(v_i) = $
for $i = 1, \ldots, n$. Then we would have $S(v_i) = w_i = T(v_i)$ for $i = 1, \ldots, n$. By part
is implies that $S = T$.*   □

To put in words Lemma 4.3, it says that a linear transformation is completely and
quely determined by where it sends a basis and for any list of vectors $w_1, \ldots, w_n \in$
there is a unique linear transformation sending a chosen basis to those vectors.
's illustrate the Lemma with an example.

ample 4.15 *Here we illustrate that if we know where a linear transformation
ds a basis, then we can determine where it sends any vector. Let $V = \mathbb{R}^2$, $B = $
0], [1, 1])$ and $T \in L(\mathbb{R}^2)$. Suppose we know that $T[1, 0] = [1, 2]$ and $T[1, 1] = $
$-4]$ and we want to determine $T[2, -3]$. Let's find the coordinates of $[2, -3]$ with
pect to the basis. One can show that*

$$[2, -3] = (5)[1, 0] + (-3)[1, 1].$$

Since T is a linear transformation, we have

$$T[2, -3] = T((5)[1, 0] + (-3)[1, 1]) = (5)T[1, 0] + (-3)T[1, 1]$$
$$= (5)[1, 2] + (-3)[3, -4] = [-4, 22].$$

ample 4.16 *Let $V = P_1$ and $W = \mathbb{R}^3$. Take the ordered basis $(v_1 = 1 - x,$
$= 1 + 2x)$ and choose $w_1 = [1, 1, 0]$ and $w_2 = [0, 1, -1]$. We will discover the
mula for the linear transformation $T \in L(V, W)$ with the property that $T(v_1) = w_1$
d $T(v_2) = w_2$. The proof of Lemma 4.3.ii provides an algorithm for how to do this.
ke any $v = a + bx \in V$. We first need to find $a_1, a_2 \in F$ such that $v = a_1v_1 + a_2v_2$,
the coordinates of v with respect to (v_1, v_2). We already know how to do this with
augmented matrix.*

$$\begin{bmatrix} 1 & 1 & | & a \\ -1 & 2 & | & b \end{bmatrix} \xrightarrow{R_1 + R_2} \begin{bmatrix} 1 & 1 & | & a \\ 0 & 3 & | & a + b \end{bmatrix} \xrightarrow{1/3R_2}$$

$$\begin{bmatrix} 1 & 1 & \bigm| & a \\ 0 & 1 & \bigm| & \frac{1}{3}(a+b) \end{bmatrix} \xrightarrow{-R_2+R_1} \begin{bmatrix} 1 & 0 & \bigm| & \frac{2}{3}a - \frac{1}{3}b \\ 0 & 1 & \bigm| & \frac{1}{3}a + \frac{1}{3}b \end{bmatrix}.$$

Hence, the coordinates are $a_1 = \frac{2}{3}a - \frac{1}{3}b$ *and* $a_2 = \frac{1}{3}a + \frac{1}{3}b$. *Now Lemma 4.3.ii defines* T *by*

$$T(a + bx) = \left(\frac{2}{3}a - \frac{1}{3}b\right)[1, 1, 0] + \left(\frac{1}{3}a + \frac{1}{3}b\right)[0, 1, -1] =$$

$$\left[\frac{2}{3}a - \frac{1}{3}b, a, -\frac{1}{3}a - \frac{1}{3}b\right].$$

Just to reassure ourselves, let's check that $T(v_1) = w_1$.

$$T(v_1) = T(1 - x) = \left[\frac{2}{3}(1) - \frac{1}{3}(-1), (1), -\frac{1}{3}(1) - \frac{1}{3}(-1)\right] = [1, 1, 0].$$

Lemma 4.3 allows us to make the following definition.

Definition 4.5 *Let* V *and* W *be vector spaces,* $B_1 = (v_1, \ldots, v_n)$ *be an ordered basis for* V, $B_2 = (w_1, \ldots, w_m)$ *be an ordered basis for* W *and* $T \in L(V, W)$. *The* **matrix representation** *of* T *with respect to* B_1 *and* B_2 *is an* $m \times n$ *matrix of the form*

$$[\ [T(v_1)]_{B_2} \ \cdots \ [T(v_n)]_{B_2}\ ] \quad \text{represented in columns.}$$

A useful notation for this matrix is the descriptive

$$[T]_{B_1}^{B_2} = [\ [T(v_1)]_{B_2} \ \cdots \ [T(v_n)]_{B_2}\ ].$$

This matrix captures, or encodes, the essential information of a linear transformation. Indeed, as we shall see, one can recover the definition of a linear transformation from this matrix. Lemma 4.3 assures us that, once we specify ordered bases, each linear transformation has its own unique matrix representation and, in addition, a matrix of the appropriate dimensions (i.e. $dim(W) \times dim(V)$), when viewed as a matrix representation of a linear transformation, is describing a specific linear transformation. It's quite remarkable that a finite set of numbers set in an array can completely describe a linear transformation. As we shall later see, there are much more remarkable things occurring with this matrix representation. Note the following shorthand notation: If V is a vector space with ordered basis B, and $T \in L(V)$, then we write $[T]_B$ for the matrix representation $[T]_B^B$.

Example 4.17 *Let's find such a matrix representation in a specific setting. Consider the linear transformation* $T \in L(\mathbb{R}^2, P_2)$ *defined by*

$$T[a, b] = (a + b) + (a - b)x + (b - a)x^2.$$

se the following ordered bases for $\mathbb{R}^2$ and P_2, respectively:

$$B_1 = ([1,2],[-2,1]) \quad and \quad B_2 = (1, 1+x, 1+x+x^2).$$

We shall compute $[T]_{B_1}^{B_2}$. First, we need to apply T to each element in B_1 (in *er).*

$$T[1,2] = 3 - x + x^2 \quad and \quad T[-2,1] = -1 - 3x + 3x^2.$$

Now we need to find the coordinates of each of the above outputs of T with respect B_2. Observe how we can find them simultaneously in a single augmented matrix. place B_2 (in their proper order) in the columns to the left of the bar and the puts (also in order) to the right of the bar and row reduce.

$$\left[\begin{array}{ccc|cc} 1 & 1 & 1 & 3 & -1 \\ 0 & 1 & 1 & -1 & -3 \\ 0 & 0 & 1 & 1 & 3 \end{array}\right] \quad which\ reduces\ to \quad \left[\begin{array}{ccc|cc} 1 & 0 & 0 & 4 & 2 \\ 0 & 1 & 0 & -2 & -6 \\ 0 & 0 & 1 & 1 & 3 \end{array}\right].$$

Hence, $[T[1,2]]_{B_2} = [4,-2,1]$ and $[T[-2,1]]_{B_2} = [2,-6,3]$. Now drop these coor- ates (again in order) into the columns of a matrix to get

$$[T]_{B_1}^{B_2} = \left[\begin{array}{cc} 4 & 2 \\ -2 & -6 \\ 1 & 3 \end{array}\right].$$

Notice that in the reduced augmented matrix above, what is to the right of the bar lready the matrix we seek.

*ample **4.18** Now keep the same setup as the previous problem and suppose that $\in L(\mathbb{R}^2, P_2)$ with*

$$[S]_{B_1}^{B_2} = \left[\begin{array}{cc} 3 & -1 \\ -4 & 3 \\ 3 & -1 \end{array}\right].$$

We illustrate the fact that the matrix representation of a linear transformation tains all necessary information about S in order to recover its definition, by finding formula which defines S. By definition of matrix representation, the columns yield

$$[S[1,2]]_{B_2} = [3,-4,3] \quad and \quad [S[-2,1]]_{B_2} = [-1,3,-1].$$

In other words,

$$S[1,2] = (3)(1) + (-4)(1+x) + (3)(1+x+x^2) = 2 - x + 3x^2,$$

$$S[-2, 1] = (-1)(1) + (3)(1 + x) + (-1)(1 + x + x^2) = 1 + 2x - x^2.$$

At this point we know where S sends a basis, so as before, we can find the formula for S. Recall that we need the coordinates of an arbitrary $[a, b]$ in terms of the basis B_1. We use an augmented matrix.

$$\begin{bmatrix} 1 & -2 & a \\ 2 & 1 & b \end{bmatrix} \text{ which reduces to } \begin{bmatrix} 1 & 0 & (1/5)a + (2/5)b \\ 0 & 1 & -(2/5)a + (1/5)b \end{bmatrix}.$$

Therefore, $[a, b]_{B_1} = [(1/5)a + (2/5)b, -(2/5)a + (1/5)b]$ and thus the formula for S is

$$S[a, b] = ((1/5)a + (2/5)b)(2 - x + 3x^2) + (-(2/5)a + (1/5)b)(1 + 2x - x^2) =$$

$$b - ax + (a + b)x^2.$$

What we have illustrated without specifically proving is the following fact:

Theorem 4.3 *Consider two finite dimensional vector spaces V and W with $dimV = n$ and $dimW = m$ and with respective bases B_1 and B_2. For any matrix $A \in M_{mn}$ there exists a unique linear transformation $T \in L(V, W)$ such that $[T]_{B_1}^{B_2} = A$.*

Now we look at some elegant and intuitive results derived from the notion of the matrix representation of a linear transformation.

Lemma 4.4 *Let U, V, W be finite dimensional vector spaces with corresponding ordered bases B_1, B_2, B_3. Then*

i. If $T \in L(U, V)$ and $u \in U$, then $[T(u)]_{B_2} = [T]_{B_1}^{B_2}[u]_{B_1}$.

ii. If 1_U is the identity map for U, then $[1_U]_{B_1} = I$.

iii. If $S, T \in L(U, V)$, then $[S + T]_{B_1}^{B_2} = [S]_{B_1}^{B_2} + [T]_{B_1}^{B_2}$.

iv. If $T \in L(U, V)$ and scalar $a \in \mathbb{R}$, then $[aT]_{B_1}^{B_2} = a[T]_{B_1}^{B_2}$.

v. If $S \in L(U, V)$ and $T \in L(V, W)$, then $[T \circ S]_{B_1}^{B_3} = [T]_{B_1}^{B_2}[S]_{B_2}^{B_3}$.

Proof 4.7 *Let $B_1 = (u_1, \ldots, u_n)$ and let $dim(V) = m$. To prove part i, define the following two linear transformations (one should really check that they are indeed linear transformations) $R, S \in L(U, \mathbb{R}^m)$:*

$$R(u) = [T(u)]_{B_2} \quad and \quad S(u) = [T]_{B_1}^{B_2}[u]_{B_1}.$$

For each u_i in the basis for U, notice that

$$S(u_i) = [T]_{B_1}^{B_2}[u_i]_{B_1} = [T]_{B_1}^{B_2}e_i,$$

the ith column of $[T]_{B_1}^{B_2}$, *which by definition is* $[T(u_i)]_{B_2} = R(u_i)$. *Since R and S agree on a basis, by Lemma 4.3.i, $R = S$. In other words $R(u) = S(u)$ for all $u \in U$. By how we defined R and S, we have proved part i.*

Part ii is easy and we leave it to the reader to prove. To prove part iii, we employ part i. For $i = 1, \ldots, n$,

$$([S]_{B_1}^{B_2} + [T]_{B_1}^{B_2})e_i = ([S]_{B_1}^{B_2} + [T]_{B_1}^{B_2})[u_i]_{B_1} = [S]_{B_1}^{B_2}[u_i]_{B_1} + [T]_{B_1}^{B_2}[u_i]_{B_1}$$

$$= [S(u_i)]_{B_2} + [T(u_i)]_{B_2} = [S(u_i) + T(u_i)]_{B_2} = [(S + T)(u_i)]_{B_2}$$

$$= [S + T]_{B_1}^{B_2}[u_i]_{B_1} = [S + T]_{B_1}^{B_2}e_i.$$

Hence, by Exercise 7 in Section 1.5, $[S + T]_{B_1}^{B_2} = [S]_{B_1}^{B_2} + [T]_{B_1}^{B_2}$.
Parts iv and v are proved in a similar manner to iii. □

We point out that part i of the lemma confirms an earlier remark concerning how linear transformations can be represented as multiplication on the left by a matrix (i.e. the matrix representation of the linear transformation). In fact, the following result follows immediately if we take the bases to be the standard ones:

Corollary 4.3 *Given any linear transformation $T \in L(\mathbb{R}^n, \mathbb{R}^m)$ there exists a unique matrix $A \in M_{mn}$ such that $T(v) = Av$ for all $v \in \mathbb{R}^n$.*

Example 4.19 *We illustrate Lemma 4.4.i with an example. Consider the linear transformation $T \in L(D_2, P_2)$ defined by*

$$T \begin{bmatrix} a & 0 \\ 0 & b \end{bmatrix} = a + bx + (a + b)x^2.$$

Consider the following bases for D_{22} and P_2, respectively:

$$B_1 = \left(\begin{bmatrix} 2 & 0 \\ 0 & 0 \end{bmatrix}, \begin{bmatrix} -1 & 0 \\ 0 & 1 \end{bmatrix} \right) \quad \text{and} \quad B_2 = (1 + x + x^2, 1 + x, x).$$

One can compute

$$[T]_{B_1}^{B_2} = \begin{bmatrix} 2 & 0 \\ 0 & -1 \\ -2 & 2 \end{bmatrix}.$$

We illustrate how one can use Lemma 4.4.i to compute $T(v)$ where

$$v = \begin{bmatrix} 2 & 0 \\ 0 & -1 \end{bmatrix}.$$

First, we compute $[v]_{B_1}$. *Notice that*

$$\begin{bmatrix} 2 & 0 \\ 0 & -1 \end{bmatrix} = (1/2)\begin{bmatrix} 2 & 0 \\ 0 & 0 \end{bmatrix} + (-1)\begin{bmatrix} -1 & 0 \\ 0 & 1 \end{bmatrix}.$$

Hence,

$$\begin{bmatrix} 2 & 0 \\ 0 & -1 \end{bmatrix}_{B_1} = [1/2, -1].$$

By Lemma 4.4.i,

$$\left[T\begin{bmatrix} 2 & 0 \\ 0 & -1 \end{bmatrix}\right]_{B_2} = \begin{bmatrix} 2 & 0 \\ 0 & -1 \\ -2 & 2 \end{bmatrix}\begin{bmatrix} 1/2 \\ -1 \end{bmatrix} = \begin{bmatrix} 1 \\ 1 \\ -3 \end{bmatrix}.$$

This means that

$$T\begin{bmatrix} 2 & 0 \\ 0 & -1 \end{bmatrix} = (1)(1 + x + x^2) + (1)(1 + x) + (-3)(x) = 2 - x + x^2.$$

Of course, a much easier way to compute this result is to use the formula.

$$T\begin{bmatrix} 2 & 0 \\ 0 & -1 \end{bmatrix} = (2) + (-1)x + ((2) + (-1))x^2 = 2 - x + x^2.$$

We point out, as the reader can probably tell, this is not a very practical method for computing the outputs of T. We gave this example merely as a way to see Lemma 4.4.i in action. Indeed, as is often the case in this text we illustrate mathematical results by way of example. This is done to help the reader grasp abstract concepts by having a concrete example.

EXERCISES

1. Consider the following ordered bases for $\mathbb{R}^3$ and P_2:

$$B_1 = (\ [1,0,0],\ [1,1,0],\ [1,1,1]\ ) \qquad B_2 = (x^2, x + x^2, 1 + x + x^2),$$

and the matrix

$$A = \begin{bmatrix} 2 & 1 & 4 \\ 3 & 1 & 1 \\ 1 & 1 & 1 \end{bmatrix}.$$

Find the formula for $T \in L(\mathbb{R}^3, P_2)$ if $[T]_{B_1}^{B_2} = A$.

2. Consider the following ordered bases for D_{22} and P_2:

$$B_1 = \left(\begin{bmatrix} 2 & 0 \\ 0 & 1 \end{bmatrix}, \begin{bmatrix} -1 & 0 \\ 0 & 1 \end{bmatrix} \right), \qquad B_2 = (x, -1 - 3x, 2 - x^2),$$

and the matrix

$$A = \begin{bmatrix} 1 & 2 \\ -2 & 0 \\ 1 & -1 \end{bmatrix}.$$

Find the formula for $T \in L(D_{22}, P_2)$ if $[T]_{B_1}^{B_2} = A$.

3. Consider the linear transformation $T \in L(\mathbb{R}^3, P_1)$ defined by $T[a, b, c] = (a+b) + (b+c)x$ and ordered bases $B_1 = ([1, 0, 2], [0, -1, 2], [1, 1, 1])$ and $B_2 = (x, 2 - x)$.

   a. Compute $[T]_{B_1}^{B_2}$.

   b. Use the fact that $[T]_{B_1}^{B_2}[v]_{B_1} = [T(v)]_{B_2}$ to compute $T[1, 2, 0]$.

4. Consider the linear transformation $T \in L(\mathbb{R}^2, P_1)$ defined by $T[a, b] = (a+b) + bx$ and ordered bases $B_1 = ([1, -1], [0, 2])$ and $B_2 = (1 + x, 1 - x)$.

   a. Compute $[T]_{B_1}^{B_2}$.

   b. Use the fact that $[T]_{B_1}^{B_2}[v]_{B_1} = [T(v)]_{B_2}$ to compute $T[1, -2]$.

   c. Let $S \in L(\mathbb{R}^2, P_1)$ and suppose $[S]_{B_1}^{B_2} = \begin{bmatrix} -1/2 & 2 \\ -1/2 & 0 \end{bmatrix}$. Find the formula which defines S.

5. Consider the linear transformation $T \in L(P_1, D_{22})$ defined by

$$T(a + bx) = \begin{bmatrix} a+b & 0 \\ 0 & b-a \end{bmatrix}$$

and ordered bases $B_1 = (1 + x, 1 - x)$ and

$$B_2 = \left(\begin{bmatrix} 1 & 0 \\ 0 & 1 \end{bmatrix}, \begin{bmatrix} -1 & 0 \\ 0 & 1 \end{bmatrix} \right).$$

   a. Compute $[T]_{B_1}^{B_2}$.

   b. Use the fact that $[T]_{B_1}^{B_2}[v]_{B_1} = [T(v)]_{B_2}$ to compute $T(2 - x)$.

c. Let $S \in L(P_1, D_{22})$ and suppose $[S]_{B_1}^{B_2} = \begin{bmatrix} 3/2 & 1/2 \\ 1/2 & -1/2 \end{bmatrix}$. Find the formula which defines S.

6. Consider the following data:

$$T \in L(U_{22}, P_1) \text{ defined by } T\begin{bmatrix} a & b \\ 0 & c \end{bmatrix} = (a+b) + (b-c)x.$$

$$B_1 = \left(\begin{bmatrix} 1 & 0 \\ 0 & 0 \end{bmatrix}, \begin{bmatrix} 1 & 0 \\ 0 & -1 \end{bmatrix}, \begin{bmatrix} 0 & -1 \\ 0 & 0 \end{bmatrix} \right) \text{ and } B_2 = (1+x, -x).$$

a. Compute $[T]_{B_1}^{B_2}$.

b. Use Lemma 4.4.i to compute $T\begin{bmatrix} 1 & 2 \\ 0 & 3 \end{bmatrix}$.

7. Consider $T \in L(U_{22}, P_1)$ defined by

$$T\begin{bmatrix} a & b \\ 0 & c \end{bmatrix} = (2a-b) + (c-a)x.$$

and the ordered bases for U_{22} and P_1, respectively,

$$B_1 = \left(\begin{bmatrix} 2 & 1 \\ 0 & 0 \end{bmatrix}, \begin{bmatrix} 0 & 1 \\ 0 & 1 \end{bmatrix}, \begin{bmatrix} -3 & 0 \\ 0 & 2 \end{bmatrix} \right) \qquad B_2 = (2+x, -5-2x).$$

a. Compute $[T]_{B_1}^{B_2}$

b. Find the formula for $S \in L(U_{22}, P_1)$ if $[S]_{B_1}^{B_2} = \begin{bmatrix} 3 & -1 & -2 \\ 1 & 0 & -1 \end{bmatrix}$.

8. Consider $T \in L(P_1, \mathbb{R}^3)$ defined by

$$T(a+bx) = [a+b, a-b, b-a].$$

and the ordered bases for P_1 and $\mathbb{R}^3$, respectively,

$$B_1 = (2-x, -1+x) \qquad B_2 = ([1,1,1], [-1,1,0], [0,1,1]).$$

a. Compute $[T]_{B_1}^{B_2}$

b. Find the formula for $S \in L(P_1, \mathbb{R}^3)$ if $[S]_{B_1}^{B_2} = \begin{bmatrix} 1 & 1 \\ 1 & 0 \\ 0 & -1 \end{bmatrix}$.

9. Let V be the set of all symmetric 2×2 matrices. Consider the following data:

$$T \in L(V, P_1) \text{ defined by } T \begin{bmatrix} a & b \\ b & c \end{bmatrix} = (a+b) + (a-c)x.$$

$$B_1 = \left(\begin{bmatrix} 1 & 0 \\ 0 & 0 \end{bmatrix}, \begin{bmatrix} 1 & 1 \\ 1 & 0 \end{bmatrix}, \begin{bmatrix} 1 & 1 \\ 1 & 1 \end{bmatrix} \right) \text{ and } B_2 = (1, 1+x).$$

a. Compute $[T]_{B_1}^{B_2}$

b. Use Lemma 4.4.i to compute $T \begin{bmatrix} 0 & 1 \\ 1 & 1 \end{bmatrix}$.

10. Consider the following data:

$$T \in L(M_{22}, P_1) \text{ defined by } T \begin{bmatrix} a & b \\ c & d \end{bmatrix} = (a+b) + (c+d)x.$$

$$B_1 = \left(\begin{bmatrix} 1 & 0 \\ 0 & 0 \end{bmatrix}, \begin{bmatrix} 1 & 1 \\ 0 & 0 \end{bmatrix}, \begin{bmatrix} 1 & 1 \\ 1 & 0 \end{bmatrix}, \begin{bmatrix} 1 & 1 \\ 1 & 1 \end{bmatrix} \right) \text{ and } B_2 = (1, 1+x).$$

a. Compute $[T]_{B_1}^{B_2}$

b. Use Lemma 4.4.i to compute $T \begin{bmatrix} 2 & 1 \\ 0 & 0 \end{bmatrix}$.

11. Let $T \in L(\mathbb{R}^3, D_{22})$ by

$$T[a, b, c] = \begin{bmatrix} a + 2c & 0 \\ 0 & 2b - c \end{bmatrix}$$

and consider bases $B_1 = ([1, -1, 0], [2, 0, 1], [0, -3, -2])$ and

$$B_2 = \left(\begin{bmatrix} 1 & 0 \\ 0 & -1 \end{bmatrix}, \begin{bmatrix} 2 & 0 \\ 0 & 0 \end{bmatrix} \right)$$

for $\mathbb{R}^3$ and D_{22} respectively.

a. Compute $[T]_{B_1}^{B_2}$

b. Use the fact that $[T(v)]_{B_2} = [T]_{B_1}^{B_2}[v]_{B_1}$ to compute $T[0, 1, -2]$.

c. Given that $S \in L(\mathbb{R}^3, D_{22})$ and

$$[S]_{B_1}^{B_2} = \begin{bmatrix} 1 & -1 & 1 \\ 0 & 1 & 1/2 \end{bmatrix},$$

find the formula for S.

12. Consider the following data:

$$V = \left\{ \begin{bmatrix} a & b \\ c & a \end{bmatrix} \mid a, b, c \in \mathbb{R} \right\},$$

$T \in L(V, P_2)$ defined by $T \begin{bmatrix} a & b \\ c & a \end{bmatrix} = (a+b-2c)+(2a+2b-4c)x+(2c-a-b)x^2,$

$$B_1 = \left(\begin{bmatrix} 1 & 1 \\ 0 & 1 \end{bmatrix}, \begin{bmatrix} 0 & 1 \\ 1 & 0 \end{bmatrix}, \begin{bmatrix} 1 & 0 \\ 1 & 1 \end{bmatrix} \right) \quad \text{and} \quad B_2 = (x, 1, x + x^2).$$

a. Show that V is indeed a vector space by showing it is a subspace of M_{22}.

b. Show that B_1 and B_2 are bases for V and P_2, respectively.

c. Compute $[T]_{B_1}^{B_2}$.

d. Use Lemma 4.4.i to compute $T \begin{bmatrix} 1 & 1 \\ -2 & 1 \end{bmatrix}$.

13. Consider $T \in L(P_2, D_{22})$ defined by

$$T(a + bx + cx^2) = \begin{bmatrix} a-b+c & 0 \\ 0 & b-a-c \end{bmatrix}.$$

and the ordered bases for P_2 and D_{22}, respectively,

$$B_1 = (\, 1+x, x-x^2, 2+x^2 \,) \qquad B_2 = \left(\begin{bmatrix} 1 & 0 \\ 0 & -1 \end{bmatrix}, \begin{bmatrix} 2 & 0 \\ 0 & -3 \end{bmatrix} \right).$$

a. Compute $[T]_{B_1}^{B_2}$.

b. Use Lemma 4.4.i to compute $T(1 - x + 2x^2)$.

c. Find the formula for $S \in L(P_2, D_{22})$ if $[S]_{B_1}^{B_2} = \begin{bmatrix} -1 & 2 & 1 \\ 1 & 0 & -2 \end{bmatrix}$.

14. Consider $T \in L(L_{22}, \mathbb{R}^2)$ defined by

$$T \begin{bmatrix} a & 0 \\ b & c \end{bmatrix} = [a + b, a + c]$$

and ordered bases

$$B_1 = \left(\begin{bmatrix} 1 & 0 \\ 1 & 0 \end{bmatrix}, \begin{bmatrix} 1 & 0 \\ 0 & 1 \end{bmatrix}, \begin{bmatrix} 0 & 0 \\ 1 & 1 \end{bmatrix} \right)$$

and $B_2 = ([1, -2], [1, -1])$.

a. Compute $[T]_{B_1}^{B_2}$.

b. Use Lemma 4.4.i to compute

$$T \begin{bmatrix} 3 & 0 \\ 0 & 1 \end{bmatrix}.$$

c. Let $S \in L(L_{22}, \mathbb{R}^2)$ and suppose

$$[S]_{B_1}^{B_2} = \begin{bmatrix} -3 & -2 & -3 \\ 5 & 3 & 4 \end{bmatrix}.$$

Find the formula which defines S.

15. Consider $T \in L(P_3, D_{22})$ defined by

$$T(a + bx + cx^2 + dx^3) = \begin{bmatrix} a + b + c & 0 \\ 0 & b - c - d \end{bmatrix}.$$

and the ordered bases for P_3 and D_{22}, respectively,

$$B_1 = (\, 1, 1+x, 1+x+x^2, 1+x+x^2+x^3 \,) \qquad B_2 = \left(\begin{bmatrix} 0 & 0 \\ 0 & 1 \end{bmatrix}, \begin{bmatrix} -1 & 0 \\ 0 & 1 \end{bmatrix} \right).$$

a. Compute $[T]_{B_1}^{B_2}$.

b. Use Lemma 4.4.i to compute $T(3 + 2x^2 + x^3)$.

c. Find the formula for $S \in L(P_3, D_{22})$ if

$$[S]_{B_1}^{B_2} = \begin{bmatrix} 1 & 0 & 1 & 0 \\ 0 & -2 & 1 & -1 \end{bmatrix}.$$

16. Consider the following data:

A linear transformation $T \in L(D_2, P_3)$ defined by

$$T\begin{bmatrix} a & 0 \\ 0 & b \end{bmatrix} = (a+b) + bx - ax^2 + (4a - 2b)x^3,$$

and ordered bases

$$B_1 = \left(\begin{bmatrix} 1 & 0 \\ 0 & -1 \end{bmatrix}, \begin{bmatrix} 2 & 0 \\ 0 & 1 \end{bmatrix} \right) \text{ and } B_2 = \left(1 + x^3, \ 1 + x, \ x + x^3, \ x^2 + x^3 \right).$$

a. Verify that B_1 is indeed a basis.

b. Compute the matrix representation $[T]_{B_1}^{B_2}$.

c. Using the fact that $[T]_{B_1}^{B_2}[v]_{B_1} = [T(v)]_{B_2}$, compute

$$T\begin{bmatrix} 1 & 0 \\ 0 & 2 \end{bmatrix}.$$

d. Find the formula defining S, given that $S : D_2 \longrightarrow P_3$ is a linear transformation with

$$[S]_{B_1}^{B_2} = \begin{bmatrix} 2 & 1 \\ -1 & 1 \\ 0 & 0 \\ 0 & 3 \end{bmatrix}.$$

17. Consider the linear transformation $T \in L(D_{22}, \mathbb{R}^2)$ defined by

$$T\begin{bmatrix} a & 0 \\ 0 & b \end{bmatrix} = [a + b, a - b]$$

and ordered bases

$$B_1 = \left(\begin{bmatrix} 1 & 0 \\ 0 & 0 \end{bmatrix}, \begin{bmatrix} 1 & 0 \\ 0 & 1 \end{bmatrix} \right)$$

and $B_2 = ([1, -2], [1, -1])$.

a. Compute $[T]_{B_1}^{B_2}$.

b. Use the fact that $[T]_{B_1}^{B_2}[v]_{B_1} = [T(v)]_{B_2}$ to compute

$$T\begin{bmatrix} -1 & 0 \\ 0 & -3 \end{bmatrix}.$$

c. Let $S \in L(D_{22}, \mathbb{R}^2)$ and suppose

$$[S]_{B_1}^{B_2} = \begin{bmatrix} -3 & -3 \\ 5 & 6 \end{bmatrix}.$$

Find the formula which defines S.

18. Let $B = (v_1, v_2, \ldots, v_n)$ be an ordered basis for a vector space V. Set $B' = (cv_1, cv_2, \ldots, cv_n)$ for some scalar c. Prove that for any linear operator $T \in L(V)$ we have $[T]_B^B = [T]_{B'}^{B'}$.

19. Suppose that $S, T \in L(V)$ for a finite dimensional vector space V. Show that if $S \circ T = 1_V$, then $T \circ S = 1_V$.

20. Referring to Exercise 15 in Section 4.2 where we considered $T \in L(V)$ for V a finite dimensional vector space with $T^2 = T$ we proved that $ker(T) \cap T(V) = \{0\}$ and $V = ker(T) + T(V)$.

    a. Let $v_1, \ldots, v_k$ be a basis for $ker(T)$ and $T(u_1), \ldots, T(u_m)$ be a basis for $T(V)$. Set $B = \{ v_1, \ldots, v_k, T(u_1), \ldots, T(u_m) \}$. Verify that B is a basis for V (look at the proof of Theorem 3.14).

    b. Describe $[T]_B$ using ordered basis B from part a.

21. Check that R, S in the proof of Lemma 4.4.i are linear transformations.

22. Prove parts ii,iv and v of Lemma 4.4.

4.4 INVERSE AND ISOMORPHISM

We now look at a special class of linear transformations which have some additional properties. Before we do this we remind the reader of some results that are valid for any functions.

4.4.1 Background

Definition 4.6 *Let X and Y be any sets and $f : X \longrightarrow Y$ be any function from X to Y. A function $g : Y \longrightarrow X$ is the **inverse** of f if $g \circ f = 1_X$ and $f \circ g = 1_Y$. In other words, for all $x \in X$ and $y \in Y$ we have*

$$(g \circ f)(x) = x \quad and \quad (f \circ g)(y) = y.$$

Example 4.20 *Here is a simple example to illustrate the definition. Let $f : \mathbb{R} \longrightarrow \mathbb{R}$ by $f(x) = 2x - 5$. Then the reader can check that f has inverse $g(x) = \frac{1}{2}(x + 5)$.*

In a sense, the inverse g of a function f *undoes* what the function f does. Below are some pertinent results concerning functions in general:

Theorem 4.4 *Let $f : X \longrightarrow Y$ be a function from a set X to a set Y.*

i. f has an inverse iff f is one-to-one and maps onto Y.

ii. If f has an inverse, then it has exactly one.

iii. If f has an inverse, then the inverse is also one-to-one and maps onto X.

iv. If f_1 has inverse g_1 and f_2 has inverse g_2, then $f_1 \circ f_2$ has an inverse, namely $g_2 \circ g_1$.

Proof 4.8 *To prove i, first we assume that f has an inverse g. We show f is one-to-one. We have to resort to the original definition of one-to-one, since f is not assumed to be a linear transformation. For $x_1, x_2 \in X$, if $f(x_1) = f(x_2)$, then $g(f(x_1)) = g(f(x_2))$, i.e. $(g \circ f)(x_1) = (g \circ f)(x_2)$. By definition of inverse, this equation reduces to $x_1 = x_2$, and we have proved that f is one-to-one. To show that f maps onto Y, take any $y \in Y$. We have to find an $x \in X$ such that $f(x) = y$. The element $x = g(y)$ does the trick, since $f(g(y)) = (f \circ g)(y) = y$.*

Now assume that f is one-to-one and maps onto Y. Define a function $g : Y \longrightarrow X$ as follows: For $y \in Y$ find the unique $x \in X$ such that $f(x) = y$ (such an x exists, since f maps onto Y, and it is unique, since f is one-to-one). We then define $g(y) = x$ (i.e. the x that we found above). We prove that this g is the inverse of f. First, for any $y \in Y$, we have $(f \circ g)(y) = f(g(y)) = f(x) = y$. Second, take any $x \in X$. Set $y = f(x)$. Note that by definition of g, we have that $g(y) = x$. Hence, $(g \circ f)(x) = g(f(x)) = g(y) = x$.

To prove ii, suppose that g_1 and g_2 are inverses of f. We will show that $g_1 = g_2$ and so f has only one inverse (if it exists). For any $y \in Y$, since f is one-to-one and maps onto Y there is a unique $x \in X$ such that $f(x) = y$. Then

$$g_1(y) = g_1(f(x)) = (g_1 \circ f)(x) = x = (g_2 \circ f)(x) = g_2(f(x)) = g_2(y).$$

We leave the proof of iii and iv as exercises. □

Because of the fact that when an inverse exists there is only one, we can assign it notation without any confusion. The inverse of f will be denoted by f^{-1}. Take note that this is simply notation and should not be taken literally as $1/f$.

4.4.2 Inverse

We now restate the definition of inverse in the context of linear transformations.

Definition 4.7 *Let V and W be vector spaces with $T \in L(V, W)$. The function $T^{-1} : W \longrightarrow V$ is the **inverse** of T, if $T^{-1} \circ T = 1_V$ and $T \circ T^{-1} = 1_W$. If T is one-to-one and maps onto W (or equivalently has an inverse), we say that T is **invertible**.*

Example 4.21 *Consider* $T \in L(P_1, D_2)$ *defined by*

$$T(a + bx) = \begin{bmatrix} a + b & 0 \\ 0 & a - b \end{bmatrix}.$$

One can show (and shortly we will present an algorithm for producing it) that

$$T^{-1} \begin{bmatrix} a & 0 \\ 0 & b \end{bmatrix} = \left(\frac{a + b}{2}\right) + \left(\frac{a - b}{2}\right) x.$$

We check, for example, that $T^{-1} \circ T = 1_{P_1}$.

$$(T^{-1} \circ T)(a + bx) = T^{-1}(T(a + bx)) = T^{-1} \begin{bmatrix} a + b & 0 \\ 0 & a - b \end{bmatrix} =$$

$$\left(\frac{(a + b) + (a - b)}{2}\right) + \left(\frac{(a + b) - (a - b)}{2}\right) x = a + bx.$$

Similarly, one can verify that $T \circ T^{-1} = 1_{D_2}$.

One of the goals of this section is to have a method for finding T^{-1}. This will be given at the end of the section. Below are some results regarding linear transformations and their inverses. The first two results parallel a result that is true for functions on finite sets having the same size.

Lemma 4.5 *Let* V *and* W *be vector spaces with* V *having finite dimension and* $T \in L(V, W)$. *Then*

    i. If T *is one-to-one with* $\dim V = \dim W$ *finite, then* T *is invertible.*

    ii. If T *maps onto* W *with* $\dim V = \dim W$ *finite, then* T *is invertible.*

    iii. If T *is invertible, then* $\dim V = \dim W$.

Proof 4.9 *To prove i, assuming* $\dim V = \dim W$ *and* T *is one-to-one (and so* $\ker(T) = \{0_V\}$), *this implies that*

$$\dim(T(V)) = \dim V - \dim(\ker(T)) = \dim V - 0 = \dim V.$$

Hence T *maps onto* V, *which implies that* T *is invertible.*

    The proof of part ii is similar and left as an exercise. To prove part iii, just consider a finite basis for V *which* T *sends to a basis for* W, *since it is both onto-to-one and maps onto* W. □

The next result makes an important connection between invertible linear transformations and their corresponding matrix representations.

Theorem 4.5 *Let V and W be finite dimensional vector spaces and B_1, B_2 be any ordered bases for V, W respectively. Let $T \in L(V, W)$. Then*

   i. If T is invertible, then so is $T^{-1} \in L(W, V)$.

   ii. If T is invertible, then $[T^{-1}]_{B_2}^{B_1} = \left([T]_{B_1}^{B_2}\right)^{-1}$.

   iii. T is invertible iff $[T]_{B_1}^{B_2}$ is invertible.

Proof 4.10 *Part i makes for a nice exercise, so we leave it to the reader. Part ii is less obvious, so we supply the proof. We prove this directly using Lemma 4.4.*

$$[T^{-1}]_{B_2}^{B_1}[T]_{B_1}^{B_2} = [T^{-1} \circ T]_{B_1} = [1_V]_{B_1} = I.$$

To prove iii, if T is invertible, then by part ii, $[T]_{B_1}^{B_2}$ is invertible. Now assume $[T]_{B_1}^{B_2}$ is invertible. To show T is one-to-one, we look at $\ker(T)$. If $T(v) = 0$, then

$$0 = [T(v)]_{B_2} = [T]_{B_1}^{B_2}[v]_{B_1}.$$

Multiplying both sides by the inverse of $[T]_{B_1}^{B_2}$ yields $0 = [v]_{B_1}$ and so $v = 0$. Hence, $\ker(T) = \{0\}$ and T is one-to-one. To show T maps onto W, notice that since $[T]_{B_1}^{B_2}$ is invertible, it must be square, and so B_1 and B_2 have the same number of vectors. Hence, $\dim(V) = \dim(W)$ and by Lemma 4.5, T maps onto W. Therefore, T is invertible. □

Theorem 4.5.ii has a nice practical application. We can use it to compute T^{-1}.

Example 4.22 *Consider $T \in L(P_1, D_2)$ defined by*

$$T(a + bx) = \begin{bmatrix} a + b & 0 \\ 0 & a - b \end{bmatrix}.$$

We first compute $[T]_{B_1}^{B_2}$. To make it easy on ourselves, we choose standard bases $B_1 = ST_1 = (1, x)$ and

$$B_2 = ST_2 = \left(\begin{bmatrix} 1 & 0 \\ 0 & 0 \end{bmatrix}, \begin{bmatrix} 0 & 0 \\ 0 & 1 \end{bmatrix} \right).$$

Then

$$T(1) = \begin{bmatrix} 1 & 0 \\ 0 & 1 \end{bmatrix} = (1) \begin{bmatrix} 1 & 0 \\ 0 & 0 \end{bmatrix} + (1) \begin{bmatrix} 0 & 0 \\ 0 & 1 \end{bmatrix}, \; and$$

$$T(x) = \begin{bmatrix} 1 & 0 \\ 0 & -1 \end{bmatrix} = (1) \begin{bmatrix} 1 & 0 \\ 0 & 0 \end{bmatrix} + (-1) \begin{bmatrix} 0 & 0 \\ 0 & 1 \end{bmatrix}.$$

Hence,

$$[T]^{ST_2}_{ST_1} = \begin{bmatrix} 1 & 1 \\ 1 & -1 \end{bmatrix}.$$

Notice that since $\left|[T]^{ST_2}_{ST_1}\right| = -2 \neq 0$, *we know that* $[T]^{ST_2}_{ST_1}$ *is invertible, and so by Theorem 4.5.iii, T is invertible. Thus we have the assurance that* T^{-1} *does indeed exist. By Theorem 4.5.ii,*

$$[T^{-1}]^{ST_1}_{ST_2} = \begin{bmatrix} 1 & 1 \\ 1 & -1 \end{bmatrix}^{-1} = \begin{bmatrix} -1/-2 & -1/-2 \\ -1/-2 & 1/-2 \end{bmatrix} = \begin{bmatrix} 1/2 & 1/2 \\ 1/2 & -1/2 \end{bmatrix}.$$

Since now we have a matrix representation for T^{-1}, *we can find the formula for* T^{-1}. *This matrix tells us that*

$$T^{-1}\begin{bmatrix} 1 & 0 \\ 0 & 0 \end{bmatrix} = (1/2)(1) + (1/2)(x) = \frac{1}{2} + \frac{1}{2}x$$

$$T^{-1}\begin{bmatrix} 0 & 0 \\ 0 & 1 \end{bmatrix} = (1/2)(1) + (-1/2)(x) = \frac{1}{2} - \frac{1}{2}x.$$

Therefore,

$$T^{-1}\begin{bmatrix} a & 0 \\ 0 & b \end{bmatrix} = a\left(\frac{1}{2} + \frac{1}{2}x\right) + b\left(\frac{1}{2} - \frac{1}{2}x\right) = \left(\frac{a+b}{2}\right) + \left(\frac{a-b}{2}\right)x.$$

Observe that having used standard bases, we avoided the step of finding the coordinates of an arbitrary vector in D_2 *with respect to* B_2.

4.4.3 Isomorphism

We now introduce the notion of isomorphism, a topic found in a discussion of any algebraic structure. Informally, two vector spaces are **isomorphic** if they are in a sense *identical* in the way they behave. In other words, give or take a relabelling of the names of the vectors, they both act in the same way with regards to their two operations.

Example 4.23 *Let's compare* $\mathbb{R}^2$ *and* P_1. *The vector* $[a, b]$ *in* $\mathbb{R}^2$ *looks a lot like the vector* $a + bx$ *in* P_1. *Furthermore, addition in both vector spaces look very similar.*

$$[a_1, b_1] + [a_2, b_2] = [a_1 + a_2, b_1 + b_2].$$
$$(a_1 + b_1x) + (a_2 + b_2x) = (a_1 + a_2) + (b_1 + b_2)x.$$

The same holds true for scalar multiplication.

$$c[a, b] = [ca, cb].$$

$$c(a + bx) = (ca) + (cb)x.$$

Indeed, replacing the set of symbols **left square bracket, comma** *and* **right square bracket** *by the symbols* $+$ *and* x *converts* $\mathbb{R}^2$ *into* P_1. *We shall see that these two vector spaces do indeed turn out to be isomorphic.*

Definition 4.8 *Let* V *and* W *be vector spaces with* $T \in L(V, W)$. *If* T *is invertible, then we call* T *an* **isomorphism**.

In order to formally prove that two vector spaces are isomorphic, we need to produce a map between the two vector spaces which is an isomorphism.

Definition 4.9 *Two vector spaces* V *and* W *are* **isomorphic**, *written* $V \simeq W$, *if there exists an isomorphism* $T \in L(V, W)$.

In other words, if we wish to show that two vector spaces are isomorphic, then we have to be inventive enough to supply the appropriate isomorphism. Many times the choice of isomorphism is obvious (or *natural*, as mathematicians like to say).

Example 4.24 *We formally show that* $\mathbb{R}^2 \simeq P_1$. *Consider the natural map* $T \in L(\mathbb{R}^2, P_1)$ *defined by*

$$T[a, b] = a + bx.$$

One can show that T *is indeed a linear transformation, and that* $\ker(T) = \{[0, 0]\}$ *and so* T *is one-to-one. Furthermore, since* $\dim \mathbb{R}^2 = \dim P_1$, *by Lemma 4.5,* T *maps onto* P_1. *All this shows that* T *is an isomorphism, and so* $\mathbb{R}^2 \simeq P_1$.

Lemma 4.6 *Let* U, V, W *be vector spaces and* $T \in L(U, V)$.

 i. $U \simeq U$.

 ii. If $U \simeq V$, *then* $V \simeq U$.

 iii. If $U \simeq V$ *and* $V \simeq W$, *then* $U \simeq W$.

 iv. If $U \simeq V$, *then* $\dim U = \dim V$.

 v. If $\dim U = \dim V$ *and* T *is one-to-one, then* $U \simeq V$.

 vi. If $\dim U = \dim V$ *and* T *maps onto* W, *then* $U \simeq V$.

Proof 4.11 *To prove i, use the isomorphism* 1_U. *To prove ii, assuming* $U \simeq V$, *we have an isomorphism* $T \in L(U, V)$. *By Theorem 4.5.i,* T^{-1} *makes* $V \simeq U$. *We leave the proof of iii as an exercise.*

 Part iv follows immediately from Lemma 4.5.iii. Parts v and vi follow from Lemma 4.5.i and ii (respectively). □

The next theorem lists some important isomorphisms of vector spaces.

Theorem 4.6 *Let V and W be finite dimensional vector spaces with $\dim(V) = n$ and $\dim(W) = m$.*

i. $V \simeq \mathbb{R}^n$.

ii. $L(V, W) \simeq M_{mn}$, and so $\dim(L(V, W)) = mn = (\dim V)(\dim W)$.

Proof 4.12 *To prove i, take any ordered basis $B_1 = (v_1, \ldots, v_n)$ for V and define the map $T : V \longrightarrow \mathbb{R}^n$ by $T(v) = [v]_{B_1}$. One can easily show that T is a linear transformation. Let's look at $\ker(T)$. If $T(v) = [0, \ldots, 0]$, then $[v]_{B_1} = [0, \ldots, 0]$ and so $v = 0v_1 + \cdots + 0v_n = 0_V$. Hence, $\ker(T) = \{0_v\}$ and T is one-to-one. Now by Lemma 4.6.v, $V \simeq \mathbb{R}^n$.*

To prove ii, take an ordered basis B_2 for W. Define the map $S : L(V, W) \longrightarrow M_{mn}$ by $S(T) = [T]_{B_1}^{B_2}$. First note that S is a linear transformation. Indeed, if $T_1, T_2 \in L(V, W)$, by Theorem 4.4.iii,

$$S(T_1 + T_2) = [T_1 + T_2]_{B_1}^{B_2} = [T_1]_{B_1}^{B_2} + [T_2]_{B_1}^{B_2} = S(T_1) + S(T_2).$$

Furthermore, if $a \in \mathbb{R}$ and $T \in L(V, W)$, by Theorem 4.4.iv,

$$S(aT) = [aT]_{B_1}^{B_2} = a[T]_{B_1}^{B_2} = aS(T).$$

Second, S is one-to-one, since matrix representation with respect to B_1 and B_2 is uniquely determined. Now by Lemma 4.6.v, $L(V, W) \simeq M_{mn}$ and $\dim(L(V, W)) = \dim(M_{mn}) = mn$. □

Corollary 4.4 *If V and W are finite dimensional vector spaces, then $V \simeq W$ iff $\dim V = \dim W$.*

Proof 4.13 *First suppose that $\dim V = \dim W = n$. By Theorem 4.6.i, $V \simeq \mathbb{R}^n$ and $W \simeq \mathbb{R}^n$ and hence $V \simeq W$. The reverse direction follows from Lemma 4.6.iv* □

The above corollary is an example of a Classification Theorem, in the sense that we have completely classified all vector spaces of finite dimension. It says that if a vector space has dimension n, then the vector space is more or less $\mathbb{R}^n$. Note also that Corollary 4.4 improves on Lemma 4.6.v and vi. There is no need anymore to exhibit an isomorphism, but rather it is enough to check that two vector spaces have the same dimension in order to conclude that they are isomorphic.

EXERCISES

1. Consider $T \in L(\mathbb{R}^2, P_1)$ defined by $T[a, b] = (2a + b) + ax$.

   a. Compute $[T]_{ST_1}^{ST_2}$ using standard bases.

b. Use part a. to explain why T^{-1} exists.

c. Use part a. to find the formula which defines T^{-1}.

2. Repeat the previous exercise for $T \in L(D_{22}, \mathbb{R}^2)$ defined by

$$T \begin{bmatrix} a & 0 \\ 0 & b \end{bmatrix} = [a+b, a-b].$$

3. Repeat the previous exercise for $T \in L(D_{22}, P_1)$ defined by

$$T \begin{bmatrix} a & 0 \\ 0 & b \end{bmatrix} = 2b - ax.$$

4. Repeat the previous exercise for $T \in L(P_1, D_{22})$ defined by

$$T(a + bx) = \begin{bmatrix} a+b & 0 \\ 0 & a-b \end{bmatrix}.$$

5. Let $T \in L(D_{22}, \mathbb{R}^2)$ defined by

$$T \begin{bmatrix} a & 0 \\ 0 & b \end{bmatrix} = [a-b, a+2b].$$

a. Use matrix representations to show that T^{-1} exists.

b. Find the formula for T^{-1} using part a.

6. Repeat the previous exercise for $T \in L(P_2, U_{22})$ defined by

$$T(a + bx + cx^2) = \begin{bmatrix} a+2c & 2a-b \\ 0 & b+3c \end{bmatrix}.$$

7. Repeat the previous exercise for $T \in L(\mathbb{R}^3, P_2)$ defined by

$$T[a, b, c] = (a+b+c) + (a-b+c)x + (a+b-c)x^2.$$

8. Consider $T \in L(P_3, U_{22})$ defined by

$$T(a + bx + cx^2 + dx^3) = \begin{bmatrix} a+b-c+d & -2a-b+c+d \\ 0 & -a+2d \end{bmatrix}.$$

and the ordered bases for P_3 and U_{22}, respectively,

$$B_1 = (\, x, 1-x^2, x+2x^3, 2 \,) \qquad B_2 = \left(\begin{bmatrix} 1 & 0 \\ 0 & 0 \end{bmatrix}, \begin{bmatrix} 2 & 1 \\ 0 & 0 \end{bmatrix}, \begin{bmatrix} 1 & 1 \\ 0 & -1 \end{bmatrix} \right).$$

a. Find ker(T) and decide whether or not T is one-to-one.

b. Find a basis for and dimension of ker(T).

c. Find $dim(T(P_3))$ and decide whether or not T maps onto U_{22}.

d. Is T invertible? (explain)

e. Compute $[T]_{B_1}^{B_2}$.

f. Use Lemma 4.4.i to compute $T(-2 + 3x + 4x^3)$.

9. Let $T \in L(P_1, \mathbb{R}^2)$ by $T(a + bx) = [2a + 2b, -a - 3b]$.

a. Use matrix representations to verify that T is invertible.

b. Using part a., find a formula for T^{-1}.

c. Verify that your answer in part b. is indeed the inverse of T (i.e. $T \circ T^{-1} = 1_{\mathbb{R}^2}$ and $T^{-1} \circ T = 1_{P_1}$).

10. Consider the following data:

$$T \in L(P_2, \mathbb{R}^3) \text{ defined by } T(a + bx + cx^2) = [a + b, a - c, a + b].$$

$$B_1 = (1 + x, x + 2x^2, -1 + x^2) \text{ and } B_2 = ([2, 3, 1], [0, -1, 0], [0, 2, -2]).$$

a. Show that $T \in L(P_2, \mathbb{R}^3)$.

b. Show that B_1 and B_2 are bases for P_2 and $\mathbb{R}^3$, respectively.

c. Compute ker(T) and use this to decide whether or not T is one-to-one.

d. Use the Linear Transformation Dimension Theorem to decide whether or not T maps onto $\mathbb{R}^3$.

e. Is T an isomorphism? (explain)

f. Calculate $[T]_{B_1}^{B_2}$.

g. Use part e. to decide whether or not $[T]_{B_1}^{B_2}$ is invertible.

11. Consider the following data:

$$T \in L(U_{22}, P_2) \text{ defined by } T \begin{bmatrix} a & b \\ 0 & c \end{bmatrix} = (a + b) + (b + c)x + (a + c)x^2.$$

$$B_1 = \left(\begin{bmatrix} 1 & 0 \\ 0 & 0 \end{bmatrix}, \begin{bmatrix} 1 & 1 \\ 0 & 0 \end{bmatrix}, \begin{bmatrix} 0 & 1 \\ 0 & 1 \end{bmatrix} \right) \text{ and } B_2 = (1, 1 + x, 1 + x^2).$$

   a. Compute $[T]_{B_1}^{B_2}$.

   b. Use part a. to decide whether or not T is invertible.

   c. Compute $\ker(T)$ and use it to decide whether or not T is one-to-one.

   d. Use Theorem 4.2 to decide whether or not T maps onto P_2.

12. Consider the following data:

$$A = \begin{bmatrix} 2 & 1 & 4 \\ 3 & 1 & 1 \\ 1 & 1 & 1 \end{bmatrix}, \quad B_1 = ([1,0,0],[1,1,0],[1,1,1]), \quad B_2 = ([0,0,1],[0,1,1],[1,1,1]).$$

Suppose that $T \in L(\mathbb{R}^3)$ and that $[T]_{B_1}^{B_2} = A$.

   a. Use Lemma 4.4.i to compute $T[3,1,0]$.

   b. Use $|A|$ to explain why A is invertible, then find A^{-1}.

   c. Find the formula for T^{-1} using Theorem 4.5.ii.

13. Use Corollary 4.4 to decide whether or not each of the following pairs of vector spaces are isomorphic:

   a. P_2 and 2×2 symmetric matrices.

   b. M_{33} and $\mathbb{R}^6$.

   c. U_{22} and P_3.

14. Let V, W be finite dimensional vectors spaces and $T \in L(V, W)$.

   a. Prove that if $dim(V) < dim(W)$, then T does not map onto W.

   b. Prove that if $dim(V) > dim(W)$, then T is not one-to-one.

   c. Prove that if $dim(V) \neq dim(W)$, then T is not an isomorphism.

15. Let V be a vector space and $T \in L(V)$. Prove that if $T^2 = 1_V$, then T is an isomorphism (note: $T^2 = T \circ T$).

16. Let $T \in L(V)$ and $S = T \circ T$. Prove that T is invertible iff S is invertible.

17. Prove that if a linear operator $T : V \longrightarrow V$ has the property that $T \circ T$ is the zero transformation (i.e. every input is mapped to 0_V), then T^{-1} does not exist.

18. Prove part ii of Lemma 4.5.

19. Prove part iii of Theorem 4.4.

20. Prove Theorem 4.5.i.

21. Verify that the map in part i of Theorem 4.6 is a linear transformation.

4.5 SIMILARITY OF MATRICES

In this section, we introduce the notion of similarity of square matrices. We show that the matrix representations of a linear operator have the special property of being similar to each other.

Definition 4.10 *A matrix $A \in M_{nn}$ is **similar** (or **conjugate**) to a matrix $B \in M_{nn}$ if there exists an invertible matrix $P \in M_{nn}$ such that $B = P^{-1}AP$.*

Example 4.25 *Consider the following matrices:*

$$A = \begin{bmatrix} 1 & 2 \\ 3 & 4 \end{bmatrix} \quad and \quad B = \begin{bmatrix} -1 & -1 \\ 4 & 6 \end{bmatrix}.$$

Notice that A is similar to B since for invertible $P = \begin{bmatrix} 1 & 1 \\ 1 & 2 \end{bmatrix}$ (invertible, since $|P| = 1 \neq 0$),

$$P^{-1}AP = \begin{bmatrix} 2 & -1 \\ -1 & 1 \end{bmatrix} \begin{bmatrix} 1 & 2 \\ 3 & 4 \end{bmatrix} \begin{bmatrix} 1 & 1 \\ 1 & 2 \end{bmatrix} = \begin{bmatrix} -1 & -1 \\ 4 & 6 \end{bmatrix} = B.$$

Lemma 4.7 *Let $A, B, C \in M_{nn}$. The following are true:*

i. A is similar to A.

ii. If A is similar to B, then B is similar to A.

iii. If A is similar to B and B is similar to C, then A is similar to C.

Proof 4.14 *To prove i, take $P = I$. To prove ii, we are assuming that there is an invertible P such that $P^{-1}AP = B$. Solving for A we have $A = PBP^{-1} = (P^{-1})^{-1}BP^{-1}$ and so the invertible matrix P^{-1} makes B similar to A. We leave the proof of iii as an exercise.* □

Note that since similarity has property ii, we can just say that A and B are similar in no particular order. Let's consider a finite dimensional vector space V and two ordered bases B and B' for V. Let $T \in L(V)$ be a linear operator. The main result in this section is to show that the matrices $[T]_B$ and $[T]_{B'}$ are related, namely they are similar.

Theorem 4.7 *Let V be a finite dimensional vector space with two ordered bases B and $B' = (v'_1, \ldots, v'_n)$. Choose any linear operator $T \in L(V)$. Then the matrix representations of T, $[T]_B$ and $[T]_{B'}$, are similar, i.e. $[T]_{B'} = P^{-1}[T]_B P$, for some invertible matrix P.*

Furthermore, the invertible matrix P which makes $[T]_B$ and $[T]_{B'}$ similar is defined as follows:

$$P = [1_V]_{B'}^B = [[v'_1]_B \quad \cdots \quad [v'_n]_B] \quad as\ columns.$$

Proof 4.15 *First observe that since 1_V is an invertible linear operator, by Theorem 4.5.iii, $P = [1_V]_{B'}^B$ is invertible. Second, notice that for any $v \in V$, by Theorem 4.4.i,*

$$P[v]_{B'} = [1_V]_{B'}^B[v]_{B'} = [1_V(v)]_B = [v]_B.$$

To complete the proof we rely heavily on Theorem 4.4.i and the observation just made. For any $i = 1, \ldots, n$,

$$P[T]_{B'}e_i = P[T]_{B'}[v_i']_{B'} = P[T(v_i')]_{B'} = [T(v_i')]_B$$

$$= [T]_B[v_i']_B = [T]_BP[v_i']_{B'} = [T]_BPe_i.$$

This shows that $P[T]_{B'} = [T]_BP$, and so $[T]_{B'} = P^{-1}[T]_BP$. □

The matrix P in the theorem above is the **change of basis** matrix from B' to B introduced in Section 3.5. The name arises from the fact that, for any $v \in V$, we have $P[v]_{B'} = [v]_B$ (as was shown in the proof above). In other words P changes the coordinates of a vector with respect to B' into coordinates of the vector with respect to B.

One can show (we leave this as an exercise) that

$$P^{-1} = [1_V]_B^{B'} = [[v_1]_{B'} \quad \cdots \quad [v_n]_{B'}],$$

where $B = (v_1, \ldots, v_n)$. In other words P^{-1} is the change of basis matrix from B to B'. There is a result which generalizes Theorem 4.7 to arbitrary linear transformations $L \in L(V, W)$ with $dimV, dimW < \infty$. We state this result without proof (although the proof is very similar to the one above).

Theorem 4.8 *Let V and W be a finite dimensional vector spaces each with two ordered bases B_1, B_1' and B_2, B_2' respectively. Choose any linear transformation $T \in L(V, W)$. Then the following relationship holds between two matrix representations of T,*

$$[T]_{B_1}^{B_2} = Q[T]_{B_1'}^{B_2'}P.$$

Where P is the (invertible) change of basis matrix from B_2 to B_2' and Q is the (invertible) change of basis matrix from B_1' to B_1.

Example 4.26 *Let $V = P_1$ with ordered bases $B = (1, 1 + x)$ and $B' = (x, 1 - x)$. Define $T \in L(P_1)$ by $T(a + bx) = a - bx$.*

Let's first compute $[T]_B$ directly.

$$T(1) = 1 = (1)(1) + (0)(1 + x) \quad \text{and} \quad T(1 + x) = 1 - x = (2)(1) + (-1)(1 + x).$$

Hence,

$$[T]_B = \begin{bmatrix} 1 & 2 \\ 0 & -1 \end{bmatrix}.$$

Now we could compute $[T]_{B'}$ the same way, but let's find it by way of Theorem 4.7 as an illustration. First, let's find P.

$$x = (-1)(1) + (1)(1 + x) \quad and \quad 1 - x = (2)(1) + (-1)(1 + x).$$

Thus,

$$P = \begin{bmatrix} -1 & 2 \\ 1 & -1 \end{bmatrix}.$$

Since $|P| = -1$, we have

$$P^{-1} = \begin{bmatrix} -1/(-1) & -2/(-1) \\ -1/(-1) & -1/(-1) \end{bmatrix} = \begin{bmatrix} 1 & 2 \\ 1 & 1 \end{bmatrix}.$$

Hence,

$$[T]_{B'} = \begin{bmatrix} 1 & 2 \\ 1 & 1 \end{bmatrix} \begin{bmatrix} 1 & 2 \\ 0 & -1 \end{bmatrix} \begin{bmatrix} -1 & 2 \\ 1 & -1 \end{bmatrix} = \begin{bmatrix} -1 & 2 \\ 0 & 1 \end{bmatrix}.$$

EXERCISES

1. Consider $T \in L(P_1)$ and bases $B = (1+x, 2x)$ and $B' = (-2, 1-x)$ and suppose that

$$[T]_B = \begin{bmatrix} 1 & -2 \\ -1 & 3 \end{bmatrix}.$$

Use the fact $[T]_{B'}$ is similar to $[T]_B$ in order to compute $[T]_{B'}$.

2. Repeat the previous exercise for $T \in L(\mathbb{R}^2)$, $B = ([1,1],[-1,2])$, $B' = ([1,1],[-1,1])$ and

$$[T]_B = \begin{bmatrix} -1 & 2 \\ 1 & 0 \end{bmatrix}.$$

3. Let $T \in L(P_1)$, $B = (x, 1-x)$, $B' = (1, -1+x)$ and suppose that

$$[T]_B = \begin{bmatrix} 1 & -1 \\ 1 & 2 \end{bmatrix}.$$

   a. Use Theorem 4.7 to compute $[T]_{B'}$.

   b. Use $[T]_B$ to find the formula which defines T.

   c. Using part b, compute $[T]_{B'}$ directly and verify your answer in part a.

4. Let $T \in L(D_{22})$ with

$$[T]_B = \begin{bmatrix} 1 & -1 \\ 1 & 2 \end{bmatrix},$$

where

$$B = \left(\begin{bmatrix} 1 & 0 \\ 0 & -2 \end{bmatrix}, \begin{bmatrix} 1 & 0 \\ 0 & 0 \end{bmatrix} \right) \text{ and } B' = \left(\begin{bmatrix} 1 & 0 \\ 0 & -1 \end{bmatrix}, \begin{bmatrix} 0 & 0 \\ 0 & -4 \end{bmatrix} \right).$$

Repeat the previous exercise.

5. Let $T \in L(U_{22})$ with

$$T \begin{bmatrix} a & b \\ 0 & c \end{bmatrix} = \begin{bmatrix} a+b & b+c \\ 0 & a+c \end{bmatrix}.$$

and consider the following bases for U_{22}:

$$B = \left(\begin{bmatrix} 1 & 0 \\ 0 & 0 \end{bmatrix}, \begin{bmatrix} 1 & -1 \\ 0 & 0 \end{bmatrix}, \begin{bmatrix} 0 & 0 \\ 0 & 2 \end{bmatrix} \right) \text{ and}$$

$$B' = \left(\begin{bmatrix} -1 & 0 \\ 0 & 1 \end{bmatrix}, \begin{bmatrix} 0 & 2 \\ 0 & 0 \end{bmatrix}, \begin{bmatrix} 0 & 0 \\ 0 & 1 \end{bmatrix} \right).$$

   a. Compute directly $[T]_B$.

   b. Using part a and Theorem 4.7, compute $[T]_{B'}$.

6. Consider the following data:

$$T \in L(D_{22}, D_{22}) \text{ defined by } T \begin{bmatrix} a & 0 \\ 0 & b \end{bmatrix} = \begin{bmatrix} a+b & 0 \\ 0 & a+b \end{bmatrix}.$$

$$B_1 = \left(\begin{bmatrix} 2 & 0 \\ 0 & 0 \end{bmatrix}, \begin{bmatrix} 2 & 0 \\ 0 & 1 \end{bmatrix} \right) \text{ and } B_2 = \left(\begin{bmatrix} 0 & 0 \\ 0 & 1 \end{bmatrix}, \begin{bmatrix} 1 & 0 \\ 0 & -1 \end{bmatrix} \right).$$

   a. Show that T is a linear transformation.

   b. Compute $[T]_{B_1}^{B_2}$.

c. Using Lemma 4.4, compute

$$T \begin{bmatrix} -2 & 0 \\ 0 & 1 \end{bmatrix}.$$

d. Suppose $S \in L(D_{22})$ and $[S]_{B_1}^{B_2} = \begin{bmatrix} 2 & 2 \\ 2 & 1 \end{bmatrix}$. Find the formula for S.

e. Compute $[T]_{B_1}$.

f. Use Theorem 4.7 to compute $[T]_{B_2}$.

7. Prove that if A is similar to B, then $|A| = |B|$.

8. Prove the following facts about the trace of a matrix:

a. For $A, B \in M_n$ that $tr(AB) = tr(BA)$.

b. Prove that if A is similar to B, then $tr(A) = tr(B)$.

9. Given the assumptions of Theorem 4.7, prove that $P^{-1} = [1_V]_B^{B'} = [[v_1]_{B'} \; \cdots \; [v_n]_{B'}]$, where $B = (v_1, \ldots, v_n)$.

10. Prove part iii of Lemma 4.7.

4.6   EIGENVALUES AND DIAGONALIZATION

In this section, we introduce some special scalars called **eigenvalues** (or **characteristic values**) which are associated with a matrix and play a crucial role in linear algebra.

Definition 4.11 *Let V be a vector space and $T \in L(V)$. If there is a $v \in V$ with $v \neq 0$, and scalar λ such that $T(v) = \lambda v$, then*

1. *The scalar λ is called an **eigenvalue** of T.*

2. *The vector v is called an **eigenvector** of T with respect to λ.*

3. *The set $E_\lambda = \{ v \in V : T(v) = \lambda v \}$ is called the **eigenspace** of T with respect to λ.*

Note that E_λ is the collection of all eigenvectors of T with respect to λ together with the zero vector.

Example 4.27 *We still need to introduce the method for finding the above objects, but we give a simple example (which we shall revisit) to illustrate the definitions. Let $T \in L(P_2)$ be defined by*

$$T(a + bx + cx^2) = (5a - 6b - 6c) + (-a + 4b + 2c)x + (3a - 6b - 4c)x^2.$$

Since $T(2 + x^2) = 4 + 2x^2 = (2)(2 + x^2)$, an eigenvalue of T is $\lambda = 2$, and $v = 2 + x^2$ is a eigenvector of T with respect to the eigenvalue 2. We shall see that

$$E_2 = \{ (2b + 2c) + bx + cx^2 \mid b, c \in \mathbb{R} \}.$$

We point out that eigenspaces are subspaces of V. Indeed, if $v_1, v_2 \in E_\lambda$, then $T(v_1) = \lambda v_1$ and $T(v_2) = \lambda v_2$. Hence,

$$T(v_1 + v_2) = T(v_1) + T(v_2) = \lambda v_1 + \lambda v_2 = \lambda(v_1 + v_2),$$

and so $v_1 + v_2 \in E_\lambda$. For any scalar a,

$$T(av_1) = aT(v_1) = a(\lambda v_1) = \lambda(av_1),$$

and so $av_1 \in E_\lambda$. Hence, we have demonstrated that E_λ is indeed a subspace of V.

Definition 4.12 *Let $A \in M_{nn}$ and define $p_A(t) = |A - tI|$. Then $p_A(t)$ is called the* **characteristic polynomial** *for A.*

Example 4.28 *Consider the matrix*

$$A = \begin{bmatrix} 5 & -6 & -6 \\ -1 & 4 & 2 \\ 3 & -6 & -4 \end{bmatrix}.$$

Then

$$p_A(t) = \left| \begin{bmatrix} 5 & -6 & -6 \\ -1 & 4 & 2 \\ 3 & -6 & -4 \end{bmatrix} - t \begin{bmatrix} 1 & 0 & 0 \\ 0 & 1 & 0 \\ 0 & 0 & 1 \end{bmatrix} \right|$$

$$= \begin{vmatrix} 5 - t & -6 & -6 \\ -1 & 4 - t & 2 \\ 3 & -6 & -4 - t \end{vmatrix} \underset{\underline{\underline{R_2 + R_3}}}{} \begin{vmatrix} 5 - t & -6 & -6 \\ -1 & 4 - t & 2 \\ 2 & -2 + t & -2 + t \end{vmatrix}$$

$$\underset{\underline{\underline{-C_2 + C_3}}}{} \begin{vmatrix} 5 - t & -6 & 0 \\ -1 & 4 - t & -2 + t \\ 2 & -2 + t & 0 \end{vmatrix} = -(-2 + t) \begin{vmatrix} 5 - t & -6 \\ 2 & -2 - t \end{vmatrix}$$

$$= -(t - 2)[(5 - t)(-2 - t) - (-6)(2)] = -(t - 2)(t^2 - 3t + 2)$$

$$= -(t - 2)(t - 2)(t - 1) = -(t - 2)^2(t - 1).$$

Observe how we used elementary row and column operations in order to get two zeros in a column and so obtain one of the factors of $p_A(t)$ as well as reduce our computation to a 2×2 determinant. For 3×3 determinants (and higher) this approach is always advisable, otherwise we would have been stuck with the task of factoring high degree polynomials. Finally, notice that $p_A(t)$ is indeed a polynomial in t.

The next theorem introduces our method for finding eigenvalues.

Theorem 4.9 *Let V be a finite dimensional vector space and $T \in L(V)$. Pick any ordered basis B for V and set $A = [T]_B$. Then the following three statements are equivalent:*

i. λ is a eigenvalue for T.

ii. $(A - \lambda I)X = 0$ has nontrivial solutions, and these non-trivial solutions correspond to the eigenvectors of T with respect to λ.

iii. $p_A(\lambda) = 0$.

Proof 4.16 *We first show that i is equivalent to ii. Let B be any basis for V. Now λ is a eigenvalue for T iff There is a non-zero $v \in V$ such that $T(v) = \lambda v$ iff $T(v) = \lambda 1_V(v)$ iff $T(v) - \lambda 1_V(v) = 0$ iff $(T - \lambda 1_V)(v) = 0$ iff $[(T - \lambda 1_V)(v)]_B = [0]_B$ iff $[T - \lambda 1_V]_B [v]_B = 0$ iff $(A - \lambda I)[v]_B = 0$ and therefore the homogeneous system $(A - \lambda I)X = 0$ has non-trivial solutions, namely the coordinates of v with respect to the ordered basis B. Furthermore, if $(A - \lambda I)X = 0$ has non-trivial solutions we can view these solutions as the coordinates of eigenvectors with respect to the ordered basis B.*

The fact that ii is equivalent to iii follows immediately from Theorem 2.21. Indeed, $(A - \lambda I)X = 0$ has non-trivial solutions iff $p_A(\lambda) = |A - \lambda I| = 0$. □

The theorem says that to find eigenvalues of a linear operator T, take any ordered basis B, form the matrix representation $A = [T]_B$, and find the roots of the characteristic polynomial $p_A(t)$. When doing this we normally take the standard basis $B = ST$ to make the computation of A easier.

Furthermore, the proof of the theorem also implies more. If $T(v) = \lambda v$ with $v \neq 0$ (i.e. v is a eigenvector), then as in the proof $(T - \lambda 1_V)(v) = 0$. Then for any ordered basis B, $[(T - \lambda 1_V)(v)]_B = [0]_B = 0$ which implies, by Lemma 4.4.i, that $[T - \lambda 1_V]_B [v]_B = 0$. As in the proof of the theorem this reduces to the equation $(A - \lambda I)[v]_B = 0$. The reverse direction holds as well and we have that v is a eigenvector for T with respect to λ iff the coordinates of v with respect to B are a solution to the homogeneous system $(A - \lambda I)X = 0$.

In other words, when we compute the solution set to the homogeneous system $(A - \lambda I)X = 0$ we are finding E_λ (at least the coordinates of vectors in E_λ with respect to B). If we choose $B = ST$, the standard basis, then finding such vectors is made even easier. This will all make better sense after the following example:

Example 4.29 *Let $T \in L(P_2)$ be defined as in the previous example by*

$$T(a + bx + cx^2) = (5a - 6b - 6c) + (-a + 4b + 2c)x + (3a - 6b - 4c)x^2.$$

We will find the characteristic polynomial, values and spaces. First we take $ST = (1, x, x^2)$ and compute $[T]_{ST}$. Since

$$T(1) = 5 - x + 3x^2, \quad T(x) = -6 + 4x - 6x^2, \quad T(x^2) = -6 + 2x - 4x^2,$$

we have that

$$[T]_{ST} = \begin{bmatrix} 5 & -6 & -6 \\ -1 & 4 & 2 \\ 3 & -6 & -4 \end{bmatrix}.$$

Call this matrix A. This is the same matrix as in the previous example. We computed the characteristic polynomial to be $p_A(t) = -(t-2)^2(t-1)$. Therefore, the eigenvalues of T are 1 and 2.

Now we compute E_1, the eigenspace for T with respect to $\lambda = 1$. As we stated in the discussion above we need to solve the homogeneous system $(A - \lambda I)X = 0$ for $\lambda = 1$. This amounts to replacing λ by 1 in the matrix $A - \lambda I$ and row-reducing (we can omit the column of zeros in the augmented matrix, since they will remain zero). Hence,

$$\begin{bmatrix} 5 - (1) & -6 & -6 \\ -1 & 4 - (1) & 2 \\ 3 & -6 & -4 - (1) \end{bmatrix} = \begin{bmatrix} 4 & -6 & -6 \\ -1 & 3 & 2 \\ 3 & -6 & -5 \end{bmatrix}.$$

This matrix reduces to

$$\begin{bmatrix} 1 & 0 & -1 \\ 0 & 1 & 1/3 \\ 0 & 0 & 0 \end{bmatrix}.$$

Thus, the solutions to the homogeneous system satisfy the relations $a - c = 0$ and $b + (1/3)c = 0$. Solving for the pivots yields $a = c$ and $b = -(1/3)c$. Therefore,

$$E_1 = \{ c - (1/3)cx + cx^2 \mid c \in F \}.$$

A basis for this subspace consists of one vector, $1 - (1/3)x + x^2$. For the reader's reassurance we compute

$$T(1 - (1/3)x + x^2) =$$

$$[5(1) - 6(-1/3) - 6(1)] + [-(1) + 4(-1/3) + 2(1)]x + [3(1) - 6(-1/3) - 4(1)]x^2 =$$

$$1 - (1/3)x + x^2 = (1)(1 - (1/3)x + x^2).$$

Now we compute the other eigenspace, E_2. Replace λ by 2 this time in $A - \lambda I$.

$$\begin{bmatrix} 5-(2) & -6 & -6 \\ -1 & 4-(2) & 2 \\ 3 & -6 & -4-(2) \end{bmatrix} = \begin{bmatrix} 3 & -6 & -6 \\ -1 & 2 & 2 \\ 3 & -6 & -6 \end{bmatrix}.$$

This matrix reduces to

$$\begin{bmatrix} 1 & -2 & -2 \\ 0 & 0 & 0 \\ 0 & 0 & 0 \end{bmatrix}.$$

Thus, the solutions to the homogeneous system satisfy the relation $a - 2b - 2c = 0$. Solving for the pivot yields $a = 2b + 2c$. Therefore,

$$E_2 = \{\ (2b + 2c) + bx + cx^2 \mid b, c \in \mathbb{R}\ \}.$$

A basis for this subspace consists of two vectors, $2 + x$ and $2 + x^2$. These vectors which form bases for the eigenspaces will play an important role for the remainder of the discussion in this section.

We wish to point out that any square matrix can be considered to have eigenvalues as well, for we can view an $n \times n$ matrix as a linear operator on $\mathbb{R}^n$ by the rule $A(v) = Av$, multiplication of A on the left of a column vector $v \in \mathbb{R}^n$.

Example 4.30 *This example deals with the case of complex eigenvalues. Consider the matrix*

$$A = \begin{bmatrix} -2 & 1 \\ -1 & -2 \end{bmatrix},$$

Note that as an operator in $L(\mathbb{C})$ the matrix representation with respect to the standard basis of the map $v \to Av$ is the matrix A. One can compute $p_A(t) = t^2 + 4t + 5$ and using the quadratic formula we find that A has eigenvalues $-2 \pm i$. Let's compute the eigenspace for $-2 + i$, namely E_{-2+i}. Replace t by $-2 + i$ this time in $A - tI$ to get

$$\begin{bmatrix} -i & 1 \\ -1 & -i \end{bmatrix} \quad \text{which reduces to} \quad \begin{bmatrix} 1 & i \\ 0 & 0 \end{bmatrix}.$$

Thus, the solutions to the homogeneous system satisfy the relation $a + bi = 0$ or $a = -bi$. Therefore,

$$E_{-2+i} = \{\ [-bi, b] \mid b \in \mathbb{C}\ \}.$$

A basis for this eigenspace consists of the vector $[-i, 1]$ *when we set* $b = 1$. *In a similar manner one can compute*

$$E_{-2-i} = \{\ [bi, b]\ |\ b \in \mathbb{C}\ \}.$$

A basis for this eigenspace consists of the vector $[i, 1]$ *when we set* $b = 1$. *In fact, one can show that for any matrix* A *with real-valued entries, eigenspaces corresponding to complex conjugate pair eigenvalues are also complex conjugates, i.e. if* $\lambda \in \mathbb{C}$ *and* $Av = \lambda v$ *for* $v \in \mathbb{C}^n$, *then* $A\overline{v} = \overline{\lambda}\overline{v}$.

Definition 4.13 *Let* $A \in M_{nn}$ *be a square matrix.*

1. *We say that* A *is* **triangularizable** *if* A *is similar to an upper triangular matrix. A linear operator* $T \in L(V)$ *is* **triangularizable** *if any (or equivalently all, by Theorem 4.7) of its matrix representations is triangularizable. Equivalently (again, by Theorem 4.7), there is an ordered basis* B *such that* $[T]_B$ *is a triangular matrix.*

2. *We say that* A *is* **diagonalizable** *if* A *is similar to a diagonal matrix. A linear operator* $T \in L(V)$ *is* **diagonalizable** *if any (or equivalently all) of its matrix representations is diagonalizable. Equivalently, there is an ordered basis* B *such that* $[T]_B$ *is a diagonal matrix.*

A result for which we will prove in the next chapter is the fact that any matrix $A \in M_{nn}$ is triangularizable. The result below is what we are after in this section.

Theorem 4.10 *Let* V *be a finite dimensional vector space and* $T \in L(V)$ *a linear operator. Let* $\lambda_1, \ldots, \lambda_n \in \mathbb{R}$ *be distinct eigenvalues of* T *with corresponding eigenvectors* $v_1, \ldots, v_n \in V$. *Then*

i. The vectors $v_1, \ldots, v_n$ *are linearly independent.*

ii. For $i \neq j$, *we have* $E_{\lambda_i} \cap E_{\lambda_j} = \{0\}$.

iii. For $i = 1, \ldots, n$, *we have* $1 \leq \dim(E_{\lambda_i}) \leq$ *the multiplicity of* λ *as a root of the characteristic polynomial.*

iv. T is diagonalizable iff $\dim(E_{\lambda_1}) + \cdots + \dim(E_{\lambda_n}) = \dim V$.

Proof 4.17 *The proof of part i is by induction on* n. *For* $n = 1$, *one non-zero vector* v_1 *is certainly linearly independent. Assume the statement is true for* k *and we show it is true for* $k + 1$. *Suppose that*

$$a_1 v_1 + \cdots + a_{k+1} v_{k+1} = 0.$$

For the case when $a_{k+1} = 0$, *the equation reduces to* $a_1 v_1 + \cdots + a_k v_k = 0$, *and by induction* $a_1 = \cdots = a_k = 0$ *and we are done. For the case when* $a_{k+1} \neq 0$ *we will get a contradiction, as desired. Solve the equation for* v_{k+1} *to get*

$$v_{k+1} = (a_1/a_{k+1})v_1 + \cdots + (a_k/a_{k+1})v_k.$$

For brevity, let's define $b_i = a_i/a_{k+1}$ for $i = 1, \ldots, k$. Thus our equation becomes

$$v_{k+1} = b_1 v_1 + \cdots + b_k v_k.$$

Let's call the equation above $*$. Applying T to both sides of equation $*$ yields

$$T(v_{k+1}) = b_1 T(v_1) + \cdots + b_k T(v_k).$$

Since $v_1, \ldots, v_{k+1}$, are eigenvectors, we have

$$\lambda_{k+1} v_{k+1} = b_1(\lambda_1 v_1) + \cdots + b_k(\lambda_k v_k).$$

Use equation $*$ to make a replacement on the left hand side to get

$$\lambda_{k+1}(b_1 v_1 + \cdots + b_k v_k) = b_1(\lambda_1 v_1) + \cdots + b_k(\lambda_k v_k).$$

Collecting like terms yields

$$b_1(\lambda_{k+1} - \lambda_1)v_1 + \cdots + b_k(\lambda_{k+1} - \lambda_k)v_k = 0.$$

By induction, $v_1, \ldots, v_k$ are linearly independent, thus

$$b_1(\lambda_{k+1} - \lambda_1) = \cdots = b_k(\lambda_{k+1} - \lambda_k) = 0.$$

Since $\lambda_1, \ldots, \lambda_{k+1}$ are distinct, it must be the case that $b_1 = \cdots = b_k = 0$. But then equation $*$ reduces to $v_{k+1} = 0$, contradicting the fact that v_{k+1} is a eigenvector.

Verifying part ii is easy and we therefore leave it as an exercise.

For part iii, since λ_i is an eigenvalue this implies there are non-zero eigenvectors in E_{λ_i} and so $\dim(E_{\lambda_i}) > 0$. Take a basis for E_{λ_i}, say $u_1, u_2, \ldots, u_k$ and extend it to $B = (u_1, u_2, \ldots, u_k, v_{k+1}, \ldots, v_n)$ a basis for V. The matrix representation of T with respect to this basis (in block form) is

$$[T]_B = \left[\begin{array}{ccccc|c} \lambda_1 & & & & & \\ & \lambda_2 & & & & 0_{n-k} \\ & & \ddots & & & \\ & & & \lambda_k & & \\ \hline & & 0_{n-k} & & & C \end{array}\right].$$

Using this matrix representation of T to compute the characteristic polynomial we get $(t - \lambda_i)^k p_C(t)$ and so λ_i is a root of the characteristic polynomial at least $k = \dim(E_{\lambda_i})$ times.

For the proof of part iv, let B_i be an ordered basis for E_i for $i = 1, \ldots, n$. One can show (see Exercise 17) that $B = B_1 \cup \cdots \cup B_n$ is a linearly independent set of vectors. Now order B so that the elements of B_1 precede elements of B_2, and so on. Now B is a basis, since the number of elements in B equals $\dim(E_{\lambda_1}) + \cdots + \dim(E_{\lambda_n}) = \dim V$ and B is a linearly independent set of vectors (see Theorem 3.10). Furthermore, since B is made up of eigenvectors $[T]_B$ will be diagonal. Indeed, $[T]_B$ will be a diagonal matrix with the eigenvalues occurring on the diagonal (see Example 4.31). The number of occurrences of each eigenvalue corresponds exactly to the dimension of its associated eigenspace. □

Example 4.31 *In our ongoing example of $T \in L(P_2)$, we found a basis for each eigenspace E_1 and E_2. Hence,*

$$\dim(E_1) + \dim(E_2) = 1 + 2 = 3 = \dim(P_2).$$

By Theorem 4.10.ii, T is diagonalizable. Furthermore, the proof of Theorem 4.10.ii shows that the ordered basis B which makes $[T]_B$ diagonal is the combined bases for E_1 and E_2, namely

$$B = (1 + \frac{1}{3}x + x^2, 2 + x, 2 + x^2).$$

Note the fact that B is comprised of eigenvectors implies

$$T(1 + \frac{1}{3}x + x^2) = (1)(1 + \frac{1}{3}x + x^2) = (1)(1 + \frac{1}{3}x + x^2) + (0)(2 + x) + (0)(2 + x^2),$$

$$T(2 + x) = (2)(2 + x) = (0)(1 + \frac{1}{3}x + x^2) + (2)(2 + x) + (0)(2 + x^2),$$

$$T(2 + x^2) = (2)(2 + x^2) = (0)(1 + \frac{1}{3}x + x^2) + (0)(2 + x) + (2)(2 + x^2).$$

Hence,

$$[T]_B = \begin{bmatrix} 1 & 0 & 0 \\ 0 & 2 & 0 \\ 0 & 0 & 2 \end{bmatrix}.$$

Some immediate corollaries to Theorem 4.10 are the following:

Corollary 4.5 *Let V be a non-trivial vector space of finite dimension n and $T \in L(V)$.*

1. *T has at most n distinct eigenvalues.*

2. *If T has exactly n distinct eigenvalues, then T is diagonalizable.*

Example 4.32 *We use the technique in the previous example to decide whether an arbitrary matrix is diagonalizable and to find a diagonal matrix to which it is similar.*

The logic goes as follows: Given a matrix $A \in M_{nn}$, consider the linear transformation $T \in L(\mathbb{R}^n)$ by $T(v) = Av$. One can check that $[T]_{ST} = A$, where $ST = ([1,0], [[0,1])$. Therefore, checking diagonalizabilty of T is also checking diagonalizability of A.

Consider the matrix

$$A = \begin{bmatrix} 1 & 4 \\ 1 & 1 \end{bmatrix}.$$

The characteristic polynomial,

$$p_A(t) = \begin{vmatrix} 1-t & 4 \\ 1 & 1-t \end{vmatrix} = (1-t)^2 - 4 =$$

$$t^2 - 2t - 3 = (t-3)(t+1).$$

Therefore, we have two eigenvalues, namely 3 and -1. Observe that we now know that A is diagonalizable by Corollary 4.5.ii. Let's find a basis for each eigenspace.

For the eigenvalue $\lambda = 3$, we row-reduce

$$\begin{bmatrix} 1-(3) & 4 \\ 1 & 1-(3) \end{bmatrix} = \begin{bmatrix} -2 & 4 \\ 1 & -2 \end{bmatrix} \quad \textit{which reduces to} \quad \begin{bmatrix} 1 & -2 \\ 0 & 0 \end{bmatrix}.$$

Therefore, $a = 2b$ and $E_3 = \{ [2b, b] \mid b \in \mathbb{R}\}$ and the basis is the single vector $[2,1]$. In a similar manner one can compute that $E_{-1} = \{ [-2b, b] \mid b \in \mathbb{R}\}$ and the basis is the single vector $[-2,1]$.

Set $B = ([2,1], [-2,1])$. As was discussed in the previous example, we know that $[T]_B = \begin{bmatrix} 3 & 0 \\ 0 & -1 \end{bmatrix}$. By Theorem 4.7, we also know that A and $[T]_B$ are similar. More specifically, $[T]_B = P^{-1}AP$ where P is the change of basis matrix from B to ST. This is easy to compute, since we write the vectors in B in terms of ST. Hence, we simply drop the vectors of B (in order) into the columns of P, i.e.

$$P = \begin{bmatrix} 2 & -2 \\ 1 & 1 \end{bmatrix}.$$

Since $|P| = 4$, we have

$$P^{-1} = \begin{bmatrix} 1/4 & 1/2 \\ -1/4 & 1/2 \end{bmatrix}.$$

Hence,

$$P^{-1}AP = \begin{bmatrix} 1/4 & 1/2 \\ -1/4 & 1/2 \end{bmatrix} \begin{bmatrix} 1 & 4 \\ 1 & 1 \end{bmatrix} \begin{bmatrix} 2 & -2 \\ 1 & 1 \end{bmatrix} =$$

$$\begin{bmatrix} 3 & 0 \\ 0 & -1 \end{bmatrix} = [T]_B.$$

Here could be a possible application of diagonalization. Suppose we needed to compute A^5. It would be quite tedious if we multiplied A by itself 5 times (and the tedium increases as the exponent increases). By the work above, $A = P[T]_B P^{-1}$, and by an exercise to follow,

$$A^5 = (P[T]_B P^{-1})^5 = P[T]_B^5 P^{-1} =$$

$$\begin{bmatrix} 2 & -2 \\ 1 & 1 \end{bmatrix} \begin{bmatrix} 3 & 0 \\ 0 & -1 \end{bmatrix}^5 \begin{bmatrix} 1/4 & 1/2 \\ -1/4 & 1/2 \end{bmatrix} =$$

$$\begin{bmatrix} 2 & -2 \\ 1 & 1 \end{bmatrix} \begin{bmatrix} (3)^5 & 0 \\ 0 & (-1)^5 \end{bmatrix} \begin{bmatrix} 1/4 & 1/2 \\ -1/4 & 1/2 \end{bmatrix} =$$

$$\begin{bmatrix} 2 & -2 \\ 1 & 1 \end{bmatrix} \begin{bmatrix} 243 & 0 \\ 0 & -1 \end{bmatrix} \begin{bmatrix} 1/4 & 1/2 \\ -1/4 & 1/2 \end{bmatrix} = \begin{bmatrix} 121 & 244 \\ 61 & 121 \end{bmatrix}.$$

Observe that no matter how high the exponent we wish to compute, we are always reduce to multiplying three matrices, a great savings in labor.

EXERCISES

1. Consider the following linear transformation $T \in L(P_2)$ defined by

$$T(a + bx + cx^2) = -a + (2b + 3c)x + (2b + c)x^2.$$

   a. Compute $[T]_{ST}$.

   b. Use part a. to verify that T is invertible.

   c. Use part a. to find the formula for T^{-1}.

   d. Find eigenvalues and spaces for T.

   e. Find a basis for and dimension of each eigenspace of T.

   f. Is there a basis B for P_2 which makes $[T]_B$ diagonal? If so, exhibit the basis and the matrix $[T]_B$.

2. Consider $T \in L(U_{22})$ defined by

$$T \begin{bmatrix} a & b \\ 0 & c \end{bmatrix} = \begin{bmatrix} a+b+c & a+b+c \\ 0 & a+b+c \end{bmatrix}.$$

a. Find the eigenvalues for T.

b. Is there enough information from part a. to decide whether or not T is diagonalizable? (explain)

c. Find the eigenspaces for T.

d. Find the basis and dimension of the eigenspaces for T.

e. Use part d. to explain why T is diagonalizable and exhibit a basis B such that $[T]_B$ is diagonal. What does $[T]_B$ equal?

3. Repeat the previous exercise for the following linear transformation $T \in L(U_{22})$:

$$T \begin{bmatrix} a & b \\ 0 & c \end{bmatrix} = \begin{bmatrix} 4a & a+b \\ 0 & -2a+c \end{bmatrix}.$$

4. Consider $T \in L(P_2)$ defined by

$$T(a + bx + cx^2) = (8a + 10b) + (-5a - 7b)x - 2cx^2.$$

a. Find the eigenvalues for T.

b. Is there enough information from part a. to decide whether or not T is diagonalizable? (explain)

c. Find the eigenspaces for T.

d. Find the basis and dimension of the eigenspaces for T.

e. Use part d. to explain why T is diagonalizable and exhibit a basis B such that $[T]_B$ is diagonal. What does $[T]_B$ equal?

f. Use $[T]_B$ in part e. and Theorem 4.5.ii to find the formula for T^{-1}.

5. Repeat the previous exercise for $T \in L(U_{22})$ defined by

$$T \begin{bmatrix} a & b \\ 0 & c \end{bmatrix} = \begin{bmatrix} a-b+c & -2b+3c \\ 0 & c \end{bmatrix}.$$

6. Consider the following linear transformation $T \in L(P_2)$ defined by

$$T(a + bx + cx^2) = 2a + (a + b + c)x + (2b - 2a)x^2.$$

a. Find eigenvalues and eigenspaces for T.

b. Find a basis for and dimension of each eigenspace of T.

c. Using part b. explain why T is diagonalizable.

d. Find a basis B for P_2 which makes $[T]_B$ diagonal and exhibit the matrix $[T]_B$.

7. Consider the following matrix:

$$A = \begin{bmatrix} -3 & -2 \\ 2 & 2 \end{bmatrix}.$$

a. Find an invertible matrix P such that $P^{-1}AP$ is diagonal.

b. Use part a. to compute A^5.

8. Repeat the previous exercise with the following matrix:

$$A = \begin{bmatrix} 2 & 1 \\ 1 & 2 \end{bmatrix}.$$

9. Repeat the previous exercise with the following matrix:

$$A = \begin{bmatrix} -7 & 0 & 5 \\ 0 & 3 & 0 \\ -10 & 0 & 8 \end{bmatrix}.$$

10. Consider the matrix $A = \begin{bmatrix} 1 & 2 \\ 3 & 2 \end{bmatrix}$

a. Find a matrix P which makes A similar to a diagonal matrix.

b. Use part b to compute A^3.

c. When would a 2×2 matrix **not** be similar to a diagonal matrix by this method?

11. Prove if A is a 2 by 2 symmetric matrix, then A is diagonalizable.

12. Prove if A is a 2 by 2 matrix, then the sum of the entries on the diagonal of A equals the sum of the eigenvalues of A and $|A|$ equals the product of the eigenvalues of A.

13. Prove that the following statement can be added to the list of equivalent statements in Theorem 3.7:

x. $\lambda = 0$ is not an eigenvalue of A.

14. Prove that $A \in M_{nn}$ and A^T have the same eigenvalues.

15. Prove that similar matrices have the same eigenvalues.

16. Prove for $A, B \in M_n$ that AB and BA have the same eigenvalues.

17. Prove that if $X_1, \ldots, X_m$ are sets each consisting of linearly independent vectors such that the following sum is a direct sum (see Section 3.1, Exercise 17):

$$\text{span}(X_1) + \text{span}(X_2) + \cdots + \text{span}(X_m),$$

then $X_1 \cup \cdots \cup X_m$ is linearly independent.

18. Given a matrix $A \in M_{nn}$, consider the linear transformation $T \in L(\mathbb{R}^n)$ by $T(v) = Av$. Prove that $[T]_{ST} = A$.

19. Prove that for $A, P \in M_{nn}$ and P invertible, we have $(PAP^{-1})^n = PA^nP^{-1}$.

20. Prove that if $\lambda \in \mathbb{C}$ and $Av = \lambda v$ for $v \in \mathbb{C}^n$, then $A\overline{v} = \overline{\lambda}\overline{v}$.

21. Consider the linear transformation defined as follows: Let $\mathcal{C}$ represent the vector space of continuous functions and $T : \mathcal{C} \to \mathcal{C}$ by

$$T(f)(x) = \int_0^x f(t)\ dt$$

Show that T has no eigenvalues.

22. Prove parts ii and iii of Theorem 4.9.

23. Prove part ii of Theorem 4.10.

24. Prove Corollary 4.5.

4.7 AXIOMATIC DETERMINANT

This section is a bit more on the theoretical side than on the application side. It can be easily skipped, but some students may find it mathematically interesting. In it we determine the axioms for which the determinant is uniquely defined. One consequence is that all our definitions of determinant in earlier sections must coincide since they all satisfy these axioms. For the definition below, recall the notation for the cartesian product of vector spaces, namely for a vector space V,

$$V^n = \underbrace{V \times V \times \cdots \times V}_{n \text{ times}} = \{\ (v_1, v_2, \ldots, v_n) : v_1, v_2, \ldots, v_n \in V\ \}.$$

Definition 4.14 *Let $V = \mathbb{R}^n$. A function $d : V^n \to \mathbb{R}$ is n-**linear** if it is linear in each of its coordinates, i.e. for any $1 \le i \le n$ and any $v_1, \ldots, v_i, v_i', \ldots, v_n \in V$ we have*

1. *$d(v_1, \ldots, v_i + v_i', \ldots, v_n) = d(v_1, \ldots, v_i, \ldots, v_n) + d(v_1, \ldots, v_i', \ldots, v_n)$ and*

2. *$d(v_1, \ldots, av_i, \ldots, v_n) = ad(v_1, \ldots, v_i, \ldots, v_n)$ for any $a \in \mathbb{R}$.*

Example 4.33 *We list here several important examples of n-linear functions.*

1. *Any linear transformation $T \in L(\mathbb{R}^n, \mathbb{R})$ is a 1-linear function (these functions are sometimes called **linear functionals**).*

2. *Any inner product on $\mathbb{R}^n$ is a 2-linear (or **bilinear**) function.*

3. *If we represent a matrix $A = (c_1, c_2, \ldots, c_n)$ as an n-tuple of its columns, then the determinant is an n-linear function.*

We list a couple results about n-linear functions which are left as exercises for the reader.

Lemma 4.8

1. *If $(v_1, v_2, \ldots, v_n)$ includes a coordinate which is the zero vector, then $d(v_1, v_2, \ldots, v_n) = 0$.*

2. *Any linear combination of n-linear functions is again n-linear.*

3. *An n-linear function on $\mathbb{R}^n$ is completely determined by its values on the inputs*

$$\left(e_{\sigma(1)}, e_{\sigma(2)}, \ldots, e_{\sigma(n)} \right),$$

where $e_1, e_2, \ldots, e_n$ is the standard basis for $\mathbb{R}^n$ and σ is any permutation of the numbers $1, 2, \ldots, n$.

Our focus in this section will be n-linear functions for which the inputs are matrices represented in columns (see Example 3 above). The goal is to determine what additional properties (i.e. axioms) this type of n-linear function needs in order to completely characterize the determinant, since there are other such n-linear functions with the very same domain, but which are not the determinant function.

Example 4.34 *For $A = [a_{ij}] \in M_{nn}$ the function defined by $d(A) = a_{11}a_{22} \cdots a_{nn}$ has the same domain as a determinant, however it is an n-linear function different from the determinant function.*

Definition 4.15 *An n-linear function d is **alternating** if $d(v_1, v_2, \ldots, v_n) = 0$ whenever some $v_i = v_j$ $(i \neq j)$.*

Lemma 4.9 *An n-linear function d is alternating iff interchanging any two vectors in the n-tuple $(v_1, v_2 \ldots, v_n)$ reverses the sign of the output of d.*

Proof 4.18 *For one direction, assume first that d is alternating. Then*

$$0 = d(v_1, \ldots, \underbrace{v_i + v_j}_{ith\ slot}, \ldots, \underbrace{v_i + v_j}_{jth\ slot}, \ldots, v_n) =$$

$$d(v_1, \ldots, v_i, \ldots, v_i, \ldots v_n) + d(v_1, \ldots, v_i, \ldots, v_j, \ldots v_n) +$$

$$d(v_1, \ldots, v_j, \ldots, v_i, \ldots v_n) + d(v_1, \ldots, v_j, \ldots, v_j, \ldots v_n) =$$

$$d(v_1, \ldots, v_i, \ldots, v_j, \ldots v_n) + d(v_1, \ldots, v_j, \ldots, v_i, \ldots v_n).$$

Therefore,

$$d(v_1, \ldots, v_i, \ldots, v_j, \ldots v_n) = -d(v_1, \ldots, v_j, \ldots, v_i, \ldots v_n).$$

For the reverse direction, consider the output

$$d(v_1, \ldots, \underbrace{v_i}_{ith\ slot}, \ldots, \underbrace{v_i}_{jth\ slot}, \ldots, v_n).$$

Interchanging the ith and jth coordinate yields the same n-tuple, however by assumption

$$d(v_1, \ldots, v_i, \ldots, v_i, \ldots v_n) = -d(v_1, \ldots, v_i, \ldots, v_i, \ldots v_n) \quad and\ so$$

$$2d(v_1, \ldots, v_i, \ldots, v_i, \ldots v_n) = 0 \quad or \quad d(v_1, \ldots, v_i, \ldots, v_i, \ldots v_n) = 0.$$

$\square$

It will be useful later on in this section to prove the following additional characterization of alternating:

Lemma 4.10 *An n-linear function d is alternating iff interchanging of any two adjacent vectors in the n-tuple $(v_1, v_2 \ldots, v_n)$ reverses the sign of the output of d.*

Proof 4.19 *One direction is an immediate consequence of Lemma 4.9. For the other direction, assume that interchanging of any two adjacent vectors in the n-tuple $(v_1, v_2 \ldots, v_n)$ changes the sign of the output of d. We shall show that interchanging any two vectors in the n-tuple $(v_1, v_2 \ldots, v_n)$ changes the sign of the output of d, which by Lemma 4.9 completes the proof. Consider the two outputs*

$$d(\ldots, \underbrace{v_i}_{ith\ slot}, \ldots, \underbrace{v_j}_{jth\ slot}, \ldots) \quad and \quad d(\ldots, \underbrace{v_j}_{ith\ slot}, \ldots, \underbrace{v_i}_{jth\ slot}, \ldots).$$

We can move from the first output to the second by a sequence of interchanges of adjacent vectors the number of which is $(j - i) + (j - i) - 1$ or $2(j - i) - 1$, so that by assumption

$$d(\ldots, \underbrace{v_j}_{ith\ slot}, \ldots, \underbrace{v_i}_{jth\ slot}, \ldots) =$$

$$(-1)^{2(j-i)-1}d(\ldots, \underbrace{v_i}_{ith\ slot}, \ldots, \underbrace{v_j}_{jth\ slot}, \ldots) =$$

$$-d(\ldots, \underbrace{v_i}_{ith\ slot}, \ldots, \underbrace{v_j}_{jth\ slot}, \ldots).$$

□

The following result states that there can be at most one n-linear alternating function with the additional property that it equals 1 on the n-tuple $(e_1, e_2, \ldots, e_n)$. This result will lead to the uniqueness of the determinant.

Theorem 4.11 *If d_1 and d_2 are two n-linear alternating function satisfying the additional property that $d_i(e_1, e_2, \ldots, e_n) = 1$ for $i = 1, 2$, then $d_1 = d_2$.*

Proof 4.20 *Set $f = d_1 - d_2$ which is an n-linear function, by part 1 of Lemma 4.8, and is certainly still alternating. By expressing each v_i as a linear combination of the standard basis $e_1, e_2, \ldots, e_n$ one can eventually express $f(v_1, v_2, \ldots, v_n) = af(e_1, e_2, \ldots, e_n)$ for some scalar $a \in \mathbb{R}$. Therefore,*

$$d_1(v_1, v_2, \ldots, v_n) - d_2(v_1, v_2, \ldots, v_n) = a[d_1(e_1, e_2, \ldots, e_n) - d_2(e_1, e_2, \ldots, e_n)] = 0.$$

Hence, $d_1(v_1, v_2, \ldots, v_n) = d_2(v_1, v_2, \ldots, v_n)$ for all n-tuples $(v_1, v_2, \ldots, v_n)$. □

Theorem 4.12 *Expanding on a row or column of a matrix to compute the determinant yields an n-linear alternating function satisfying the additional property that its value on I_n equals 1.*

Proof 4.21 *The proof is done by induction on n. For $n = 1$, since the determinant is the identity map, it is certainly 1-linear, alternating (vacuously) and its value on $I_1 = [1]$ equals 1. For $n > 1$, let's first reestablish our notation for expanding on row i for $A \in M_{nn}$:*

$$|A|_n = \sum_{j=1}^{n}(-1)^{i+j}a_{ij}|A_{ij}|_{n-1}.$$

Since $|*|_n$ is a linear combination of $|*|_{n-1}$'s and by induction the determinant function on the A_{ij}'s is $(n-1)$-linear, it follows from part 1 of Lemma 4.8 that the determinant on A is n-linear. We now show it is alternating. By Lemma 4.10, it's enough to consider a matrix A with two adjacent equal columns, say column k and

$k + 1$. Then A_{ij} will have two equal columns for $j \neq k, k+1$, $a_{ik} = a_{i,k+1}$ and $A_{ik} = A_{i,k+1}$. Therefore, using this observation and by induction

$$|A|_n = (-1)^{i+k} a_{ik} |A_{ik}|_{n-1} + (-1)^{i+k+1} a_{i,k+1} |A_{i,k+1}|_{n-1} = 0.$$

Finally, we show that $|I_n|_n = 1$. Set $A = I_n$. Notice for $i = j$ that $A_{ij} = I_{n-1}$ and for $i \neq j$ that A_{ij} will have a zero column. Therefore, using these observations and by induction

$$|I_n|_n = |I_{n-1}|_{n-1} = 1.$$

A similar argument can be used to prove the result for expanding on a column. □

Corollary 4.6 *The determinant function obtained by expanding on any row or column is the unique n-linear alternating function on M_{nn} satisfying the additional property $|I_n| = 1$.*

Proof 4.22 *We have already seen in the proof of Theorem 4.12 that the determinant function is n-linear and alternating on matrices $A = (c_1, c_2, \ldots, c_n)$ viewed as an n-tuple of column vectors and satisfies $|(e_1, e_2, \ldots, e_n)| = 1$. By Theorem 4.11, it is the only such function.* □

EXERCISES

1. Verify that each of the following functions are n-linear:

   a. Any linear transformation $T \in L(\mathbb{R}^n, \mathbb{R})$ is a 1-linear function.

   b. Any inner product on $\mathbb{R}^n$ is a 2-linear (or **bilinear**) function.

   c. If we represent a matrix $A = (c_1, c_2, \ldots, c_n)$ as an n-tuple of its columns, then the determinant is an n-linear function.

   d. For $A = [a_{ij}] \in M_{nn}$ the function defined by $d(A) = a_{11} a_{22} \cdots a_{nn}$ is an n-linear function. Illustrate why d is different from the determinant function.

2. Show that the following functions are not n-linear (give a counterexample):

   a. $d : V^n \to \mathbb{R}$ by $d(v_1, v_2, \ldots, v_n) = a$ for some fixed scalar a.

   b. For $A = [a_{ij}] \in M_{nn}$ the function defined by $d(A) = a_{11}^2 a_{22} \cdots a_{nn}$

3. Prove that if d is n-linear and alternating on a vector space V, then for any set of linearly dependent vectors $v_1, v_2, \ldots, v_n$ in V we have $d(v_1, v_2, \ldots, v_n) = 0$.

4. Prove Lemma 4.8.

5. Prove Theorem 4.12 for expanding on a column.

4.8 QUOTIENT VECTOR SPACE

If one is dealing with any algebraic structure, there is always a notion of a quotient structure. The reader should already have some good experience with equivalence relations and classes, otherwise it would be strongly recommended to study or review these concepts. In the first subsection of this section, we provide the reader with all the necessary review if needed. Quotient structures are important in the study of algebraic structures. Some important applications are the ability to equate algebraic structure via an isomorphism to a quotient structure. One important application we will see in this section is the First Isomorphism Theorem. Quotient structures can be useful in induction proofs where one has a notion of measuring size. In the case of a vector space, it is dimension which can be used for measuring size and therefore allows us to prove things by induction. One important application which is at the end of this section is the fact that every matrix is triagularizable over the complex numbers.

We begin with a review of equivalence relations which the reader may skip if they are already comfortable with this concept.

4.8.1 Equivalence Relations

The notion of a relation on a set is important in many fields of mathematics. We shall see many applications of a particular type of relation (called an equivalence relation) in this text. We start by defining a relation and then narrow things down to an equivalence relation.

Definition 4.16 *A **relation** $\sim$ on a set A is simply any subset of the cartesian product $A \times A$. If $(a, b) \in \sim$ we instead write $a \sim b$ and we say a relates to b.*

Example 4.35 *Here, we list a number of examples including several that you have already seen in this text.*

1. *Let $A = \{a, b, c, d\}$ and set $\sim = \{(a, b), (b, b), (c, d)\}$. For instance, according to our definition of $\sim$, we have $c \sim d$ or c relates to d.*

2. *Let $A = \mathbb{Z}$ and $\sim$ be $<$. In other words, $(n, m) \in \sim$ or $n \sim m$ exactly when $n < m$.*

3. *Set $A = \mathcal{P}(\mathbb{Z})$ which represents all the subsets of $\mathbb{Z}$ (called the **power set** of $\mathbb{Z}$). Let $\sim$ be $\subseteq$, i.e. subset. In other words, two subsets X and Y of $\mathbb{Z}$ will relate exactly when $X \subseteq Y$.*

4. *Take any set A and let $\sim$ be equality, i.e. $a \sim b$ exactly when $a = b$. In other words $\sim = \{(a, a) : a \in A\}$.*

5. *Let $f : A \to B$ be a function from a set A to another set B. Define a relation on A as follows: $a \sim b$ iff $f(a) = f(b)$.*

6. Let $A = \mathbb{Z}$ and define $\sim$ as follows: $n \sim m$ iff There exists an integer k such that $m = nk$. On says that n **divides** m and we write $n|m$. For instance, $(3, -15) \in \sim$ since 3 divides -15 because $-15 = 3(-5)$.

7. Define a relation on the set $\mathbb{Z}$ as follows: Fix a positive integer n and define $m \sim k$ iff $n|(m - k)$. This relation is called **congruence modulo** n and in place of $m \sim k$ we typically write $m \equiv_n k$ or $m \equiv k \pmod{n}$.

8. Define a relation on $\mathbb{Q}$ as follows: $\frac{a}{b} \sim \frac{c}{d}$ iff $ad = bc$. So for instance $(\frac{1}{2}, \frac{-3}{-6}) \in \sim$ or $\frac{1}{2} \sim \frac{-3}{-6}$.

9. Let A be the set of all $m \times n$ matrices (for a fixed m and n) and for two such matrices A and B define $A \sim B$ iff A and B are elementarily equivalent, i.e. if B can be obtained by applying a finite number of elementary row operations to A.

10. Let A be the set of all $n \times n$ matrices and for two such matrices A and B define $A \sim B$ iff There is an invertible matrix P such that $B = P^{-1}AP$. Recall, one says that A is **similar to** B.

11. Define a relation $\sim$ on the set of all vector spaces as follows: Two vector spaces V and W relate iff they are isomorphic. For instance, $(\mathbb{R}^3, P_2) \in \sim$ or $\mathbb{R}^3 \sim P_2$, since 3-tuples are isomorphic to polynomials of degree 2 or less (because they have the same dimension).

There are various properties that one may wish to investigate in regards to a relation. We list a few below.

Definition 4.17 Let $\sim$ be a relation on a set A. We say $\sim$ is

1. **reflexive** if for all $a \in A$, we have $a \sim a$.

2. **symmetric** if for all $a, b \in A$, we have $a \sim b$ implies $b \sim a$.

3. **transitive** if for all $a, b, c \in A$, we have $a \sim b$ and $b \sim c$ implies $a \sim c$.

4. **irreflexive** if for all $a \in A$, we have $a \not\sim a$.

5. **anti-symmetric** if for all $a, b \in A$, we have $a \sim b$ and $b \sim a$ implies $a = b$.

Example 4.36 The reader may wish to prove which properties are satisfied by each of the examples presented above. For instance, $<$ on $\mathbb{Z}$ is irreflexive, anti-symmetric (vacuously) and transitive. The relation $\subseteq$ on $\mathcal{P}(\mathbb{Z})$ is reflexive, anti-symmetric and transitive. Congruence $\equiv_n$ on $\mathbb{Z}$ is reflexive, symmetric and transitive.

Some examples of types of relations that are of particular importance in mathematics are the following:

Definition 4.18 Let $\sim$ be a relation on a set A. We say that $\sim$ is

1. *a* **partial ordering** *of A if it is reflexive, anti-symmetric and transitive.*

2. *an* **equivalence relation** *on A if it is reflexive, symmetric and transitive.*

3. *a* **function** *on A if whenever a ∼ b and a ∼ c, then it must be that b = c.*

Example 4.37 *We see now that ⊆ is a partial ordering on $\mathcal{P}(\mathbb{Z})$ and $\equiv_n$ is an equivalence relation on $\mathbb{Z}$. Note that if we restrict the relation* **divides** *to a relation on positive integers, then it becomes a partial ordering on positive integers.*

The focus of our discussion for the remainder of this subsection is equivalence relations. Let's list here the examples introduced above which are equivalence relations. The reader should take the time to prove that they are indeed equivalence relations.

Example 4.38 *Here are some equivalence relations which are not specifically related to linear algebra.*

1. *Take any set A and let ∼ be equality, i.e. a ∼ b exactly when a = b.*

2. *Let $f : A \to B$ be a function from a set A to another set B. Define a relation on A as follows: a ∼ b iff f(a) = f(b).*

3. *Define a relation on the set $\mathbb{Z}$ as follows: Fix a positive integer n and define m ∼ k iff m ≡ k (mod n).*

4. *Define a relation on $\mathbb{Q}$ as follows: $\frac{a}{b} \sim \frac{c}{d}$ iff ad = bc.*

Here are some equivalence relations specific to linear algebra.

1. *Matrix equivalence is an equivalence relation on the set of m × n matrices, i.e. for two m × n matrices A and B define A ∼ B iff there exist a finite number of elementary row operations which change A into B.*

2. *Matrix similarity is an equivalence relation n × n matrices, i.e. for two n × n matrices A and B define A ∼ B iff there is an invertible matrix P such that $B = P^{-1}AP$.*

3. *Isomorphism is an equivalence relation on the set of vector spaces, i.e. two vector spaces V and W relate iff they are isomorphic.*

As we have stated already, for the remainder of this subsection we will be assuming that ∼ is an equivalence relation and as such we typically use the notation ≡ in place of ∼.

Definition 4.19 *Let ≡ be an equivalence relation on a set A and a ∈ A. The* **equivalence class** *of a* **with respect to** *≡, written*

$$[a]_\equiv = \{\, b \in A \;:\; a \equiv b\}.$$

The element a is sometimes called the **representative** *of the class $[a]_\equiv$. The collection of all equivalence classes of A with respect to $\equiv$, in other words $\{\, [a]_\equiv\, :\, a \in A \,\}$, is denoted by $A/\equiv$ and is called the* **quotient set** *of A.*

At times we will simply write $[a]$ in place of $[a]_\equiv$ when the equivalence relation is understood and we may simply call $[a]$ the **class of** a for brevity. Some other notation for an equivalence class which the reader will encounter in the next subsection is $\overline{a}$ in place of $[a]$.

Example 4.39 *Let's compute some equivalence classes for the examples already presented.*

1. *The equivalence classes for equality on a set A are singleton sets, i.e. $[a] = \{a\}$, since no other element besides a relates to a.*

2. *For the equivalence relation we defined on $\mathbb{Q}$ an equivalence class represents all the different ways we can represent a particular fraction. For instance, the equivalence class*

$$\left[\frac{1}{2}\right] = \left\{\frac{1}{2}, \frac{-3}{-6}, \frac{12}{24}, \cdots\right\}.$$

3. *Consider the equivalence relation congruence modulo 3 on $\mathbb{Z}$. There are exactly three distinct equivalence classes. Each class contains integers which when divided by 3 yield the same remainder.*

$$[0] = \{0, \pm 3, \pm 6, \ldots\}$$

$$[1] = \{\ldots, -8, -5, -2, 1, 4, 7, \ldots\}$$

$$[2] = \{\ldots, -7, -4, -1, 2, 5, 8, \ldots\}$$

One can view equivalence relations as a generalization of equality. Each class in a sense contains all the elements of a set which we view as being the same. Just consider the example of the equivalence class of $\frac{1}{2}$. We view $\frac{1}{2}$ and $\frac{-3}{-6}$ as essentially being the same even though symbolically the look very different. Equivalence classes are simply a formal way of equating things which we wish to view as being equal. Consider also isomorphic vector spaces. For the sake of study of vector spaces we might as well group isomorphic types as being equal since isomorphic types are alike in all ways in terms of properties they share, except that the objects are named and look differently.

We now prove a result which uncovers the essential properties of an equivalence relation.

Lemma 4.11 *Let $\equiv$ be an equivalence relation on a set A.*

1. *For all $a \in A$ we have $a \in [a]$.*

2. *For all $a, b \in A$ we have $[a] = [b]$ iff $a \equiv b$.*

3. *For all $a, b \in A$ either $[a] = [b]$ or $[a] \cap [b] = \emptyset$.*

Proof 4.23 *The first part follows immediately from the reflexive property. For the second part, assume first that $[a] = [b]$. Now, since $a \in [a]$ we have $a \in [b]$ and so by definition and symmetry $a \equiv b$. Now assume that $a \equiv b$. Using transitivity and symmetry, notice that $c \in [a]$ iff $a \equiv c$ iff $c \equiv b$ iff $c \in [b]$ and so $[a] = [b]$. For the last part, either $[a] = [b]$ or $[a] \neq [b]$. In the latter case we show that $[a]$ and $[b]$ are disjoint, thus proving the result. Indeed, we prove this by proving the contrapositive, for if $[a] \cap [b] \neq \emptyset$ then there is a $c \in [a] \cap [b]$. Then $c \in [a]$ and $c \in [b]$ and so $c \equiv a$ and $c \equiv b$. Using symmetry and transitivity we have $a \equiv b$ and so by the second part $[a] = [b]$.* □

Notice that the second part of the lemma says that any element of a class can represent that class, i.e. if $b \in [a]$ then $[b] = [a]$. The first and third part of the lemma says that equivalence classes partition the set A into a union of disjoint sets.

Example 4.40 *Let's consider what equivalence classes look like in the case of our linear algebraic examples of an equivalence relation.*

1. *Consider the equivalence relation on M_{mn} matrix equivalence. Each equivalence class can be represented by a reduced row-echelon form.*

2. *Consider the equivalence relation isomorphism on finite dimensional vector spaces. Each equivalence class can be represented by $\mathbb{R}^n$, i.e. the quotient space is*

$$\{[\mathbb{R}^n] \mid n = 0, 1, 2, \ldots\},$$

if we allow $\mathbb{R}^0$ to be the set $\{0\}$.

3. *Consider the equivalence relation similarity on $n \times n$ diagonalizable matrices. Each equivalence class can be represented by a diagonal matrix.*

Let's formally define the notion of a partition of a set.

Definition 4.20 *Let A be a non-empty set and $\mathcal{P}$ be a family of non-empty subsets of A. We say $\mathcal{P}$ is a **partition** of A or **partitions** A if*

1. *For all $a \in A$ there is an $X \in \mathcal{P}$ such that $a \in X$.*

2. *For all $X, Y \in \mathcal{P}$ distinct we have $X \cap Y = \emptyset$.*

According to this formal definition, we see from the lemma that $A/\equiv$ is a partition of A. One can think of a partition of a set as a puzzle where each puzzle piece is an element of the partition and when you put all the puzzle pieces together you get the set A.

Example 4.41 *Consider the earlier example of congruence modulo 3 an equivalence relation on $\mathbb{Z}$. The partition into equivalence classes, namely $\mathbb{Z}/\equiv_3$, consists of three puzzle pieces, namely $[0]$, $[1]$ and $[2]$. These three classes are pairwise disjoint and their union is all of $\mathbb{Z}$.*

The reader may wish to verify the following observations: Given a partition $\mathcal{P}$ of a set A, define the relation $a \sim b$ iff There is an $X \in \mathcal{P}$ such that $a, b \in X$. Then $\sim$ defines an equivalence relation whose equivalence classes consisting of precisely the elements of $\mathcal{P}$. Conversely, if one starts with an equivalence relation $\equiv$ on a set A and forms the partition into equivalence classes and then defines an equivalence relation on this partition as we just did above, then we wind up with the same equivalence relation we began with.

4.8.2 Introduction to Quotient Spaces

We begin our discussion by introducing a relation on vectors in a vector space. Let V be a vector space with subspace U. For $v, w \in V$ define $v \sim w$ iff $v - w \in U$. One can readily check that this defines an equivalence relation. Indeed, to verify the reflexive property, for any $v \in V$ we have $v \sim v$, since $v - v = 0 \in U$. To verify the symmetric property, for $v, w \in V$ if $v \sim w$ then $v - w \in U$, but then $-(v - w) \in U$ and so $w - v \in U$ which implies that $w \sim v$. To verify the transitive property, for $u, v, w \in V$ if $u \sim v$ and $v \sim w$, then $u - v, v - w \in U$ and so $(u - v) + (v - w) \in U$ or $u - w \in U$ and so $u \sim w$.

Now consider the equivalence classes for this equivalence relation, namely $\bar{v} = \{ w \in V : w \sim v \}$ where $v \in V$. Notice that

$$w \in \bar{v} \text{ iff } w \sim v \text{ iff } w - v \in U \text{ iff } w = v + u, \text{ some } u \in U.$$

Hence, $\bar{v} = \{ v + u : u \in U \}$ which we will denote by $v + U$. We call this a **coset of V modulo U**. We will denote the collection of all equivalence classes (or cosets of V modulo U) by the notation V/U which we call the **quotient space of V modulo U**. We will therefore move away from the notation $\bar{v}$ in favor of the more descriptive notation $v + U$. Given a coset $v + U$, the element v is called a **representative** of that coset. The use of the indefinite article "a" is justified, because of the more general fact that any element of an equivalence class can represent that class.

Example 4.42 *Consider the vector space $V = \mathbb{R}^3$ and the subspace $U = \text{span}(\hat{\imath}, \hat{\jmath})$ which generates all the points in the xy-plane. Take $v = [0, 0, c]$ for some scalar $c \in \mathbb{R}$. Then one can see that $v + U$ generates all the points on the plane $z = c$ and therefore V/U is the collection of all planes parallel to the xy-plane. Furthermore, as we shall see more formally later in the section, V/U is isomorphic to $\text{span}(\hat{k})$, the z-axis, and $\dim(V/U) = 1 = \dim(V) - \dim(U)$.*

We leave the following properties as exercises:

Lemma 4.12 (Basic Properties) *If U is a subspace of a vector space V and $v, w \in V$, then*

1. $0 + U = U$.

2. $v + U = w + U$ *iff* $v - w \in U$.

3. $w \in v + U$ *implies* $w + U = v + U$, *thus any element in a coset can represent that coset.*

Our immediate goal is to make V/U into a vector space, so we need to define a vector addition and a scalar multiplication. The obvious definitions are, for $v, w \in V$ and scalar $a \in \mathbb{R}$,

$$(v + U) + (w + U) = (v + w) + U \quad \text{and} \quad a(v + U) = (av) + U.$$

One can easily go on to check that cosets together with these two operations satisfy the axioms of a vector space (left as an exercise). Thus, perhaps the reader believes there is no more to say and the vector space has been established, however there is a potential problem with these definitions since they are defined in terms of representatives and any element of a coset can represent that coset. Hence, we need to check that coset addition and scalar multiplication are *well-defined*. In other words, no matter which representatives we use for cosets when we add or multiply by a scalar using the definitions above, we always arrive at the same coset. Therefore, we need to establish two things:

Lemma 4.13 *Let U be a subspace of a vector space V.*

1. *If $v + U = v' + U$ and $w + U = w' + U$, then $(v + w) + U = (v' + w') + U$, and*

2. *If $a \in \mathbb{R}$ and $v + U = v' + U$, then $(av) + U = (av') + U$.*

Proof 4.24 *We will establish the first and leave the second as an exercise. Thus, if $v + U = v' + U$ and $w + U = w' + U$, then $v - v', w - w' \in U$ and so $(v - v') + (w - w') \in U$ or $(v + w) - (v' + w') \in U$ which means $(v + w) + U = (v' + w') + U$, which establishes the first item.* □

Hence, because of Lemma 4.13, we have established the following result:

Theorem 4.13 *V/U together with the two vector space operations on cosets defined above forms a vector space.*

We introduce an important linear transformation which relates to quotient spaces.

Definition 4.21 *Let U be a subspace of a vector space V and define the function $\nu : V \to V/U$ by $\nu(v) = v + U$. This function is called the **canonical map**.*

We leave it as an exercise for the reader to verify that ν is a linear transformation which maps onto V/U.

We have now arrived at another dimension theorem which relates to quotient spaces. It allows us to easily compute the dimension of a quotient space under certain conditions as described in the result below.

Theorem 4.14 (Quotient Space Dimension Theorem) *For U be a subspace of a vector space V, V is finite dimensional iff both U and V/U are finite dimensional. Furthermore, in this case, $\dim(V/U) = \dim(V) - \dim(U)$.*

Proof 4.25 *First, assuming $\dim(V) < \infty$, by the Subspace Dimension Theorem, $\dim(U) < \infty$. Consider the canonical map $\nu : V \to V/U$. Since ν maps onto V/U we know that the image of any basis for V is a set of generators for V/U, and so $\dim(V/U) \leq \dim(V) < \infty$. For the reverse direction, assuming $\dim(U) < \infty$ and $\dim(V/U) < \infty$, we shall prove that $\dim(V/U) = \dim(V) - \dim(U)$ and thus as a byproduct obtain the fact that $\dim(V) < \infty$. First, let's dispense with the two extreme cases. First, if $U = V$ then V/U is the trivial subspace with only the zero vector $U = 0 + U$ (exercise). Therefore, $\dim(V/U) = 0 = \dim(V) - \dim(U)$. Second, if $U = \{0\}$, then V/U is isomorphic to V (exercise) and therefore have the same finite dimension. Hence,*

$$\dim(V/U) = \dim(V) = \dim(V) - 0 = \dim(V) - \dim(U).$$

From now on we assume that U is a proper non-trivial subspace of V and select a basis for U, say $u_1, u_2, \ldots, u_k$. Since, $\dim(V/U) < \infty$ it also has a finite basis, say $v_1 + U, v_2 + U, \ldots, v_m + U$.

Claim: $u_1, u_2, \ldots, u_k, v_1, v_2, \ldots, v_m$ *forms a basis for V.*

First we show that $u_1, u_2, \ldots, u_k, v_1, v_2, \ldots, v_m$ span V. Take any $v \in V$ and consider the coset $v + U \in V/U$. Then for some scalars $b_1, b_2, \ldots, b_m \in \mathbb{R}$, we can write

$$v + U = b_1(v_1 + U) + b_2(v_2 + U) + \cdots + b_m(v_m + U).$$

But then $v + U = (b_1v_1 + b_2v_2 + \cdots b_mv_m) + U$ and so $v - (b_1v_1 + b_2v_2 + \cdots b_mv_m) \in U$ and as such there exists scalars $a_1, a_2, \ldots, a_k \in \mathbb{R}$ such that

$$v - (b_1v_1 + b_2v_2 + \cdots b_mv_m) = a_1u_1 + a_2u_2 + \cdots a_ku_k,$$

and so $v = a_1u_1 + a_2u_2 + \cdots a_ku_k + b_1v_1 + b_2v_2 + \cdots b_mv_m$ as desired. To show $u_1, u_2, \ldots, u_k, v_1, v_2, \ldots, v_m$ are linearly independent, suppose for some scalars $a_1, a_2, \ldots, a_k, b_1, b_2, \ldots, b_m$ that $a_1u_1 + a_2u_2 + \cdots a_ku_k + b_1v_1 + b_2v_2 + \cdots b_mv_m = 0$. Then $b_1v_1 + b_2v_2 + \cdots b_mv_m = -(a_1u_1 + a_2u_2 + \cdots a_ku_k) \in U$. Therefore,

$$U = (b_1v_1 + b_2v_2 + \cdots b_mv_m) + U = b_1(v_1 + U) + b_2(v_2 + U) + \cdots + b_m(v_m + U).$$

Since, $v_1 + U, v_2 + U, \ldots, v_m + U$ are linearly independent, $b_1 = b_2 = \cdots = b_m = 0$.

But then $a_1 u_1 + a_2 u_2 + \cdots a_k u_k = 0$ and since $u_1, u_2, \ldots, u_k$ are linearly independent, $a_1 = a_2 = \cdots = a_k = 0$, which completes the proof of the claim.

Therefore, by the claim, $\dim(V/U) = m = (k+m) - k = \dim(V) - \dim(U)$. $\qquad\square$

Example 4.43 We can use ideas from this proof in order to construct a basis for a given quotient space. Consider the vector space $V = \mathbb{R}^4$ with subspace $U = \{[a + b, a, a - b, b] : a, b \in \mathbb{R}\}$. A basis for U is $u_1 = [1, 1, 1, 0]$ and $u_2 = [0, 1, -1, 1]$. Using our algorthm presented in Section 3.5, let's extend U to a basis for V by row reducing

$$
\begin{bmatrix}
1 & 0 & 1 & 0 & 0 & 0 \\
1 & 1 & 0 & 1 & 0 & 0 \\
1 & -1 & 0 & 0 & 1 & 0 \\
0 & 1 & 0 & 0 & 0 & 1
\end{bmatrix}
\quad \text{which reduces to} \quad
\begin{bmatrix}
1 & 0 & 0 & 0 & 1 & 1 \\
0 & 1 & 0 & 0 & 0 & 1 \\
0 & 0 & 1 & 0 & -1 & -1 \\
0 & 0 & 0 & 1 & -1 & -2
\end{bmatrix}.
$$

Therefore, a basis extension would be u_1, u_2, e_1, e_2. We claim that a basis for V/U is $e_1 + U$ and $e_2 + U$. To see this, notice that any $v \in V$ can be written in terms of this basis, i.e. $v = a_1 u_1 + a_2 u_2 + a_3 e_1 + a_4 e_2$ for some $a_i \in \mathbb{R}$. Then

$$
v + U = [(a_1 u_1) + U] + [(a_2 u_2) + U] + [(a_3 e_1) + U] + [(a_4 e_2) + U].
$$

But since $a_1 u_1, a_2 u_2 \in U$ and U is the zero vector in V/U, we have

$$
v + U = U + U + a_3(e_1 + U) + a_4(e_2 + U) = a_3(e_1 + U) + a_4(e_2 + U).
$$

Thus $e_1 + U$ and $e_2 + U$ span V/U with $\dim(V/U) = 4 - 2 = 2$. Therefore, $e_1 + U$ and $e_2 + U$ form a basis for V/U.

The next result will be useful in allowing us to equate a vector spaces with a quotient space via an isomorphism.

Theorem 4.15 (First Isomorphism Theorem) If V and W are vector spaces and $T \in L(V, W)$, then $V/\ker T \simeq T(V)$.

Proof 4.26 Set $K = \ker T$ and consider the map $S : V/K \to T(V)$ by $S(v + K) = T(v)$. Since the domain of this function is equivalence classes, we must first ensure that S is a well-defined map. In other words, two representations of the same coset get sent to the same place by S. Thus, if $v + K = w + K$ we need to conclude that $S(v + K) = S(w + K)$. We will do this via a series of iff statements.

$$
v + K = w + K \quad \text{iff} \quad v - w \in K \quad \text{iff} \quad T(v - w) = 0_W
$$

$$
\text{iff} \quad T(v) = T(w) \quad \text{iff} \quad S(v + K) = S(w + K).
$$

Notice that reading the iff statements from right to left also proves that S is a one-to-one map. The fact that S maps onto $T(V)$ is self-evident. So it remains to show that S is a linear transformation so that S is an isomorphism, thus proving the result. To this end

$$S[(v+K)+(w+K)] = S[(v+w)+K] = T(v+w) = T(v)+T(w) = S(v+K)+S(w+K).$$

and $\;S[a(v+K)] = S[(av)+K] = T(av) = aT(v) = aS(v+K).$

Hence, S is an isomorphism making V/K isomorphic to $T(V)$. □

4.8.3   Applications of Quotient Spaces

We illustrate in the following examples how quotient spaces can be used. Although some of these results can be proved without quotient spaces, one readily sees the mathematical elegance in which quotient spaces can be employed.

Example 4.44 *We can quickly reprove the Linear Transformation Dimension Theorem. By the First Isomorphism Theorem we know $V/\ker T \simeq T(V)$ and being isomorphic implies they have the same dimension, i.e. $\dim(V/\ker T) = T(V)$, but then by the Quotient Space Dimension Theorem, $\dim(V) - \dim(\ker T) = \dim T(V)$ or equivalently, $\dim(V) = \dim(\ker T) + \dim T(V)$.* □

Example 4.45 *We can reprove the Product Dimension Theorem for a cartesian product of vector spaces $V_1 \times V_2 \times \cdots \times V_n$ by induction on n. The result holds trivially for $n = 1$, so for $n > 1$ consider the linear transformation (one needs to check it is)*

$$T : V_1 \times V_2 \times \cdots \times V_n \to V_1 \times V_2 \times \cdots \times V_{n-1} \;\; by \;\; T[v_1, v_2, \ldots, v_n] = [v_1, v_2, \ldots, v_{n-1}].$$

*T is sometimes called a **projection map**. It's easy to see that T maps onto $V_1 \times V_2 \times \cdots \times V_{n-1}$ and*

$$\ker T = \{[0, 0, \ldots, 0, v_n] \;:\; v_n \in V_n\}.$$

and so (another easy thing to check) we have that $\ker T \simeq V_n$. Therefore, by the Linear Transformation Dimension Theorem just reverified,

$$\dim(V_1 \times V_2 \times \cdots \times V_n) = \dim(V_n) + \dim(V_1 \times V_2 \times \cdots \times V_{n-1}).$$

But then, by induction,

$$\dim(V_1 \times V_2 \times \cdots \times V_n) = \dim(V_n) + \dim(V_1) + \dim(V_2) + \cdots + \dim(V_{n-1}).$$

Rearranging, we have our result that

$$\dim(V_1 \times V_2 \times \cdots \times V_n) = \dim(V_1) + \dim(V_2) + \cdots + \dim(V_n).$$

□

The final application is a new result not reproved which employs quotient spaces.

Theorem 4.16 *Every square matrix is triangularizable over the complex numbers, i.e. given a matrix $A \in M_{nn}(\mathbb{C})$ there exists an invertible matrix $P \in M_{nn}(\mathbb{C})$ and a upper triangular matrix $B \in M_{nn}(\mathbb{C})$ such that $P^{-1}AP = B$.*

Proof 4.27 *Let T be the familiar linear operator on $\mathbb{C}^n$ defined by $T(v) = Av$. We will show that T (and therefore A) is triangularizable. The bulk of the proof relies on the following claim:*

Claim: *For any linear operator $T \in L(\mathbb{C}^n)$, there exists a ordered basis $B = (v_1, \ldots, v_n)$ for $\mathbb{C}^n$ such that $T(v_i) \in \text{span}(v_1, \ldots, v_i)$ for $i = 1, 2, \ldots, n$ (such a basis is called a **fan basis**).*

We prove the Claim by induction on n. For $n = 1$, $T \in L(\mathbb{C})$ and $\mathbb{C} = \text{span}(v_1)$ so evidentally, $T(v_1) \in \text{span}(v_1)$. For $n > 1$, first set $A = [T]_{ST}$ and consider the characteristic polynomial $p_A(t)$. Since every polynomial has a root in $\mathbb{C}$, set λ_1 to be a root of $p_A(t)$, an eigenvalue of T with respect to some eigenvector v_1. Set $U = \text{span}(v_1)$. Certainly, $T(U) \subseteq U$. Indeed, for any $av_1 \in U$, we have

$$T(av_1) = aT(v_1) = a\lambda_1 v_1 \in \text{span}(v_1) = U.$$

Now consider the map $S : V/U \to V/U$ by $S(v+U) = T(v)+U$. Since the domain of S is a quotient space, we need to check that it is well-defined. To that end, suppose that $v + U = w + U$, then $v - w \in U$, but then $T(v - w) \in U$ (see above) which means that $T(v) - T(w) \in U$, and so $T(v) + U = T(w) + U$, i.e. $S(v) = S(w)$. We leave it as an exercise to show that S is a linear transformation.

Now $\dim(V/U) = \dim(V) - \dim(U) = n - 1$, so by induction S has a fan basis, i.e. there is a basis for V/U, say $v_2 + U, \ldots, v_n + U$, such that

$$S(v_i + U) \in \text{span}(v_2 + U, \ldots, v_i + U) \quad \text{for} \quad i = 2, \ldots, n.$$

Equivalently, for $i = 2, \ldots, n$ and some scalars $a_2, \ldots, a_i \in \mathbb{C}$,

$$S(v_i + U) = a_2(v_2 + U) + \cdots + a_i(v_n + U) = (a_2 v_2 + \cdots + a_i v_i) + U.$$

Therefore, by definition of S, $T(v_i) + U = (a_2 v_2 + \cdots + a_i v_i) + U$ and so $T(v_i) - (a_2 v_2 + \cdots + a_i v_i) \in U$ which means $T(v_i) - (a_2 v_2 + \cdots + a_i v_i) = a_1 v_1$ for some $a_1 \in \mathbb{C}$ and so

$$T(v_i) = a_1 v_1 + \cdots + a_i v_i \in \text{span}(v_1, \ldots, v_i).$$

Hence, the Claim is proved. Having shown this Claim, it's easy to see that $[T]_B$ is upper triangular. Furthermore, $[T]_{ST} = A$. Therefore, by similarity of matrices the theorem is proved. □

EXERCISES

1. For each of the following vector spaces V with subspace U, find a basis for V/U.

   a. $V = \mathbb{R}^4$ and $U = \{ [a, -a, -b, b] : a, b \in \mathbb{R} \}$.

   b. $V = P_3$ and $U = \{ a + bx + (a + b)x^2 - bx^3 : a, b \in \mathbb{R} \}$.

   c. $V = M_{22}$ and $U = D_{22}$.

2. Given positive integers m and n with $m < n$, determine what the quotient space $\mathbb{R}^n/\mathbb{R}^m$ is isomorphic to.

3. Verify the Basic Properties listed in Lemma 4.12.

4. Verify that V/U together with the well-defined operations for coset addition and scalar multiplication satisfies the axioms of a vector space.

5. Verify that scalar multiplication as defined for V/U is well-defined as stated in Lemma 4.13.

6. Verify that $\nu : V \to V/U$ by $\nu(v) = v + U$ is a linear transformation which maps onto V/U.

7. Prove for any vector space V that $V/\{0\} \simeq V$ and $V/V \simeq \{0\}$, the trivial vector space.

8. Prove that if $u_1, \ldots, u_k$ is a basis for a subspace U of a finite dimensional vector space V and we extend this basis to a basis for V, say $u_1, \ldots, u_k, v_{k+1}, \ldots, v_n$, then $v_{k+1} + U, \ldots, v_n + U$ forms a basis for V/U.

9. Prove that the map S defined in Theorem 4.16 is a linear transformation.

4.9   DUAL VECTOR SPACE

To every vector space we can associate a dual vector space which is a useful structure to have around in order to prove some important results in linear algebra.

Definition 4.22 *Let V be a vector space.*

1. *The **dual space** of V, written $V^* = L(V, \mathbb{R})$, i.e. linear transformations from V to the real numbers.*

2. *An element $\phi \in V^*$ is called a **linear functional**.*

Example 4.46 *We list here several important linear functionals.*

1. *The **trace** of a matrix is a linear functional in M_{nn}^*, since $tr(A+B) = tr(A) + tr(B)$ for any square matrices A, B and $tr(aA) = atr(A)$ for any scalar $a \in \mathbb{R}$ and square matrix A.*

2. *Any* **evaluation** *transformation is a linear functional in $\mathcal{F}$. Recall, for any scalar $a \in \mathbb{R}$ and $f \in \mathcal{F}$ the evaluation transformation takes f to $f(a)$.*

3. *Define the linear functional $\phi_i \in (\mathbb{R}^n)^*$ by $\phi_i[a_1, a_2, \ldots, a_n] = a_i$, for a fixed i, $1 \leq i \leq n$. We call this map the ith* **coordinate** *functional.*

4. *Let $v_1, v_2, \ldots, v_n$ be a basis for a vector space V and fix an i, $1 \leq i \leq n$. Define the $\chi_i \in V^*$ to be the unique linear functional having the following property:*

$$\chi_i(v_j) = \begin{cases} 0, & i \neq j \\ 1, & i = j \end{cases}$$

We call this the **characteristic** *functional.*

Example 4.47 *We give a concrete example of the characteristic functionals for $V = P_2$. Consider the following basis for P_2; $v_1 = 1$, $v_2 = 1 + x$, $v_3 = 1 + x + x^2$. Let's find the formula for χ_2. We will use Lemma 4.3.ii to construct it. Take any $p = a + bx + cx^2 \in P_2$. First, we find the coordinates of p with respect to $B = (v_1, v_2, v_3)$. Create the augmented matrix*

$$\begin{bmatrix} 1 & 1 & 1 & a \\ 0 & 1 & 1 & b \\ 0 & 0 & 1 & c \end{bmatrix} \quad \text{which reduces to} \quad \begin{bmatrix} 1 & 0 & 0 & a-b \\ 0 & 1 & 0 & b-c \\ 0 & 0 & 1 & c \end{bmatrix}$$

Therefore,

$$\chi_2(a + bx + cx^2) = (a-b)(0) + (b-c)(1) + (c)(0) = b - c.$$

Similarly, $\chi_1(a + bx + cx^2) = a - b$ and $\chi_3(a + bx + cx^2) = c$.

These characteristic functionals are important as is illustrated in the following result:

Lemma 4.14 *Let V be a finite dimensional vector space. Then $\dim V^* = \dim V$. Furthermore, if $v_1, v_2, \ldots, v_n$ forms a basis for V then $\chi_1, \chi_2, \ldots, \chi_n$ forms a basis for V^* (called the* **dual** *basis).*

Proof 4.28 *Using Theorem 4.6.ii,*

$$\dim(V^*) = \dim(L(V, \mathbb{R})) = \dim(V)\dim(\mathbb{R}) = \dim V.$$

Hence, to prove $\chi_1, \chi_2, \ldots, \chi_n$ forms a basis for V^*, it's enough to show they are linearly independent. Therefore, suppose $a_1\chi_1 + a_2\chi_2 + \cdots + a_n\chi_n = 0$. Then for $1 \leq j \leq n$ we have $(a_1\chi_1 + a_2\chi_2 + \cdots + a_n\chi_n)(v_j) = 0$ or equivalently $a_1\chi_1(v_j) + a_2\chi_2(v_j) + \cdots + a_n\chi_n(v_j) = 0$. But by the definition of characteristic functionals this reduces to $a_j = 0$, for $1 \leq j \leq n$. □

Even though we did not have to show that $\chi_1, \chi_2, \ldots, \chi_n$ span V^*, it is worthwhile to see how they do this. Take any $\phi \in V^*$. We show that

$$\phi = \phi(v_1)\chi_1 + \phi(v_2)\chi_2 + \cdots + \phi(v_n)\chi_n.$$

Using Lemma 4.3.i, it's enough to show that ϕ and $\phi(v_1)\chi_1 + \phi(v_2)\chi_2 + \cdots + \phi(v_n)\chi_n$ agree on the basis $v_1, v_2, \ldots, v_n$. For $1 \leq j \leq n$,

$$[\phi(v_1)\chi_1 + \phi(v_2)\chi_2 + \cdots + \phi(v_n)\chi_n](v_j)$$
$$= \phi(v_1)\chi_1(v_j) + \phi(v_2)\chi_2(v_j) + \cdots + \phi(v_n)\chi_n(v_j) = \phi(v_j)\chi_j(v_j) = \phi(v_j).$$

Example 4.48 *Let's illustrate this discourse on span using the example we created earlier. With $V = P_2$ and basis $1, 1+x, 1+x+x^2$ we computed the dual basis χ_1, χ_2, χ_3 where*

$$\chi_1(a + bx + cx^2) = a - b, \quad \chi_2(a + bx + cx^2) = b - c, \quad \chi_3(a + bx + cx^2) = c.$$

Consider the linear functional $\phi \in (P_2)^*$ defined by $\phi(a + bx + cx^2) = a + b + c$. Notice that when we input the basis for P_2, we get

$$\phi(1) = 1, \quad \phi(1 + x) = 2 \quad and \quad \phi(1 + x + x^2) = 3.$$

But then

$$\phi(1)\chi_1 + \phi(1 + x)\chi_2 + \phi(1 + x + x^2)\chi_3 = \chi_1 + 2\chi_2 + 3\chi_3 \quad and \ so$$

$$[\phi(1)\chi_1 + \phi(1+x)\chi_2 + \phi(1+x+x^2)\chi_3](a + bx + cx^2) = [\chi_1 + 2\chi_2 + 3\chi_3](a + bx + cx^2)$$

$$= \chi_1(a + bx + cx^2) + 2\chi_2(a + bx + cx^2) + 3\chi_3(a + bx + cx^2)$$

$$= (a - b) + 2(b - c) + 3c = a + b + c = \phi(a + bx + cx^2).$$

We now shift our attention to linear transformations on dual vector spaces. For every linear transformation between two vector spaces we can associate another linear transformation between the corresponding dual spaces.

Definition 4.23 *Let V and W be vector spaces and $T \in L(V, W)$. The **transpose** of T, written T^*, is a linear transformation from W^* to V^*, i.e. $T^* \in L(W^*, V^*)$, defined by $T^*(\phi) = \phi \circ T$.*

Several remarks are in order. First, T^* is indeed a linear transformation, since it is a composition of two linear transformations. Secondly, T^* maps into V^*, since $T^*(\phi) = \phi \circ T$ takes as an input an element of V and outputs a real number. Finally, the kernel of T^* is the collection of all linear functionals in W^* which send the image of T to 0_W, since

$$\phi \in \ker T^* \text{ iff } T^*(\phi) = 0 \text{ iff } \phi \circ T = 0$$

$$\text{iff } (\phi \circ T)(v) = 0_W, \ \forall v \in V \text{ iff } \phi(T(v)) = 0_W, \ \forall v \in V.$$

The next result explains the reason why T^* is called a transpose.

Theorem 4.17 *Let V and W be finite dimensional vector spaces with bases B_1 and B_2 respectively. Denote B_1^* and B_2^* as the dual bases for V^* and W^* respectively. Then*

$$[T^*]_{B_2^*}^{B_1^*} = \left([T]_{B_1}^{B_2}\right)^T.$$

Proof 4.29 *Set $B_1 = \{v_1, v_2, \ldots, v_n\}$ and $B_2 = \{w_1, w_2, \ldots, w_m\}$. Set $B_1^* = \{\phi_1, \phi_2, \ldots, \phi_n\}$ and $B_2^* = \{\psi_1, \psi_2, \ldots, \psi_m\}$. Let $A = [T]_{B_1}^{B_2}$, where $A = [a_{ij}]$. Note that for $1 \leq j \leq m$, as we saw in remarks after Lemma 4.14,*

$$T^*(\psi_j) = \psi_j \circ T = \sum_{i=1}^{n} (\psi_j \circ T)(v_i)\psi_i.$$

Therefore, the jth column of $[T^*]_{B_2^*}^{B_1^*}$ is the vector

$$[(\psi_j \circ T)(v_1), \ (\psi_j \circ T)(v_2), \ \ldots, (\psi_j \circ T)(v_n)].$$

In particular, the ijth entry in $[T^*]_{B_2^*}^{B_1^*}$ is

$$(\psi_j \circ T)(v_i) = \psi_j(T(v_i)) = \psi_j\left(\sum_{k=1}^{m} a_{ki}w_k\right) = \sum_{k=1}^{m} a_{ki}\psi_j(w_k) = a_{ji}\psi_j(w_j) = a_{ji},$$

which is, of course, the ijth entry in $\left([T]_{B_1}^{B_2}\right)^T$. □

Example 4.49 *Let's illustrate Theorem 4.17 with an example. Take $V = P_1$ with ordered basis $B_1 = (1, \ 1 + x)$, $W = U_{22}$ with basis $B_2 = (E_{11}, \ E_{11} + E_{12}, \ E_{11} + E_{12} + E_{22})$ and consider the linear transformation*

$$T(a + bx) = \begin{bmatrix} a & b \\ 0 & a + b \end{bmatrix}.$$

First, we compute $[T]_{B_1}^{B_2}$. Now

$$T(1) = \begin{bmatrix} 1 & 0 \\ 0 & 1 \end{bmatrix} \quad and \quad T(1+x) = \begin{bmatrix} 1 & 1 \\ 0 & 2 \end{bmatrix}.$$

We row reduce

$$\left[\begin{array}{ccc|cc} 1 & 1 & 1 & 1 & 1 \\ 0 & 1 & 1 & 0 & 1 \\ 0 & 0 & 1 & 1 & 2 \end{array}\right] \quad which \ reduces \ to \quad \left[\begin{array}{ccc|cc} 1 & 0 & 0 & 1 & 0 \\ 0 & 1 & 0 & -1 & -1 \\ 0 & 0 & 1 & 1 & 2 \end{array}\right]$$

Therefore,

$$[T]_{B_1}^{B_2} = \begin{bmatrix} 1 & 0 \\ -1 & -1 \\ 1 & 2 \end{bmatrix}.$$

Before we can compute $[T^*]_{B_2^*}^{B_1^*}$ we first need to derive the dual bases B_1^* and B_2^*. To find B_1^* we first find the coordinates of an arbitrary $a + bx \in P_1$ with respect to B_1. So we create the augmented matrix

$$\left[\begin{array}{cc|c} 1 & 1 & a \\ 0 & 1 & b \end{array}\right] \quad which \ reduces \ to \quad \left[\begin{array}{cc|c} 1 & 0 & a-b \\ 0 & 1 & b \end{array}\right]$$

Therefore,

$$\phi_1(a + bx) = (a - b)(1) + (b)(0) = a - b \quad and \quad \phi_2(a + bx) = (a - b)(0) + (b)(1) = b.$$

Now we find B_2^* by first finding the coordinates of an arbitrary $\begin{bmatrix} a & b \\ 0 & c \end{bmatrix} \in U_{22}$ with respect to B_2. So we create the augmented matrix

$$\left[\begin{array}{ccc|c} 1 & 1 & 1 & a \\ 0 & 1 & 1 & b \\ 0 & 0 & 1 & c \end{array}\right] \quad which \ reduces \ to \quad \left[\begin{array}{ccc|c} 1 & 0 & 0 & a-b \\ 0 & 1 & 0 & b-c \\ 0 & 0 & 1 & c \end{array}\right]$$

Therefore,

$$\psi_1 \begin{bmatrix} a & b \\ 0 & c \end{bmatrix} = (a-b)(1) + (b-c)(0) + (c)(0) = a - b,$$

$$\psi_2 \begin{bmatrix} a & b \\ 0 & c \end{bmatrix} = (a-b)(0) + (b-c)(1) + (c)(0) = b - c,$$

$$\psi_3 \begin{bmatrix} a & b \\ 0 & c \end{bmatrix} = (a-b)(0) + (b-c)(0) + (c)(1) = c.$$

We again use the remarks after Lemma 4.14 to find coordinates each $T^*(\psi_i)$ ($1 \leq i \leq 3$) with respect to B_1^*.

$$T^*(\psi_1) = [(T^*(\psi_1))(1), \ (T^*(\psi_1))(1+x)] =$$

$$[\psi_1(T(1)), \ \psi_1(T(1+x))] = \left[\psi_1 \begin{bmatrix} 1 & 0 \\ 0 & 1 \end{bmatrix}, \psi_1 \begin{bmatrix} 1 & 1 \\ 0 & 2 \end{bmatrix} \right] = [1, 0].$$

$$T^*(\psi_2) = [(T^*(\psi_2))(1), \ (T^*(\psi_2))(1+x)]$$

$$= [\psi_2(T(1)), \ \psi_2(T(1+x))] = \left[\psi_2 \begin{bmatrix} 1 & 0 \\ 0 & 1 \end{bmatrix}, \psi_2 \begin{bmatrix} 1 & 1 \\ 0 & 2 \end{bmatrix} \right] = [-1, -1].$$

$$T^*(\psi_3) = [(T^*(\psi_3))(1), \ (T^*(\psi_3))(1+x)]$$

$$= [\psi_3(T(1)), \ \psi_3(T(1+x))] = \left[\psi_3 \begin{bmatrix} 1 & 0 \\ 0 & 1 \end{bmatrix}, \psi_3 \begin{bmatrix} 1 & 1 \\ 0 & 2 \end{bmatrix} \right] = [1, 2].$$

Therefore,

$$[T^*]_{B_2^*}^{B_1^*} = \begin{bmatrix} 1 & -1 & 1 \\ 0 & -1 & 2 \end{bmatrix} = \left([T]_{B_1}^{B_2} \right)^T.$$

Definition 4.24 *Let U be a subspace of a vector space V. The **annihilator** of U in V, written*

$$U^\circ = \{ \ \phi \in V^* \ : \ \phi(u) = 0, \ \forall u \in U \ \}.$$

Example 4.50 *We give several examples of direct computation of the annihilator of a subspace.*

1. *We have seen for $T \in L(V, W)$ that $\ker T^* = T(V)^\circ$.*

2. *Let $V = \mathbb{R}^2$ and $U = \text{span}([1, 1])$. Note that $\phi \in V^*$ is in U° exactly when $\phi[1, 1] = 0$. Extend $[1, 1]$ to a basis for V by adding the vector $[1, 0]$. Find the coordinates of $[a, b] \in V$ with respect to $B = ([1, 0], [1, 1])$ by row reducing*

$$\begin{bmatrix} 1 & 1 & | & a \\ 0 & 1 & | & b \end{bmatrix} \quad \text{which reduces to} \quad \begin{bmatrix} 1 & 0 & | & a - b \\ 0 & 1 & | & b \end{bmatrix}$$

Therefore,

$$\phi[a, b] = (a - b)\phi[1, 0] + b\phi[1, 1] = c(a - b).$$

If we define $\phi_1 \in V^*$ by $\phi_1[a, b] = a - b$, then we see that $U^\circ = \text{span}(\phi_1)$.

3. Let $V = P_2$ with $U = \text{span}(1 + x)$. Note that $\phi \in V^*$ is in U° exactly when $\phi(1 + x) = 0$. Extend $1 + x$ to a basis for V by adding the vector $1, x, x^2$ and row reducing

$$\begin{bmatrix} 1 & 1 & 0 & 0 \\ 1 & 0 & 1 & 0 \\ 0 & 0 & 0 & 1 \end{bmatrix} \quad \text{which reduces to} \quad \begin{bmatrix} 1 & 0 & 1 & 0 \\ 0 & 1 & -1 & 0 \\ 0 & 0 & 0 & 1 \end{bmatrix}.$$

Thus, by identifying the pivots of the reduced row-echelon form we see that the extended basis is $B = (1+x, 1, x^2)$. Therefore, $\phi \in U^\circ$ if $\phi(1+x) = 0$ while $\phi(1)$ and $\phi(x^2)$ can equal whatever we like, say $\phi(1) = d \in \mathbb{R}$ and $\phi(x^2) = e \in \mathbb{R}$. Find the coordinates of $a + bx + cx^2 \in V$ with respect to $B = (1 + x, 1, x^2)$ by row reducing

$$\begin{bmatrix} 1 & 1 & 0 & a \\ 1 & 0 & 0 & b \\ 0 & 0 & 1 & c \end{bmatrix} \quad \text{which reduces to} \quad \begin{bmatrix} 1 & 0 & 0 & b \\ 0 & 1 & 0 & a - b \\ 0 & 0 & 1 & c \end{bmatrix}$$

Therefore,

$$\phi(a + bx + cx^2) = b\phi(1 + x) + (a - b)\phi(1) + c\phi(x^2) = d(a - b) + ec.$$

If we define $\phi_1 \in V^*$ by $\phi_1(a + bx + cx^2) = a - b$ and $\phi_2(a + bx + cx^2) = c$, then we see that $U^\circ = \text{span}(\phi_1, \phi_2)$.

Notice that in all these examples above there is a relationship between the dimension of U, V and U° which leads to the following result:

Theorem 4.18 (Annihilator Dimension Theorem) *If U is a subspace of a finite dimensional vector space, then*

$$\dim U^\circ = \dim V - \dim U.$$

Proof 4.30 *Consider the inclusion linear transformation $T \in L(U, V)$ defined by $T(u) = u$ for all $u \in U$. Then $T^* \in L(V^*, U^*)$ is defined by $T^*(\phi) = \phi \circ T$ which is the map ϕ restricted to U. First note that $\ker T^* = T(U)^\circ = U^\circ$. Second T^* maps onto U^*, since for any $\phi \in U^*$, define $\psi \in V^*$ to be the linear functional that agrees with ϕ on U and equals 0 elsewhere (in the usual way by extending a basis for U to a basis for V). Now by Theorem 4.2, Theorem 4.14 and our work above,*

$$\dim V = \dim V^* = \dim(\ker T^*) + \dim(T^*(V^*)) = \dim U^\circ + \dim U^* = \dim U^\circ + \dim U.$$

□

We can use this result to derive further useful results.

Theorem 4.19 *Let V, W be vector spaces and $T \in L(V, W)$. Then*

1. $\dim T(V) = \dim T^*(W^*)$

2. *T maps onto W iff T^* is one-to-one*

3. *T is one-to-one iff T^* maps onto V^**

Proof 4.31 *To prove the first item, using earlier results on dimension,*

$$\dim T^*(W^*) = \dim W^* - \dim(\ker T^*) = \dim W - \dim T(V)^\circ =$$

$$\dim W - (\dim W - \dim T(V)) = \dim T(V).$$

The second and third item follow from the first and are left as exercises. □

There is nothing to stop us from taking the dual of the dual and this is what we do now. We shall see that in a sense, i.e. in the sense of isomorphism, this returns us back to the original vector space.

Definition 4.25 *For vector spaces V, W and $T \in L(V, W)$*

1. *the* **bidual** *of V, written*

$$V^{**} = (V^*)^* = L(V^*, \mathbb{R}) = L(L(V, \mathbb{R}), \mathbb{R}).$$

2. *the* **bitranspose** *of T, written*

$$T^{**} = (T^*)^* \in L(V^{**}, W^{**}) \quad \text{and is defined by} \quad T^{**}(v^{**}) = v^{**} \circ T^*.$$

3. *For $v \in V$, the notation $v^{**} \in V^{**}$ will be given more meaning, namely for a fixed $v \in V$ and any $\phi \in V^*$, we define $v^{**}(\phi) = \phi(v)$.*

4. *We then define the map $\Phi : V \to V^{**}$ by $\Phi(v) = v^{**}$ called the* **canonical map**.

Note that for $0 \in V$ the element $0^{**} \in V^{**}$ is the zero map from V^* to $\mathbb{R}$ (this short proof is left as an exercise).

From our earlier work we already know that when the dimension of V is finite, then $\dim V^{**} = \dim V^* = \dim V$, but we can say more about the relation between V and V^{**},

Theorem 4.20 *If $\dim(V)$ is finite, then the canonical map Φ is a vector space isomorphism thus making $V^{**} \cong V$.*

Proof 4.32 *First we show that Φ is a linear transformation.*

1. *For $v_1, v_2 \in V$, $\Phi(v_1 + v_2) = \Phi(v_1) + \Phi(v_2)$.*

   This is equivalent to showing $(v_1 + v_2)^{**} = v_1^{**} + v_2^{**}$ and to see this, for any $\phi \in V^*$,

   $$(v_1 + v_2)^{**}(\phi) = \phi(v_1 + v_2) = \phi(v_1) + \phi(v_2) = v_1^{**}(\phi) + v_2^{**}(\phi).$$

2. *For $v \in V$ and $a \in \mathbb{R}$, $\Phi(av) = a\Phi(v)$.*

   This is equivalent to showing $(av)^{**} = av^{**}$ and to see this, for any $\phi \in V^*$,

   $$(av)^{**}(\phi) = \phi(av) = a\phi(v) = av^{**}(\phi).$$

Thus, Φ is a linear transformation. Since $\dim V = \dim V^{**}$ it's enough to show that Φ is one-to-one by showing that the kernel of Φ is trivial. To show this we will rely on a result which we leave as an exercise, namely $v \in V$ and $\phi(v) = 0$ for all $\phi \in V^*$ iff $v = 0$. Assuming this result

$$v \in \ker\Phi \;\; iff \;\; \Phi(v) = 0^{**} \;\; iff \;\; v^{**} = 0^{**} \;\; iff \;\; v^{**}(\phi) = 0, \; \forall \phi \in V^*$$

$$iff \;\; \phi(v) = 0, \; \forall \phi \in V^* \;\; iff \;\; v = 0.$$

□

EXERCISES

1. Verify that the coordinate functional is indeed a linear functional.

2. Verify that the characteristic functional is indeed a linear functional.

3. For each of the following vector spaces and bases compute the corresponding dual basis:

   a. $V = \mathbb{R}^3$ with basis $B = ([1, 1, 0], [1, 0, 1], [0, 1, 1])$.

   b. $V = P_3$ with basis $B = (1 - x, x - x^2, 1 + x + x^2)$.

   c. $V = M_{22}$ with basis

   $$B = \left(\begin{bmatrix} 1 & 0 \\ 0 & 0 \end{bmatrix}, \begin{bmatrix} 1 & 1 \\ 0 & 0 \end{bmatrix}, \begin{bmatrix} 1 & 1 \\ 1 & 0 \end{bmatrix}, \begin{bmatrix} 1 & 1 \\ 1 & 1 \end{bmatrix} \right)$$

4. For each of the following vector spaces, bases and linear transformations, illustrate Theorem 4.17. Note that often the solutions to Exercise 3 can assist.

   a. $V = \mathbb{R}^3$ with basis $B_1 = ([1, 1, 0], [1, 0, 1], [0, 1, 1])$ and $W = P_3$ with basis $B_2 = (1 - x, x - x^2, 1 + x + x^2)$.

b. $V = D_{22}$ with basis $B_1 = (E_{11},\ E_{11} + E_{22})$ and $W = M_{22}$ with basis

$$B_2 = \left(\begin{bmatrix} 1 & 0 \\ 0 & 0 \end{bmatrix}, \begin{bmatrix} 1 & 1 \\ 0 & 0 \end{bmatrix}, \begin{bmatrix} 1 & 1 \\ 1 & 0 \end{bmatrix}, \begin{bmatrix} 1 & 1 \\ 1 & 1 \end{bmatrix} \right)$$

5. Compute the annihilator of a subspace of a vector space in each of the following settings:

a. $U = span\left(\begin{bmatrix} 1 & 0 \\ 0 & 2 \end{bmatrix} \right)$ in $V = D_{22}$.

b. $U = span\left(\begin{bmatrix} 1 & 0 \\ 0 & 2 \end{bmatrix} \right)$ in $V = U_{22}$.

c. $U = span\left(\begin{bmatrix} 1 & 0 \\ 0 & 2 \end{bmatrix} \right)$ in $V = M_{22}$.

6. When we use standard bases for $\mathbb{R}^n$, P_n and M_{mn}, what is another name for the characteristic functionals? Justify your answer.

7. Prove for a given subspace U of a vector space V that U° is a subspace of V^*.

8. Prove that if $U \subseteq W$ are subspaces of a vector space V, then $W^\circ \subseteq U^\circ$.

9. Prove that if U is a subspace of a finite dimensional vector space V and $v \in V - U$, then there exists a $\phi \in U^\circ$ such that $\phi(v) \neq 0$.

10. Let U, W be subspaces of a vector space V. Prove that $U = W$ iff $U^\circ = W^\circ$.

11. For $0 \in V$ prove that the element $0^{**} \in V^{**}$ is the zero map from V^* to $\mathbb{R}$.

12. Given a finite dimensional vector space V, if $v \in V$ and $\phi(v) = 0$ for all $\phi \in V^*$, then $v = 0$.

    (hint: for the non-trivial direction make use of a basis for V and the corresponding dual basis for V^*)

13. Prove the last two parts of Theorem 4.19.

CHAPTER 5

Inner Product Spaces

IN THIS CHAPTER, real inner product space are covered which generalizes dot product. Orthogonality, the Gram-Schmidt Process, and best approximation are introduced. In Section 5.1, a real inner product space is defined and examples are given. In Section 5.2, perpendicular vectors are defined and a method for producing an orthonormal basis from any given basis is given. The annihilator of a set is given which will be used in later sections of the chapter. In Section 5.3, orthogonal matrices are defined and all 2×2 orthogonal matrices are classified as rotations or reflections of two dimensional space. Section 5.4, is another numerical methods application called the QR factorization. In Section 5.5, is it shown how any square matrix is similar to a triangular matrix over the complex numbers. In Section 5.6, the notions of component and projection are reintroduced as well as the orthogonal projection of a vector onto a subspace. These play a crucial role in best approximation. In Section 5.7, it is shown that any symmetric matrix can be diagonalized by eigenvectors. In Section 5.8, the singular value decomposition is presented. In Section 5.9, the technique of best approximation is applied to overdetermined systems of linear equations, a best fitting polynomial, a best fitting hyperplane, underdetermined systems of linear equations and to continuous functions.

5.1 DEFINITION, EXAMPLES, AND PROPERTIES

A (Real) Inner Product Space is simply a generalization of the vector space $\mathbb{R}^n$ together with its dot product. Our goal is to generalize the definition of dot product and associate it with the other classic examples of a vector space. First, we begin with the definition:

Definition 5.1 *A* **(real) inner product space** *is a vector space V together with a binary map (called the* **inner product***) from $V \times V$ to $\mathbb{R}$ sending (v, w) to $(v|w)$ satisfying the following properties:*

i. $(v|v) > 0$, *for $0 \neq v \in V$.*

ii. $(v|w) = (w|v)$, *for all $v, w \in V$.*

iii. $(u + v|w) = (u|w) + (v|w)$, *for all $u, v, w \in V$.*

DOI: 10.1201/9781003217794-5

iv. $(av|w) = a(v|w)$, *for all* $v, w \in V$ *and* $a \in \mathbb{R}$.

If $\dim(V) < \infty$, *then* V *is called a* **Euclidean space**.

There are names for these properties. Property i says that $(*|*)$ is **positive definite**, ii is the **symmetric** property, while iii and iv combined say that $(*|*)$ is **bilinear**. An inner product is called **non-degenerate** if $(v|w) = 0$ for all $w \in V$, then $v = 0$. It is left as an exercise to show that an inner product is necessarily non-degenerate.

Example 5.1 *Listed below are some standard examples of real inner product spaces.*

1. *Let* $V = \mathbb{R}^n$ *and define*

$$([a_1, \ldots, a_n]|[b_1, \ldots, b_n]) = a_1 b_1 + \cdots + a_n b_n.$$

   This is the standard dot product we have seen in Section 1.2. In that section we proved that the dot product satisfied properties i–iv.

2. *Let* $V = \mathcal{C}[a, b]$ *(continuous real-valued functions on the interval* $[a, b]$*) and define* $(f|g) = \int_a^b f(x)g(x) \, dx$. *One can easily check that properties i–iv are satisfied here.*

3. *Let* $V = M_n$ *and define* $(A|B) = tr(B^T A)$. *Again, one can check that properties i–iv are satisfied.*

Example 5.2 *Let* $V = \mathbb{R}^2$ *and define*

$$([a_1, a_2]|[b_1, b_2]) = a_1 b_1 - a_1 b_2 - a_2 b_1 + 4 a_2 b_2.$$

With a little patience, one can easily verify properties ii–iv, however it is **not** *an inner product, since it does not satisfy property i.*

Example 5.3 *Let* $V = \{0\}$, *the trivial vector space, with* $(0|0) = 0$. *This is aptly called the* **trivial** *inner product space. It is easy to check that properties i–iv are satisfied in this case.*

Example 5.4 *Let* W *be an inner product space with inner product* $(*|*)$ *and suppose* $T \in L(V, W)$ *for another vector space* V. *Define the binary map* $(u|v)_V = (T(u)|T(v))$ *for* $u, v \in V$. *One can verify that this maps makes* V *into an inner product space.*

Listed below are some properties of an inner product space V as a consequence of its definition.

iii'. $(w|u + v) = (w|u) + (w|v)$, for all $u, v, w \in V$.

iv'. $(w|av) = a(w|v)$, for all $v, w \in V$ and $a \in \mathbb{R}$.

v. $(v|0) = (0|v) = 0$ for all $v \in V$.

Properties iii' and iv' follow immediately from property ii and property v follows from iv' and ii.

Definition 5.2 *Let V be an inner product space with inner product $(*|*)$. The* **norm** *(or* **length,** *or* **magnitude***) of $v \in V$, written*

$$|v| = \sqrt{(v|v)}.$$

A **unit** *vector v has the property that $|v| = 1$. The* **distance** *between two vectors $v, w \in V$ is defined to be $|v - w|$.*

Example 5.5 *Referring to the inner products defined above,*

1. $|[2, -3, 1]| = \sqrt{(2)^2 + (-3)^2 + (1)^2} = \sqrt{14}.$

2. *For $[a, b] = [0, 1]$, $|e^x| = \sqrt{\int_0^1 e^{2x} \, dx} = \sqrt{\frac{1}{2}(e^2 - 1)}.$*

3. *For $A = \begin{bmatrix} 2 & 1 \\ -1 & 1 \end{bmatrix}$,*

$$|A| = \sqrt{tr(A^T A)} = \sqrt{tr\left(\begin{bmatrix} 5 & 1 \\ 1 & 2 \end{bmatrix}\right)} = \sqrt{7}.$$

One can check that $|A|^2$ is simply the sum of the squares of the entries in A.

Notice that there is no notational confusion between norm and absolute value, since the context (i.e. scalar or vector) dictates what $| * |$ signifies (however, $|A|$ can be confused with determinant).

As in Section 1.2, the vector $\frac{1}{|v|} v$ is a unit vector called the **normalization** of the vector $v \in V$ where v is necessarily non-zero.

We summarize some properties pertaining to the above definitions.

Lemma 5.1 *Let V be an inner product space with inner product $(*|*)$ with $v, w \in V$ and $a \in \mathbb{R}$. Then*

1. $|v| \geq 0$. *Furthermore, $|v| = 0$ iff $v = 0$.*

2. $|av| = |a||v|.$

3. *(Polarization Identity) $(v|w) = \frac{1}{4}(|v + w|^2 - |v - w|^2).$*

4. *(Parallelogram Law) $|v + w|^2 + |v - w|^2 = 2|v|^2 + 2|w|^2.$*

5. *(Cauchy-Schwarz Inequality) $|(v|w)| \leq |v||w|.$*

6. *(Triangle Inequality)* $|v + w| \leq |v| + |w|$.

7. *(Right Triangle Law)* $(v|w) = 0$ *iff* $|v + w|^2 = |v|^2 + |w|^2$.

Proof 5.1 *To prove 1, use properties i and v. To prove 2, notice that*

$$|av|^2 = (av|av) = a^2(v|v) = a^2|v|^2.$$

Now take the square root of both sides to get the desired result. We can prove 3 and 4 at the same time. Note that, using the bilinear and symmetric property,

$$|v + w|^2 = (v + w|v + w) = (v|v) + (v|w) + (w|v) + (w|w) = |v|^2 + |w|^2 + 2(v|w).$$

Let's call the above equation $*$. Replacing w by $-w$ in $*$ yields

$$|v - w|^2 = |v|^2 + |-w|^2 + 2(v|-w) = |v|^2 + |w|^2 - 2(v|w).$$

Let's call this second equation $**$. Adding the two equations $*$ and $**$ yields 4, while subtracting $**$ from $*$ and multiplying the result by $\frac{1}{4}$ yields 3.

To prove 5, we first need to show that we have the result for unit vectors, i.e. for $v, w \in V$ unit vectors, we have

$$|(v|w)| \leq 1.$$

Indeed, using $*$,

$$0 \leq |v + w|^2 = |v|^2 + |w|^2 + 2(v|w) = 2 + 2(v|w),$$

which simplifies to $(v|w) \geq -1$. Similarly, using $**$,

$$0 \leq |v - w|^2 = |v|^2 + |w|^2 - 2(v|w) = 2 - 2(v|w),$$

which simplifies to $(v|w) \leq 1$. Putting these two results together yields

$$-1 \leq (v|w) \leq 1 \qquad or \qquad |(v|w)| \leq 1.$$

Now we can prove 5 for arbitrary $v, w \in V$. Since $v/|v|$ and $w/|w|$ are unit vectors, we have $|(v/|v|, w/|w|)| \leq 1$. Then

$$\frac{(v|w)}{|v||w|} \leq 1, \qquad and \ so \qquad (v|w) \leq |v||w|.$$

To prove 6, using $*$ again and what we just proved,

$$|v + w|^2 = |v|^2 + |w|^2 + 2(v|w) \leq |v|^2 + |w|^2 + 2|v||w| = (|v| + |w|)^2.$$

Now take the square root of both sides of the above inequality to get the desired result.

*Part 7 follows easily from equation *.* □

EXERCISES

1. Compute the inner product in each of the following examples:

   a. In Example 5.1.1, compute $(u|v)$, where $u = [-1, 2, 4]$ and $v = [-2, 0, 5]$.

   b. In Example 5.1.2, compute $(f|g)$, where $f(x) = x$, $g(x) = x^2$ and $[a, b] = [1, 2]$.

   c. In Example 5.1.3, compute $(A|B)$, where

   $$A = \begin{bmatrix} 1 & -2 \\ 3 & 0 \end{bmatrix} \text{ and } B = \begin{bmatrix} 0 & -1 \\ 2 & -3 \end{bmatrix}.$$

   d. In Example 5.2, compute $(u|v)$, where $u = [1, -2]$ and $v = [3, -1]$.

   e. In Example 5.4, let $W = \mathbb{R}^2$ with the inner product from Example 2, and let $T \in L(\mathbb{R}^2)$ be defined by $T[a, b] = [a + b, a - b]$ (so $V = \mathbb{R}^2$). Compute $(u|v)_V$, where $u = [1, -2]$ and $v = [3, -1]$.

2. Compute the magnitude of each vector:

   a. In Example 5.1.1, compute $||[-1, 2, 0]||$.

   b. In Example 5.1.2, compute $|\sec x|$ on $[a, b] = [0, \pi/4]$.

   c. In Example 5.1.3, compute $\left\Vert \begin{bmatrix} 1 & 2 \\ -1 & 0 \end{bmatrix} \right\Vert$.

   d. In Example 5.4, let $W = \mathbb{R}^2$ with the inner product from Example 2, and let $T \in L(\mathbb{R}^2)$ be defined by $T[a, b] = [a + b, a - b]$. Compute $||[1, -2]||_V$.

3. Prove that each of the following examples is an inner product space:

   a. Example 5.1.2.

   b. Example 5.1.3.

   c. Example 5.4.

4. Decide whether or not each of the following functions is an inner product.

   a. For $V = \mathbb{R}^2$ define $([a_1, b_1] | [a_2, b_2]) = |a_1 b_1| + |a_2 b_2|$.

   b. For $V = P$ (polynomials) define $(p|q) = \int_0^\infty p(x)q(x)e^{-x}\, dx$.

5. Prove that Example 5.2 satisfies properties ii - iv, but **not** property i.

6. Verify Properties iii', iv' and v using Properties i–iv.

7. Prove that for each $0 \neq v \in V$ an inner product space, $\frac{1}{|v|}v$ is a unit vector.

8. Fix a vector $v \in V$ and define a map $v^* : V \to \mathbb{R}$ by $v^*(w) = (w|v)$ for all $w \in V$. Prove that v^* is a linear transformation.

9. Prove that for an inner product space V and $w \in V$, if $(v|w) = 0$ for all $v \in V$, then $w = 0$.

10. Prove Lemma 5.1.1 and 5.1.7.

5.2 ORTHOGONAL AND ORTHONORMAL

The geometric notion of two vectors being perpendicular (or orthogonal) can be extended from $\mathbb{R}^n$ with dot product to arbitrary inner product spaces. We saw in Section 1.2 that two vectors in $\mathbb{R}^n$ are perpendicular iff their dot product is zero. This is exactly how we extend this notion.

Definition 5.3 *Let V be an inner product space with inner product $(*|*)$. Two vectors $v, w \in V$ are **orthogonal**, written $v \perp w$, if $(v|w) = 0$.*

Example 5.6 *Referring to the inner products defined in Section 5.1,*

1. *(Example 5.1.1)* $[1, -2, -1] \perp [1, 1, -1]$.

2. *(Example 5.1.2)* $\sin x \perp \cos x$ *on* $[a, b] = [0, \pi]$.

3. *(Example 5.1.3)* $\begin{bmatrix} 2 & 1 \\ 1 & -1 \end{bmatrix} \perp \begin{bmatrix} 1 & -2 \\ 2 & 2 \end{bmatrix}$.

Below are listed some simple properties of orthogonality which we leave as an exercise.

Lemma 5.2 *Let V be an inner product space with inner product $(*|*)$ and $v, w \in V$. Then*

i. $v \perp w$ iff $w \perp v$.

ii. $v \perp v$ iff $v = 0$.

iii. If $v_1, \ldots, v_n \in V$ and $v \perp v_i$ for $i = 1, \ldots, n$, then $v \perp (a_1 v_1 + \cdots + a_n v_n)$ for all $a_1, \ldots, a_n \in \mathbb{R}$.

iv. If $v_1, \ldots, v_n \in V$ are non-zero and pairwise orthogonal (i.e. $v_i \perp v_j$ for all $i \neq j$), then $v_1, \ldots, v_n$ are linearly independent.

Proof 5.2 *Parts i-iii are left as exercises. To prove iv, we suppose for $a_1, \ldots, a_n \in \mathbb{R}$ that $a_1 v_1 + \cdots + a_n v_n = 0$. Then, for all $j = 1, \ldots, n$,*

$$0 = (0|v_j) = \left(\sum_{i=1}^{n} a_i v_i \middle| v_j \right) = \sum_{i=1}^{n} a_i(v_i|v_j) = a_j(v_j|v_j) = a_j|v_j|^2.$$

Since $v_j \neq 0$, it must be that $a_j = 0$ (for all $j = 1, \ldots, n$). □

Definition 5.4 *Let V be an inner product space.*

1. *A set of vectors in V is **orthogonal** if they are pairwise orthogonal.*

2. *A set of vectors in V is **orthonormal** if they are orthogonal and of unit length (i.e. all unit vectors).*

Example 5.7 *The standard basis for $\mathbb{R}^n$ and M_{nn} are orthonormal in their respective inner product spaces (Examples 5.1.1 and 5.1.2).*

The next theorem is very useful, for it gives a constructive proof of the existence of an orthonormal basis for any Euclidean space. This algorithm for constructing the orthonormal basis is called the **Gram-Schmidt Process**.

Theorem 5.1 *If V is a non-trivial Euclidean space, then V has an orthonormal basis.*

Proof 5.3 *Start with any basis for V, say $v_1, \ldots, v_n$. We will first we find an orthogonal basis for V which we shall call $w_1, \ldots, w_n$.*

Set $w_1 = v_1$. Inductively assume that we have found non-zero orthogonal vectors $w_1, \ldots, w_k$ (for $1 \leq k < n$) with $\mathrm{span}(w_1, \ldots, w_k) = \mathrm{span}(v_1, \ldots, v_k)$. Then

$$w_{k+1} = v_{k+1} - \sum_{i=1}^{k} \frac{(v_{k+1}|w_i)}{|w_i|^2} w_i.$$

First, note that $w_{k+1} \neq 0$, otherwise because of how w_{k+1} is defined it would follow that $v_{k+1} \in \mathrm{span}(w_1, \ldots, w_k) = \mathrm{span}(v_1, \ldots, v_k)$, contradicting the linear independence of $v_1, \ldots, v_n$. Second, $w_{k+1} \perp w_j$ for $j = 1, \ldots, k$, since

$$(w_j|w_{k+1}) = (w_j|v_{k+1}) - \sum_{i=1}^{k} \frac{(v_{k+1}|w_i)}{|w_i|^2}(w_j|w_i)$$

$$= (w_j|v_{k+1}) - \frac{(v_{k+1}|w_j)}{|w_j|^2}(w_j|w_j) = (w_j|v_{k+1}) - (v_{k+1}|w_j) = 0.$$

Finally, $\mathrm{span}(w_1, \ldots, w_{k+1}) = \mathrm{span}(v_1, \ldots, v_{k+1})$, since $w_{k+1} \in \mathrm{span}(w_1, \ldots, w_k, v_{k+1}) = \mathrm{span}(v_1, \ldots, v_k, v_{k+1})$ and $v_{k+1} \in \mathrm{span}(w_1, \ldots, w_{k+1})$, since

$$v_{k+1} = \sum_{i=1}^{k} \frac{(v_{k+1}|w_i)}{|w_i|^2} w_i + w_{k+1}.$$

Now to get an orthonormal basis for V simply normalize the vectors $w_1, \ldots, w_n$.

□

We want to reiterate again that the proof of the existence of an orthonormal basis is indeed constructive. As such, in the examples below we will apply this algorithm for constructing an orthonormal basis.

Example 5.8 *Let's start we a basic example in a familiar setting. Set $V = \mathbb{R}^3$, standard Euclidean space. Now the standard basis is already an orthonormal basis, so let's begin with a basis which is not orthonormal, such as $v_1 = [1, 0, 0]$, $v_2 = [1, 1, 0]$, $v_3 = [1, 1, 1]$. Set $w_1 = v_1 = [1, 0, 0]$ and*

$$w_2 = v_2 - \left(\frac{v_2 \cdot w_1}{w_1 \cdot w_1} \right) w_1 = [1, 1, 0] - \left(\frac{[1, 1, 0] \cdot [1, 0, 0]}{[1, 0, 0] \cdot [1, 0, 0]} \right) [1, 0, 0]$$

$$= [1, 1, 0] - \left(\frac{1}{1} \right) [1, 0, 0] = [0, 1, 0].$$

$$w_3 = v_3 - \left(\frac{v_3 \cdot w_1}{w_1 \cdot w_1} \right) w_1 - \left(\frac{v_3 \cdot w_2}{w_2 \cdot w_2} \right) w_2$$

$$= [1, 1, 1] - \left(\frac{[1, 1, 1] \cdot [1, 0, 0]}{[1, 0, 0] \cdot [1, 0, 0]} \right) [1, 0, 0] - \left(\frac{[1, 1, 1] \cdot [0, 1, 0]}{[0, 1, 0] \cdot [0, 1, 0]} \right) [0, 1, 0]$$

$$= [1, 1, 1] - [1, 0, 0] - [0, 1, 0] = [0, 0, 1].$$

Normally, we would then normalize vectors w_1, w_2 and w_3, however in this example they are already normalized since they are the standard orthonormal basis for $\mathbb{R}^3$.

Example 5.9 *Take $V = P_2$ with $(p|q) = \int_0^1 p(x)q(x) \, dx$. We start with the standard basis $v_1 = 1$, $v_2 = x$, $v_3 = x^2$. Set $w_1 = 1$.*

$$w_2 = v_2 - \frac{(v_2|w_1)}{(w_1|w_1)} w_1 = x - \frac{(x|1)}{(1|1)} \cdot 1 = x - \frac{1/2}{1} = -\frac{1}{2} + x.$$

$$w_3 = v_3 - \frac{(v_3|w_1)}{(w_1|w_1)} w_1 - \frac{(v_3|w_2)}{(w_2|w_2)} w_2 = x^2 - \frac{(x^2|1)}{(1|1)} - \frac{(x^2| -\frac{1}{2} + x)}{(-\frac{1}{2} + x| -\frac{1}{2} + x)} \cdot (-\frac{1}{2} + x)$$

$$= x^2 - \frac{1}{3} + x - \frac{1/12}{1/12}(-\frac{1}{2} + x) = \frac{1}{6} + x^2.$$

As stated in the proof above, we now need to normalize w_1, w_2 and w_3.

$$|w_1| = \sqrt{(w_1|w_1)} = \sqrt{\int_0^1 dx} = 1.$$

$$|w_2| = \sqrt{(w_2|w_2)} = \sqrt{\int_0^1 (-\frac{1}{2} + x)^2 \, dx} = 1/\sqrt{12}.$$

$$|w_3| = \sqrt{(w_3|w_3)} = \sqrt{\int_0^1 (\frac{1}{6} + x^2)^2 \, dx} = \sqrt{61/180}.$$

Hence, the orthonormal basis is then $w_1/|w_1|$, $w_2/|w_2|$, $w_3/|w_3|$. More precisely, it is

$$1, -\sqrt{3} + \sqrt{12}x, \ \sqrt{5} + \sqrt{\frac{180}{61}}x^2.$$

The next result is easy to verify, so we leave it as an exercise. What the lemma says basically is that any finite dimensional inner product space is essentially Euclidean space with the standard dot product.

Lemma 5.3 *Let $B = (u_1, \ldots, u_n)$ be an ordered orthonormal basis for a Euclidean space V. For $u, v \in V$, suppose that $[u]_B = [a_1, \ldots, a_n]$ and $[v]_B = [b_1, \ldots, b_n]$. Then $(u|v) = a_1 b_1 + \cdots + a_n b_n$.*

Example 5.10 *Let's illustrate Lemma 5.3. Consider the example above in which we found an orthonormal basis for P_2, namely,*

$$B = \left(1, -\sqrt{3} + \sqrt{12}x, \ \sqrt{5} + \sqrt{\frac{180}{61}}x^2\right).$$

We shall compute $(1|x)$ in two different ways. On the one hand, $(1|x) = \int_0^1 x \, dx = \frac{1}{2}$. On the other hand, according to Lemma 5.3, first we need to find the coordinates of 1 and x with respect to B. Clearly, $[1]_B = [1, 0, 0]$. To get $[x]_B$ we row reduce

$$\begin{bmatrix} 1 & -\sqrt{3} & \sqrt{5} & 0 \\ 0 & \sqrt{12} & 0 & 1 \\ 0 & 0 & \sqrt{\frac{180}{61}} & 0 \end{bmatrix} \quad \text{which reduces to} \quad \begin{bmatrix} 1 & 0 & 0 & \frac{1}{2} \\ 0 & 1 & 0 & \frac{\sqrt{3}}{6} \\ 0 & 0 & 1 & 0 \end{bmatrix},$$

so that $[x]_B = [1/2, \sqrt{3}/6, 0]$. Therefore, $(1|x) = (1)(1/2) + (0)(\sqrt{3}/6) + (0)(0) = \frac{1}{2}$, as expected.

Definition 5.5 *Let X be a subset of an inner product space V. We define*

$$X^\perp = \{ \, v \in V \mid (v|u) = 0 \text{ for all } u \in X \, \},$$

*called the **annihilator** of X in V.*

Example 5.11 *Let's compute a simple annihilator to get some intuition for the idea. Consider the standard inner product space on $\mathbb{R}^3$ and set $X = \{\hat{\imath}, \hat{\jmath}\}$. We compute $X^\perp$. Note that $v = [a, b, c] \in X^\perp$ iff $v \cdot \hat{\imath} = 0$ and $v \cdot \hat{\jmath} = 0$. These two equations simplify to $a = 0$ and $b = 0$. Hence,*

$$X^\perp = \{[0, 0, c] \mid c \in \mathbb{R}\} = \text{span}(\hat{k}).$$

Now this makes sense, since $\text{span}(\hat{\imath}, \hat{\jmath})$ is the xy-plane and the collection of vectors perpendicular to the xy-plane is the z-axis.

Example 5.12 *Now let's compute an annihilator in a less familiar setting. Consider the inner product space M_{22} with inner product $(A|B) = \text{tr}(B^T A)$. Let $X = \{A, B\}$, where*

$$A = \begin{bmatrix} 1 & 1 \\ 0 & 0 \end{bmatrix} \quad and \quad B = \begin{bmatrix} 0 & 0 \\ 1 & -1 \end{bmatrix}.$$

$C = \begin{bmatrix} a & b \\ c & d \end{bmatrix} \in X^\perp$ *iff* $(C|A) = 0$ *and* $(C|B) = 0$, *i.e.* $\text{tr}(A^T C) = 0$ *and* $\text{tr}(B^T C) = 0$. *Equivalently,*

$$\text{tr}\left(\begin{bmatrix} 0 & 1 \\ 0 & 1 \end{bmatrix}\begin{bmatrix} a & b \\ c & d \end{bmatrix}\right) = 0 \quad and \quad \text{tr}\left(\begin{bmatrix} -1 & 0 \\ 1 & 0 \end{bmatrix}\begin{bmatrix} a & b \\ c & d \end{bmatrix}\right) = 0.$$

Equivalently,

$$\text{tr}\left(\begin{bmatrix} c & d \\ c & d \end{bmatrix}\right) = 0 \quad and \quad \text{tr}\left(\begin{bmatrix} -a & -b \\ a & b \end{bmatrix}\right) = 0.$$

Equivalently, $c + d = 0$ and $-a + b = 0$. Therefore, $d = -c$ and $b = a$ with a and c arbitrary. Hence,

$$X^\perp = \left\{\begin{bmatrix} a & a \\ c & -c \end{bmatrix} : a, c \in \mathbb{R}\right\}.$$

We will write the short-hand $X^{\perp\perp}$ to represent $(X^\perp)^\perp$, $X^{\perp\perp\perp}$ for $\left((X^\perp)^\perp\right)^\perp$, etc. We now list some properties of the annihilator.

Lemma 5.4 *If X and Y are subsets of an inner product space V, then*

  i. $X^\perp$ is a subspace of V.

  ii. $X \subseteq X^{\perp\perp}$.

  iii. If $X \subseteq Y$, then $Y^\perp \subseteq X^\perp$.

  iv. $X^{\perp\perp\perp} = X^\perp$.

v. $X \cap X^\perp \subseteq \{\,0\,\}$.

vi. $X^\perp = \mathrm{span}(X)^\perp$.

Proof 5.4 *Part i is left as an exercise. To show part ii, take any $u \in X$. We know that for any $v \in X^\perp$ that $(u|v) = 0$. But this means that $u \in (X^\perp)^\perp$, and so $X \subseteq X^{\perp\perp}$. We leave part iii as an exercise. Part iv follows from parts ii and iii. To prove part v, assuming X is non-empty, take a $v \in X \cap X^\perp$. Since $v \in X$ and $v \in X^\perp$, this implies that $(v|v) = 0$. For an inner product, this means that $v = 0$. To prove vi, since $X \subseteq \mathrm{span}(X)$, by part iii, $\mathrm{span}(X)^\perp \subseteq X^\perp$. For the reverse inclusion, take a $v \in X^\perp$. We show that for any $u \in \mathrm{span}(X)$ that $(v|u) = 0$, and so $v \in \mathrm{span}(X)^\perp$ and we're done. To do this, express $u = a_1 u_1 + \cdots + a_k u_k$ for some $a_1, \ldots, a_k \in \mathbb{R}$ and $u_1, \ldots, u_k \in X$. Then*

$$(v|u) = (v|a_1 u_1 + \cdots + a_k u_k) = a_1(v|u_1) + \cdots + a_k(v|u_k)$$

$$= a_1 \cdot 0 + \cdots + a_k \cdot 0 = 0.$$

Hence, $v \in \mathrm{span}(X)^\perp$. □

The following result will be useful later in the chapter:

Lemma 5.5 *For $A \in M_{mn}$,*

i. $(\mathrm{rowsp}(A))^\perp = \mathrm{nullsp}(A)$.

ii. $(\mathrm{colsp}(A))^\perp = \mathrm{nullsp}(A^T)$.

Proof 5.5 *Let $r_1, \ldots, r_m$ be the rows of A and $c_1, \ldots, c_n$ the columns. In order to prove part i, we first take $v \in \mathrm{rowsp}(A)^\perp$ and show it is in $\mathrm{nullsp}(A)$. Since $v \in \mathrm{rowsp}(A)^\perp$, this means that $v \cdot u = 0$ for any $u \in \mathrm{rowsp}(A)$. In particular, $v \cdot r_i = 0$ for all $i = 1, \ldots, m$. Now, by definition of matrix multiplication,*

$$Av = \begin{bmatrix} r_1 \cdot v \\ \vdots \\ r_m \cdot v \end{bmatrix} = \begin{bmatrix} 0 \\ \vdots \\ 0 \end{bmatrix} = 0.$$

Hence, $v \in \mathrm{nullsp}(A)$. Now take $v \in \mathrm{nullsp}(A)$. As above, we have $v \cdot r_i = 0$ for all $i = 1, \ldots, m$. Therefore $v \in \{r_1, \ldots, r_m\}^\perp$. Since $\mathrm{rowsp}(A) = \mathrm{span}(r_1, \ldots, r_m)$, by Lemma 5.4.vi, $v \in \mathrm{rowsp}(A)^\perp$.

For part ii, replace A by A^T in the statement of part i to get $\mathrm{rowsp}(A^T)^\perp = \mathrm{nullsp}(A^T)$ and since $\mathrm{rowsp}(A^T) = \mathit{collsp}(A)$, the result follows. □

EXERCISES

1. Verify that each pair of vector mentioned in Example 5.6 at the beginning of this section are indeed perpendicular.

2. Consider the inner product space $\mathbb{R}^3$ in Example 5.1.1.

   a. Find an orthonormal basis for $\mathbb{R}^3$ starting with the basis $[1,1,0], [1,0,1], [0,1,1]$.

   b. Compute $([2,1,1] | [1,1,2])$ in two ways: Directly from the definition of the inner product and then by using Lemma 5.3 and part a.

3. Consider the inner product space $\mathcal{C}[0,6]$ in Example 5.1.2.

   a. Find an orthonormal basis for P_2 starting with the basis $1, 1+x, 1+x^2$.

   b. Compute $(2+x | -x+x^2)$ in two ways: Directly from the definition of the inner product and then by using Lemma 5.3 and part a.

4. Consider the inner product on U_{22} in Example 5.1.3.

   a. Find an orthonormal basis for $U_2(\mathbb{R})$ starting with the following basis:

   $$\begin{bmatrix} 1 & 0 \\ 0 & 0 \end{bmatrix}, \begin{bmatrix} 0 & 1 \\ 0 & 1 \end{bmatrix}, \begin{bmatrix} 1 & 0 \\ 0 & 1 \end{bmatrix}$$

   b. Compute $\left(\begin{bmatrix} 0 & 1 \\ 0 & 0 \end{bmatrix} \middle| \begin{bmatrix} 2 & 1 \\ 0 & 2 \end{bmatrix} \right)$ in two ways: Directly from the definition of the inner product and then by using Lemma 5.3 and part a.

5. Consider the inner product on U_{22} in Example 5.1.3. Find an orthonormal basis for $U_2(\mathbb{R})$ starting with the following basis:

   $$\begin{bmatrix} 1 & 1 \\ 0 & 0 \end{bmatrix}, \begin{bmatrix} 0 & 1 \\ 0 & 1 \end{bmatrix}, \begin{bmatrix} 1 & 0 \\ 0 & 1 \end{bmatrix}$$

6. For each of the following inner product spaces V (with inner product defined in Example 5.1) and subset X of V, compute $X^{\perp}$:

   a. $V = \mathbb{R}^3$ and $X = \{ [1,3,-1], [2,2,1] \}$.

   b. $V = P_2$ and $X = \{ 1, x \}$ with $[a,b] = [0,1]$.

   c. $V = P_2$ and $X = \{ 1+x, 1+x^2 \}$ with $[a,b] = [0,1]$.

   d. $V = M_{22}$ and $X = \left\{ \begin{bmatrix} 1 & -1 \\ 1 & 0 \end{bmatrix}, \begin{bmatrix} 2 & 1 \\ 0 & -2 \end{bmatrix} \right\}$.

7. Prove that the standard basis $e_1, e_2, \ldots, e_n$ for $\mathbb{R}^n$ with the standard dot product is an orthonormal basis.

8. Prove that the standard basis $\{ E_{ij} \mid 1 \le i, j \le n \}$ for M_{nn} with the inner product defined in Example 5.1.3 is an orthonormal basis.

9. Prove each of the following statement for an inner product space V with inner product $(\ |\ )$.

   a. $(u + v) \perp (u - v)$ iff $|u| = |v|$.

   b. $u \perp v$ iff $|u| \le |u + av|$ for all $a \in \mathbb{R}$.

10. Prove Lemma 5.3.

11. Prove Lemma 5.2.i–iii.

12. Prove Lemma 5.4.i, iii and iv.

5.3 ORTHOGONAL MATRICES

In this section we study a class of square matrices and their corresponding mappings with a special property we call *orthogonal*. A special case of these mappings and matrices occurs in $\mathbb{R}^2$ in which the mappings turn out to be either rotations or reflections of $\mathbb{R}^2$.

5.3.1 Definition and Results

Definition 5.6 *A matrix $A \in M_{nn}$ is* **orthogonal** *if $A^T A = I$.*

Example 5.13 *Here we list several examples of orthogonal matrices.*

1. $A = \begin{bmatrix} \sqrt{3}/2 & -1/2 \\ 1/2 & \sqrt{3}/2 \end{bmatrix}$ *is orthogonal, since*

$$\begin{bmatrix} \sqrt{3}/2 & -1/2 \\ 1/2 & \sqrt{3}/2 \end{bmatrix}^T \begin{bmatrix} \sqrt{3}/2 & -1/2 \\ 1/2 & \sqrt{3}/2 \end{bmatrix}$$

$$= \begin{bmatrix} \sqrt{3}/2 & 1/2 \\ -1/2 & \sqrt{3}/2 \end{bmatrix} \begin{bmatrix} \sqrt{3}/2 & -1/2 \\ 1/2 & \sqrt{3}/2 \end{bmatrix} = \begin{bmatrix} 1 & 0 \\ 0 & 1 \end{bmatrix}.$$

We shall see that A represents a $30°$ rotation of the xy-plane.

2. $B = \begin{bmatrix} 1/2 & \sqrt{3}/2 \\ \sqrt{3}/2 & -1/2 \end{bmatrix}$ *is orthogonal, since*

$$\begin{bmatrix} 1/2 & \sqrt{3}/2 \\ \sqrt{3}/2 & -1/2 \end{bmatrix}^T \begin{bmatrix} 1/2 & \sqrt{3}/2 \\ \sqrt{3}/2 & -1/2 \end{bmatrix}$$

$$= \begin{bmatrix} 1/2 & \sqrt{3}/2 \\ \sqrt{3}/2 & -1/2 \end{bmatrix} \begin{bmatrix} 1/2 & \sqrt{3}/2 \\ \sqrt{3}/2 & -1/2 \end{bmatrix} = \begin{bmatrix} 1 & 0 \\ 0 & 1 \end{bmatrix}.$$

We shall see that B represents a reflection of the xy-plane across a line through the origin which makes a 30° angle with the positive x-axis.

By the end of this section we will classify all 2×2 real orthogonal matrices. We remark that $A^T A = I$ is equivalent to $AA^T = I$ (since A^T is the inverse of A). Furthermore, matrix orthogonality is equivalent to the rows or columns of the matrix forming an orthonormal basis for $\mathbb{R}^n$ (consider matrix multiplication in terms of inner product). We leave the following results as an exercise:

Theorem 5.2 *For $A \in M_{nn}$ the following are equivalent:*

  i. A is orthogonal.

  ii. A is invertible with inverse A^T.

  iii. A^T is orthogonal.

  iv. The rows (or columns) of A form an orthonormal basis for $\mathbb{R}^n$.

Definition 5.7 *Let V be a Euclidean space and $T \in L(V)$*

  *1. T is **orthogonal** if $(T(v)|T(w)) = (v|w)$ for all $v, w \in V$.*

  *2. T is an **isometry** if $|T(v)| = |v|$ for all $v \in V$.*

The definitions orthogonal and isometry will be shown below to be equivalent. We say that isometries *preserve length*. From this one can easily show that they also preserve the distance between vectors and the angle between vectors.

Theorem 5.3 *Let V be a Euclidean space and $T \in L(V)$.*

  1. T is orthogonal iff $(T(v_i)|T(v_j)) = (v_i|v_j)$ for some basis $v_1, \dots, v_n$ and all $1 \le i, j \le n$.

  2. T is orthogonal iff T is an isometry.

Proof 5.6 *The first item is left as an exercise. To prove the second item, assume first that T is orthogonal. Then for all $v \in V$ we have*

$$|T(v)| = \sqrt{(T(v)|T(v))} = \sqrt{(v|v)} = |v|.$$

Now assume that $|T(v)| = |v|$ for all $v \in V$. Then by Lemma 5.1.iii,

$$4(T(v)|T(w)) = |T(v) + T(w)|^2 - |T(v) - T(w)|^2$$

$$= |T(v + w)|^2 - |T(v - w)|^2 = |v + w|^2 - |v - w|^2 = 4(v|w).$$

$\square$

We now make the connection between orthogonal matrices and orthogonal linear operators on $\mathbb{R}^n$ (this can in fact be generalized to any Euclidean space).

Theorem 5.4 *Consider the Euclidean space $\mathbb{R}^n$ with the standard dot product. For any $A \in M_{nn}$ define $T \in L(\mathbb{R}^n)$ by $T(v) = Av$. Then A is orthogonal iff T is orthogonal.*

Proof 5.7 *First observe that for any matrix A and $v, w \in \mathbb{R}^n$,*

$$Av \cdot Aw = (Aw)^T Av = (A^T Aw)^T v = (A^T Aw) \cdot v = v \cdot (A^T Aw).$$

Assuming that A is orthogonal, using the above observation notice that

$$T(v) \cdot T(w) = Av \cdot Aw = v \cdot (A^T Aw) = v \cdot w.$$

Hence, T is orthogonal. Now assume that T is orthogonal, i.e. $v \cdot w = Av \cdot Aw$ for all $v, w \in \mathbb{R}^n$. Then by the first observation, for all $v, w \in \mathbb{R}^n$, $v \cdot w = Av \cdot Aw = v \cdot A^T Aw$ or equivalently $v \cdot (w - A^T Aw) = 0$ or equivalently $v \cdot (I - A^T A)w = 0$. Now, by Exercise 9 in Section 5.1, this implies that $(I - A^T A)w = 0$. Since this last statement is true for all $w \in \mathbb{R}^n$, by Exercise 7 in Section 1.5, $I - A^T A = 0$ and so $A^T A = I$. $\qquad\square$

Corollary 5.1 *A linear operator $T \in L(\mathbb{R}^n)$ is orthogonal iff $[T]_{ST}$ is orthogonal.*

Proof 5.8 *This follows immediately from the fact that $T(v) = [T]_{ST}\, v$.* $\qquad\square$

5.3.2 Application: Rotations and Reflections

We are ready to prove a classification theorem. We will classify 2×2 real orthogonal mappings and matrices. Let $T \in L(\mathbb{R}^2)$ be orthogonal. Set $T(\hat{\imath}) = [a, b]$ and $T(\hat{\jmath}) = [c, d]$. Then, using the fact that T is also an isometry,

$$1 = |\hat{\imath}| = |T(\hat{\imath})| = |[a, b]| = a^2 + b^2.$$

Similarly, $c^2 + d^2 = 1$. Since $\hat{\imath} \perp \hat{\jmath}$ and T is orthogonal, it follows that $T(\hat{\imath}) \perp T(\hat{\jmath})$ and so $ac + bd = 0$. Then

$$0 = (ac+bd)^2 = a^2c^2+2abcd+b^2d^2 = a^2c^2+2ac(-ac)+(1-a^2)(1-c^2) = 1-a^2-c^2.$$

Hence, $a^2 + c^2 = 1$. In a similar manner one can deduce that $b^2 + d^2 = 1$. But then $d^2 = 1 - c^2 = a^2$ and so $d = \pm a$. Similarly, one can show that $c = \pm b$.

Assume first that we are in the case that $d = a$. Then $a(c+b) = 0$, so either $a = 0$ (in which case $d = 0$ and so $c = \pm 1$) or $c = -b$.

In the case that $d = -a$. Then $a(c-b) = 0$, so either $a = 0$ (in which case $d = 0$ and so $c = \pm 1$) or $c = b$.

We can summarize these results as follows:

$$\text{Either} \quad [T]_{ST} = \begin{bmatrix} a & -b \\ b & a \end{bmatrix} \quad \text{or} \quad [T]_{ST} = \begin{bmatrix} a & b \\ b & -a \end{bmatrix}.$$

Notice that a matrix of either form listed above is orthogonal (check that $A^T A = I$) and so (by Corollary 5.1) we have the complete list of all orthogonal linear operators on $\mathbb{R}^2$. Notice also that

$$\begin{vmatrix} a & -b \\ b & a \end{vmatrix} = 1 \quad \text{while} \quad \begin{vmatrix} a & b \\ b & -a \end{vmatrix} = -1,$$

since $a^2 + b^2 = 1$. In addition $a^2 + b^2 = 1$ implies that $[a, b]$ lies on the unit circle, and so we can find $0 \le \theta < 2\pi$ such that $a = \cos\theta$ and $b = \sin\theta$. Hence,

$$\text{Either} \quad [T]_{ST} = \begin{bmatrix} \cos\theta & -\sin\theta \\ \sin\theta & \cos\theta \end{bmatrix} \quad \text{or} \quad [T]_{ST} = \begin{bmatrix} \cos\theta & \sin\theta \\ \sin\theta & -\cos\theta \end{bmatrix}.$$

Let's call these matrices Type I and Type II (respectively). We will now obtain a geometric interpretation of each of these two operators. We show now that the Type I matrices correspond to a counter-clockwise rotation of $\mathbb{R}^2$ through an angle of θ. Indeed, for any $v = [a, b] \in \mathbb{R}^2$, the smaller angle between v and $T(v)$ is

$$\cos^{-1}\left(\frac{v \cdot T(v)}{|v||T(v)|}\right) = \cos^{-1}\left(\frac{v \cdot T(v)}{|v|^2}\right)$$

$$= \cos^{-1}\left(\frac{[a, b] \cdot \begin{bmatrix} \cos\theta & -\sin\theta \\ \sin\theta & \cos\theta \end{bmatrix}\begin{bmatrix} a \\ b \end{bmatrix}}{|v|^2}\right)$$

$$= \cos^{-1}\left(\frac{[a, b] \cdot [a\cos\theta - b\sin\theta, a\sin\theta + b\cos\theta]}{a^2 + b^2}\right)$$

$$= \cos^{-1}\left(\frac{(a^2 + b^2)\cos\theta}{a^2 + b^2}\right) = \cos^{-1}(\cos\theta) = \begin{cases} \theta, & \text{if } 0 \le \theta \le \pi \\ 2\pi - \theta, & \text{if } \pi < \theta < 2\pi \end{cases}.$$

We remark that in Appendix B we show that the matrix representation of any linear operator which rotates $\mathbb{R}^2$ counter-clockwise through an angle of θ has this very form

$$\begin{bmatrix} \cos\theta & -\sin\theta \\ \sin\theta & \cos\theta \end{bmatrix}.$$

We show now that the matrices of Type II correspond to a reflection of $\mathbb{R}^2$ across the line through the origin which makes an angle of $\theta/2$ with the positive x-axis. Indeed, such a line through the origin has slope $\sin(\theta/2)/\cos(\theta/2)$. We appeal to the result proved in Appendix B which states that any linear operator which reflects $\mathbb{R}^2$ across the line $y = mx$ has the form

$$\begin{bmatrix} \frac{1-m^2}{m^2+1} & \frac{2m}{m^2+1} \\ -\frac{2m}{m^2+1} & \frac{1-m^2}{m^2+1} \end{bmatrix}.$$

Notice that

$$\frac{1-m^2}{m^2+1} = \frac{1 - \left(\frac{\sin(\theta/2)}{\cos(\theta/2)}\right)^2}{\left(\frac{\sin(\theta/2)}{\cos(\theta/2)}\right)^2 + 1} = \frac{1 - \left(\frac{\sin(\theta/2)}{\cos(\theta/2)}\right)^2}{\left(\frac{\sin(\theta/2)}{\cos(\theta/2)}\right)^2 + 1} \cdot \frac{\cos^2(\theta/2)}{\cos^2(\theta/2)}$$

$$= \frac{\cos^2(\theta/2) - \sin^2(\theta/2)}{\sin^2(\theta/2) + \cos^2(\theta/2)} = \frac{\cos\theta}{1} = \cos\theta.$$

Similarly, one can show that $\frac{2m}{m^2+1} = \sin\theta$. Hence, we have shown that the matrices of Type II are of the same form as the reflections across the line $y = mx$.

Example 5.14 *Let's apply these rotations and reflections to a specific graphic, namely the letter T (see Figure 5.1).*

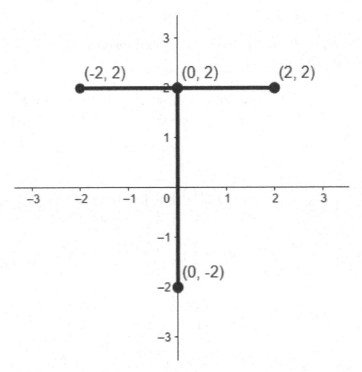

Figure 5.1 The Letter T in the xy-plane.

1. *We will first rotate the letter T through an angle of 30°. What we will do is rotate the four indicated points on the graphic and then connect the dots. First, we compute the appropriate operator*

$$[T]_{ST} = \begin{bmatrix} \cos 30° & -\sin 30° \\ \sin 30° & \cos 30° \end{bmatrix} = \begin{bmatrix} \sqrt{3}/2 & -1/2 \\ 1/2 & \sqrt{3}/2 \end{bmatrix}.$$

Therefore,

$$T[2,2] = \begin{bmatrix} \sqrt{3}/2 & -1/2 \\ 1/2 & \sqrt{3}/2 \end{bmatrix} \begin{bmatrix} 2 \\ 2 \end{bmatrix} = \begin{bmatrix} \sqrt{3}-1 \\ 1+\sqrt{3} \end{bmatrix} \approx [0.7, 2.7].$$

$$T[-2,2] = \begin{bmatrix} \sqrt{3}/2 & -1/2 \\ 1/2 & \sqrt{3}/2 \end{bmatrix} \begin{bmatrix} -2 \\ 2 \end{bmatrix} = \begin{bmatrix} -\sqrt{3}-1 \\ -1+\sqrt{3} \end{bmatrix} \approx [-2.7, -0.3].$$

$$T[0,2] = \begin{bmatrix} \sqrt{3}/2 & -1/2 \\ 1/2 & \sqrt{3}/2 \end{bmatrix} \begin{bmatrix} 0 \\ 2 \end{bmatrix} = \begin{bmatrix} -1 \\ \sqrt{3} \end{bmatrix} \approx [-1, 1.7].$$

$$T[0,-2] = \begin{bmatrix} \sqrt{3}/2 & -1/2 \\ 1/2 & \sqrt{3}/2 \end{bmatrix} \begin{bmatrix} 0 \\ -2 \end{bmatrix} = \begin{bmatrix} 1 \\ -\sqrt{3} \end{bmatrix} \approx [1, -1.7].$$

See Figure 5.2 for the resulting rotated image.

2. *Now let's reflect the letter T across line making an angle of 30° with the positive x-axis. Again we compute*

$$[T]_{ST} = \begin{bmatrix} \cos 60° & \sin 60° \\ \sin 60° & -\cos 60° \end{bmatrix} = \begin{bmatrix} 1/2 & \sqrt{3}/2 \\ \sqrt{3}/2 & -1/2 \end{bmatrix}.$$

Therefore,

$$T[2,2] = \begin{bmatrix} 1/2 & \sqrt{3}/2 \\ \sqrt{3}/2 & -1/2 \end{bmatrix} \begin{bmatrix} 2 \\ 2 \end{bmatrix} = \begin{bmatrix} 1+\sqrt{3} \\ \sqrt{3}-1 \end{bmatrix} \approx [2.7, 0.7].$$

$$T[-2,2] = \begin{bmatrix} 1/2 & \sqrt{3}/2 \\ \sqrt{3}/2 & -1/2 \end{bmatrix} \begin{bmatrix} -2 \\ 2 \end{bmatrix} = \begin{bmatrix} -1+\sqrt{3} \\ -1-\sqrt{3} \end{bmatrix} \approx [0.7, -2.7].$$

$$T[0,2] = \begin{bmatrix} 1/2 & \sqrt{3}/2 \\ \sqrt{3}/2 & -1/2 \end{bmatrix} \begin{bmatrix} 0 \\ 2 \end{bmatrix} = \begin{bmatrix} \sqrt{3} \\ -1 \end{bmatrix} \approx [1.7, -1].$$

$$T[0,-2] = \begin{bmatrix} 1/2 & \sqrt{3}/2 \\ \sqrt{3}/2 & -1/2 \end{bmatrix} \begin{bmatrix} 0 \\ -2 \end{bmatrix} = \begin{bmatrix} -\sqrt{3} \\ 1 \end{bmatrix} \approx [-1.7, 1].$$

See Figure 5.3 for the resulting reflected image.

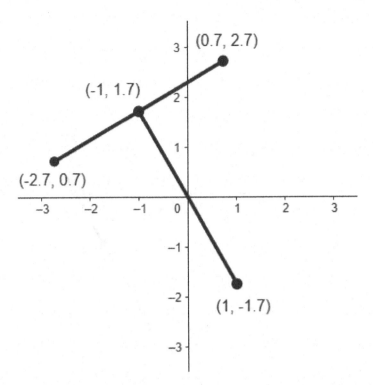

Figure 5.2 The Letter T rotated counter-clockwise 30° in the xy-plane.

We remark that one can show that a 3×3 orthogonal matrix A is one of three types: Either $|A| = 1$ and A corresponds to a rotation of $\mathbb{R}^n$ about a line through the origin, or $|A| = -1$ and A corresponds either to a reflection across a hyper-plane through the origin or a rotation about a line through the origin followed by a reflection across a hyper-plane through the origin perpendicular to that line (called a *rotoreflection*). As the dimension gets higher the classification becomes more complicated.

EXERCISES

1. Following all the steps in Example 5.14 of this section compute the following transformations of the letter T:

   a. Rotation of 120°.

   b. Reflection through 60°.

2. Let V be any inner product space. Prove that if $T \in L(V)$ is orthogonal, then $v \perp w$ implies that $T(v) \perp T(w)$ for all $v, w \in V$.

3. Let V be any inner product space. Prove that if $T \in L(V)$ is orthogonal, then for all $v, w \in V$ we have that the smaller angle between $T(v)$ and $T(w)$ equals the smaller angle between v and w.

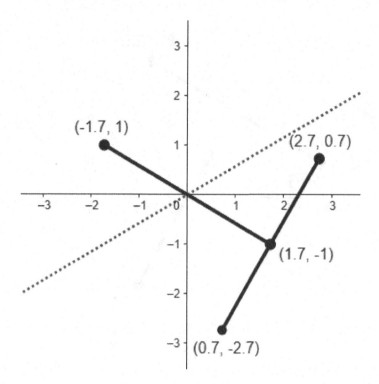

Figure 5.3 The Letter T reflected across a line making an angle of 30° with the positive x-axis.

4. Let V be any inner product space. Prove that if $T \in L(V)$ is orthogonal, then for all $v, w \in V$ we have that the distance between $T(v)$ and $T(w)$ equals the distance between v and w.

5. Prove part a of Theorem 5.3.

6. Prove Theorem 5.2.

5.4 APPLICATION: QR FACTORIZATION

We have already seen how matrix factorizations/decompositions play an important role in linear algebra in Section 2.11. In this section we present another factorization which is useful in numerical linear algebra called the QR factorization. As with the LU factorization, the QR factorization can also be used to computationally speed up solving linear systems, inverting a matrix or computing its determinant. Furthermore, it can be used in finding eigenvalues of a matrix by an iterative algorithm called the QR algorithm.

Example 5.15 *Below we express $A = QR$, where Q is an orthogonal matrix and R is upper triangular. Shortly, we will see the algorithm for performing this factorization.*

$$A = \begin{bmatrix} 1 & 1 & -1 \\ 1 & -1 & 1 \\ 1 & 1 & 1 \end{bmatrix} = \begin{bmatrix} 1/\sqrt{3} & 1/\sqrt{6} & -1/\sqrt{2} \\ 1/\sqrt{3} & -2/\sqrt{6} & 0 \\ 1/\sqrt{3} & 1/\sqrt{6} & 1/\sqrt{2} \end{bmatrix} \begin{bmatrix} \sqrt{3} & 1/\sqrt{3} & 1/\sqrt{3} \\ 0 & 2\sqrt{6}/3 & -2/\sqrt{6} \\ 0 & 0 & \sqrt{2} \end{bmatrix}.$$

For the moment we will assume A is an $n \times n$ invertible matrix with columns $v_1, v_2, \ldots, v_n$. By Theorem 3.11, $v_1, v_2, \ldots, v_n$ forms a basis for $\mathbb{R}^n$. Using the Gram-Schmidt process we can derive an orthogonal basis $w_1, w_2, \ldots, w_n$ and by normalizing this orthogonal basis we form an orthonormal basis $u_1, u_2, \ldots, u_n$. We now need a particular result which relates these three bases.

Lemma 5.6 *Consider the standard dot product for $\mathbb{R}^n$. For the three bases described above the following is true:*

1. *For $1 \le k \le n$, $w_k = |w_k| u_k$*

2. *$v_1 = |w_1| u_1$.*

3. *For $k > 1$,*

$$v_{k+1} = |w_{k+1}| u_{k+1} + \sum_{i=1}^{k} (v_{k+1} \cdot u_i) u_i.$$

Proof 5.9 *For the first statement, notice that*

$$|w_k| u_k = |w_k| \frac{w_k}{|w_k|} = w_k.$$

For the second statement, as defined in the Gram-Schmidt process, $w_1 = v_1$ and the result follows from the first statement. For the third statement, recall that in the Gram-Schmidt process, for $1 \le k < n$,

$$w_{k+1} = v_{k+1} - \sum_{i=1}^{k} \frac{v_{k+1} \cdot w_i}{|w_i|^2} w_i.$$

Solving for v_{k+1}, we have

$$v_{k+1} = w_{k+1} + \sum_{i=1}^{k} \frac{v_{k+1} \cdot w_i}{|w_i|^2} w_i$$

$$= w_{k+1} + \sum_{i=1}^{k} \left(\frac{v_{k+1} \cdot |w_i| u_i}{|w_i|} \right) \frac{w_i}{|w_i|} = w_{k+1} + \sum_{i=1}^{k} (v_{k+1} \cdot u_i) u_i.$$

$\square$

We can now present the algorithm for the QR factorization. Recall that A is an $n \times n$ invertible matrix with columns $v_1, v_2, \ldots, v_n$. Define Q to be the matrix whose columns are $u_1, u_2, \ldots, u_n$. By Theorem 5.2.iv, Q is an orthogonal matrix. Now define another matrix $R = [r_{ij}]$ as follows: For $1 \leq i, j \leq n$,

$$r_{ij} = \begin{cases} u_i \cdot v_j & i < j \\ |w_i|, & i = j \\ 0, & i > j \end{cases}$$

By its very definition, R is upper triangular, but furthermore $A = QR$ which is the QR factorization. We now verify that $A = QR$.

Theorem 5.5 *Consider A, Q and R as defined above. Then $A = QR$.*

Proof 5.10 *Let $R = [c_1 \ c_2 \ \cdots \ c_n]$ represented in columns. By Lemma 1.1, $QR = [Qc_1 \ Qc_2 \ \cdots \ Qc_n]$ represented in columns. By Lemmas 1.2 and 5.6.2,*

$$Qc_1 = \sum_{i=1}^{n} r_{i1} u_i = r_{11} u_1 = |w_1| u_1 = v_1.$$

For $1 < j \leq n$, by Lemmas 1.2 and 5.6.3,

$$Qc_j = \sum_{i=1}^{n} r_{ij} u_i = \sum_{i=1}^{j} r_{ij} u_i = \sum_{i=1}^{j-1} r_{ij} u_i + r_{jj} u_j = \sum_{i=1}^{j-1} (u_i \cdot v_j) u_i + |w_j| u_j = v_j.$$

This proves that $QR = [v_1 \ v_2 \ \cdots \ v_n] = A$. $\qquad\square$

Example 5.16 *We will perform the QR factorization algorithm on Example 5.15 where*

$$A = \begin{bmatrix} 1 & 1 & -1 \\ 1 & -1 & 1 \\ 1 & 1 & 1 \end{bmatrix}.$$

According to the notation used in this section,

$$v_1 = [1, 1, 1], \ v_2 = [1, -1, 1], \ v_3 = [-1, 1, 1].$$

Now we use the Gram-Schmidt process to generate an orthogonal basis. One can show that this basis is

$$w_1 = [1, 1, 1], \ w_2 = [2/3, -4/3, 2/3], \ w_3 = [-1, 0, 1].$$

We then normalize w_1, w_2, w_3 to obtain

$$u_1 = [1/\sqrt{3}, 1/\sqrt{3}, 1/\sqrt{3}], \ u_2 = [1/\sqrt{6}, -2/\sqrt{6}, 1/\sqrt{6}], \ u_3 = [-1/\sqrt{2}, 0, 1/\sqrt{2}].$$

Therefore,

$$Q = \begin{bmatrix} 1/\sqrt{3} & 1/\sqrt{6} & -1/\sqrt{2} \\ 1/\sqrt{3} & -2/\sqrt{6} & 0 \\ 1/\sqrt{3} & 1/\sqrt{6} & 1/\sqrt{2} \end{bmatrix}.$$

In order to obtain the matrix R, we need to compute

$$r_{11} = |w_1| = \sqrt{3}, \quad r_{12} = u_1 \cdot v_2 = 1/\sqrt{3}, \quad r_{13} = u_1 \cdot v_3 = 1/\sqrt{3},$$

$$r_{22} = |w_2| = 2\sqrt{6}/3, \quad r_{23} = u_2 \cdot v_3 = -2/\sqrt{6}, \quad r_{33} = |w_3| = \sqrt{2}.$$

Therefore,

$$R = \begin{bmatrix} \sqrt{3} & 1/\sqrt{3} & 1/\sqrt{3} \\ 0 & 2\sqrt{6}/3 & -2/\sqrt{6} \\ 0 & 0 & \sqrt{2} \end{bmatrix}.$$

Having introduced the QR factorization we now show how it can be useful in several numerical methods. First, we use it to solve square linear systems with a unique solution. Here is the general algorithm. Given a linear system $AX = B$ with QR factorization $A = QR$, since Q is invertible with inverse Q^T we can solve the equivalent system $RX = Q^T B$ by back substitution.

Example 5.17 *Consider the linear system*

$$\begin{cases} x + y - z &= -2 \\ x - y + z &= 4 \\ x + y + z &= 0 \end{cases}.$$

The coefficient matrix for this system is the matrix A from Example 5.15. We have already found the QR factorization with

$$Q = \begin{bmatrix} 1/\sqrt{3} & 1/\sqrt{6} & -1/\sqrt{2} \\ 1/\sqrt{3} & -2/\sqrt{6} & 0 \\ 1/\sqrt{3} & 1/\sqrt{6} & 1/\sqrt{2} \end{bmatrix} \quad and \quad R = \begin{bmatrix} \sqrt{3} & 1/\sqrt{3} & 1/\sqrt{3} \\ 0 & 2\sqrt{6}/3 & -2/\sqrt{6} \\ 0 & 0 & \sqrt{2} \end{bmatrix}.$$

Following the algorithm we first compute

$$Q^T B = \begin{bmatrix} 1/\sqrt{3} & 1/\sqrt{3} & 1/\sqrt{3} \\ 1/\sqrt{6} & -2/\sqrt{6} & 1/\sqrt{6} \\ -1/\sqrt{2} & 0 & 1/\sqrt{2} \end{bmatrix} \begin{bmatrix} -2 \\ 4 \\ 0 \end{bmatrix} = \begin{bmatrix} 2/\sqrt{3} \\ -10/\sqrt{6} \\ \sqrt{2} \end{bmatrix}.$$

Now we solve the system $RX = Q^T B$, i.e.

$$\begin{cases} \sqrt{3}x + (1/\sqrt{3})y + (1/\sqrt{3})z &= 2/\sqrt{3} \\ (2\sqrt{6}/3)y - (2/\sqrt{6})z &= -10/\sqrt{6} \\ \sqrt{2}z &= \sqrt{2} \end{cases},$$

or equivalently (although this step is not necessary),

$$\begin{cases} 3x + y + z &= 2 \\ 2y - z &= -5 \\ z &= 1 \end{cases}.$$

Back substituting, we have $y = (z - 5)/2 = -2$ and $x = (2 - y - z)/3 = 1$. Hence, the solution to the original system is $(1, -2, 1)$.

We can also nearly compute the determinant with the QR factorization, for if $A = QR$, then $|A| = |Q||R| = \pm|R|$ where $|R|$ can be obtained by multiplying together its diagonal entries. Hence, we can obtain the absolute value of the determinant of a matrix.

Example 5.18 *Consider again the matrix A from Example 5.15 with QR factorization*

$$Q = \begin{bmatrix} 1/\sqrt{3} & 1/\sqrt{6} & -1/\sqrt{2} \\ 1/\sqrt{3} & -2/\sqrt{6} & 0 \\ 1/\sqrt{3} & 1/\sqrt{6} & 1/\sqrt{2} \end{bmatrix} \quad and \quad R = \begin{bmatrix} \sqrt{3} & 1/\sqrt{3} & 1/\sqrt{3} \\ 0 & 2\sqrt{6}/3 & -2/\sqrt{6} \\ 0 & 0 & \sqrt{2} \end{bmatrix}.$$

Then $|A| = \pm|R| = \pm(\sqrt{3})(2\sqrt{6}/3)(\sqrt{2}) = \pm 4$. Indeed, one can compute that $|A| = -4$.

We can use the QR factorization to find the inverse of a matrix. Indeed, if $A = QR$, then $A^{-1} = R^{-1}Q^{-1} = R^{-1}Q^T$. Therefore, we are reduced to finding the inverse of R which is an easier problem since it is upper triangular, thus making the Gaussian elimination quicker.

Example 5.19 *Consider again the matrix A from Example 5.15 with QR factorization*

$$Q = \begin{bmatrix} 1/\sqrt{3} & 1/\sqrt{6} & -1/\sqrt{2} \\ 1/\sqrt{3} & -2/\sqrt{6} & 0 \\ 1/\sqrt{3} & 1/\sqrt{6} & 1/\sqrt{2} \end{bmatrix} \quad and \quad R = \begin{bmatrix} \sqrt{3} & 1/\sqrt{3} & 1/\sqrt{3} \\ 0 & 2\sqrt{6}/3 & -2/\sqrt{6} \\ 0 & 0 & \sqrt{2} \end{bmatrix}.$$

To find R^{-1} we row reduce

$$\begin{bmatrix} \sqrt{3} & 1/\sqrt{3} & 1/\sqrt{3} & 1 & 0 & 0 \\ 0 & 2\sqrt{6}/3 & -2/\sqrt{6} & 0 & 1 & 0 \\ 0 & 0 & \sqrt{2} & 0 & 0 & 1 \end{bmatrix} \; to \; \begin{bmatrix} 1 & 0 & 0 & \sqrt{3} & -\sqrt{6}/12 & -\sqrt{2}/4 \\ 0 & 1 & 0 & 0 & \sqrt{6}/4 & \sqrt{2}/4 \\ 0 & 0 & 1 & 0 & 0 & 1/\sqrt{2} \end{bmatrix}.$$

Therefore,

$$A^{-1} = \begin{bmatrix} \sqrt{3} & -\sqrt{6}/12 & -\sqrt{2}/4 \\ 0 & \sqrt{6}/4 & \sqrt{2}/4 \\ 0 & 0 & 1/\sqrt{2} \end{bmatrix} \begin{bmatrix} 1/\sqrt{3} & 1/\sqrt{3} & 1/\sqrt{3} \\ 1/\sqrt{6} & -2/\sqrt{6} & 1/\sqrt{6} \\ -1/\sqrt{2} & 0 & 1/\sqrt{2} \end{bmatrix}$$

$$= \begin{bmatrix} 1/2 & 1/2 & 0 \\ 0 & -1/2 & 1/2 \\ -1/2 & 0 & 1/2 \end{bmatrix}.$$

There's nothing stopping us from extending the QR factorization to rectangular matrices assuming the columns of the matrix are linearly independent. The very same procedure will work in this case. A nice application for the case of rectangular matrices is computing the pseudo-inverse. Hence, if we have a quicker way to compute a pseudo-inverse, then we in turn have quicker ways to solve all the applications of least squares gone over in Section 5.9. If A is an $m \times n$ matrix with linearly independent columns, we first write $A = QR$ with Q an $m \times n$ with orthonormal columns and R an $n \times n$ upper triangular matrix as defined earlier. Then

$$A^\dagger = (A^T A)^{-1} A^T = (R^T Q^T Q R)^{-1} R^T Q^T$$

$$= (R^T R)^{-1} R^T Q^T = R^{-1} (R^T)^{-1} R^T Q^T = R^{-1} Q^T.$$

Notice that we obtained the same result for a square invertible matrix.

Example 5.20 *Consider the following matrix with linearly independent columns:*

$$A = \begin{bmatrix} 1 & 1 & 1 \\ -1 & 1 & 1 \\ 1 & -1 & 1 \\ -1 & 1 & -1 \end{bmatrix}.$$

According to the QR factorization, we start with

$$v_1 = [1, -1, 1, -1], \; v_2 = [1, 1, -1, 1], \; v_3 = [-1, 1, 1, -1].$$

Now we use the Gram-Schmidt process to generate an orthogonal basis. One can show that this basis is

$$w_1 = [1, -1, 1, -1], \quad w_2 = [3/2, 1/2, -1/2, 1/2], \quad w_3 = [0, 4/3, 2/3, -2/3].$$

We then normalize w_1, w_2, w_3 to obtain

$$u_1 = [1/2, -1/2, 1/2, -1/2], \quad u_2 = [\sqrt{3}/2, \sqrt{3}/6, -\sqrt{3}/6, \sqrt{3}/6],$$
$$u_3 = [0, \sqrt{6}/3, \sqrt{6}/6, -\sqrt{6}/6].$$

Therefore,

$$Q = \begin{bmatrix} 1/2 & \sqrt{3}/2 & 0 \\ -1/2 & \sqrt{3}/6 & \sqrt{6}/3 \\ 1/2 & -\sqrt{3}/6 & \sqrt{6}/6 \\ -1/2 & \sqrt{3}/6 & -\sqrt{6}/6 \end{bmatrix}.$$

In order to obtain the matrix R, we need to compute

$$r_{11} = |w_1| = 2, \quad r_{12} = u_1 \cdot v_2 = -1, \quad r_{13} = u_1 \cdot v_3 = 0,$$

$$r_{22} = |w_2| = \sqrt{3}, \quad r_{23} = u_2 \cdot v_3 = -2\sqrt{3}/3, \quad r_{33} = |w_3| = 2\sqrt{6}/3.$$

Therefore,

$$R = \begin{bmatrix} 2 & -1 & 0 \\ 0 & \sqrt{3} & -2\sqrt{3}/3 \\ 0 & 0 & 2\sqrt{6}/3 \end{bmatrix}.$$

One can then compute

$$R^{-1} = \begin{bmatrix} 1/2 & \sqrt{3}/6 & \sqrt{6}/12 \\ 0 & \sqrt{3}/3 & \sqrt{6}/6 \\ 0 & 0 & \sqrt{6}/4 \end{bmatrix}.$$

Therefore,

$$A^\dagger = \begin{bmatrix} 1/2 & \sqrt{3}/6 & \sqrt{6}/12 \\ 0 & \sqrt{3}/3 & \sqrt{6}/6 \\ 0 & 0 & \sqrt{6}/4 \end{bmatrix} \begin{bmatrix} 1/2 & -1/2 & 1/2 & -1/2 \\ \sqrt{3}/2 & \sqrt{3}/6 & -\sqrt{3}/6 & \sqrt{3}/6 \\ 0 & \sqrt{6}/3 & \sqrt{6}/6 & -\sqrt{6}/6 \end{bmatrix}$$

$$= \begin{bmatrix} 1/2 & 0 & 1/4 & -1/4 \\ 1/2 & 1/2 & 0 & 0 \\ 0 & 1/2 & 1/4 & -1/4 \end{bmatrix}.$$

The reader can easily verify this answer by computing $(A^T A)^{-1} A^T$ directly.

EXERCISES

1. If possible, compute the QR factorization for each matrix.

   a. $A = \begin{bmatrix} 1 & 2 \\ 3 & 4 \end{bmatrix}$

   b. $B = \begin{bmatrix} 1 & -1 & 1 \\ 4 & 2 & 5 \\ 3 & 1 & 4 \end{bmatrix}$

   c. $C = \begin{bmatrix} 2 & -4 & 0 \\ 1 & -2 & 1 \\ 1 & 0 & 2 \end{bmatrix}$

   d. $D = \begin{bmatrix} 1 & 2 & 1 & -1 \\ 3 & 3 & 0 & 1 \\ 2 & 7 & -1 & 1 \\ -2 & 1 & 1 & 0 \end{bmatrix}$

2. Solve each system of equations by using the QR factorization found in Exercise 1.

   a. $\begin{cases} x + 2y & = & 1 \\ 3x + 4y & = & -2 \end{cases}$

   b. $\begin{cases} x - y + z & = & 1 \\ 4x + 2y + 5z & = & -1 \\ 3x + y + 4z & = & 0 \end{cases}$

   c. $\begin{cases} x_1 + 2x_2 + x_3 - x_4 & = & 2 \\ 3x_1 + 3x_2 + x_4 & = & -1 \\ 2x_1 + 7x_2 - x_3 + x_4 & = & 0 \\ -2x_1 + x_2 + x_3 & = & 1 \end{cases}$

3. If possible, compute the determinant of each matrix using the QR factorization found in Exercise 1.

4. If possible, find the inverse of each matrix using the QR factorization found in Exercise 1.

5. Compute the pseudo-inverse of the following matrix using QR factorization:

$$A = \begin{bmatrix} -1 & -1 & 1 \\ 1 & 3 & 3 \\ -1 & -1 & 5 \\ 1 & 3 & 7 \end{bmatrix}$$

6. (See Section 5.9) Use Exercise 5 to find a best approximation to the following overdetermined system:

$$\begin{cases} -x - y + z & = & 2 \\ x + 3y + 3z & = & -2 \\ -x - y + 5z & = & 0 \\ x + 3y + 7z & = & 4 \end{cases}$$

5.5   SCHUR TRIANGULARIZATION THEOREM

In this section, we prove an important result called the Schur Triangularization Theorem which states that any matrix over the complex numbers is similar to an upper triangular matrix via a unitary matrix. Therefore, in this section we allow the scalars of our vector space to be the complex numbers. We remind the reader about some properties of complex numbers.

1. If $z = a + bi$ is a complex number, then the **conjugate**, written $\bar{z} = a - bi$. We extend this notation to complex values vectors as $\bar{v}$ in which we take the conjugate of each of the components of v. In other words if $w = [z_1, z_2, \ldots, z_n]$, then $\overline{w} = [\overline{z_1}, \overline{z_2}, \ldots, \overline{z_n}]$. Note that if $w = [a_1 + b_1 i, a_2 + b_2 i, \ldots, a_n + b_n i] \in \mathbb{C}^n$ we can write $w = u + vi$ where $u = [a_1, a_2, \ldots, a_n] \in \mathbb{R}^n$ and $v = [b_1, b_2, \ldots, b_n] \in \mathbb{R}^n$.

2. The **magnitude** (or **norm**) of a complex number $z = a + bi$, written $|z| = \sqrt{a^2 + b^2}$. Note that $z\bar{z} = |z|^2 \geq 0$ with $z\bar{z} = 0$ iff $z = 0$. In a similar manner, for a complex vector $w = [z_1, z_2, \ldots, z_n] \in \mathbb{C}^n$,

$$w \cdot \overline{w} = z_1 \overline{z_1} + z_2 \overline{z_2} + \cdots + z_n \overline{z_n} \geq 0 \quad \text{with} \quad w \cdot \overline{w} = 0 \quad \text{iff} \quad w = 0$$

3. If z is complex with $\bar{z} = z$, then z is in fact a real number.

4. For complex numbers $z_1, z_2 \in \mathbb{C}$, $\overline{z_1 z_2} = \overline{z_1} \, \overline{z_2}$. A similar property holds for matrix multiplication and dot product, i.e. for complex entry matrices A and B, we have $\overline{AB} = \overline{A} \, \overline{B}$.

We present now some special families of complex valued matrices.

Definition 5.8 *Let $A = [a_{ij}]$ be a matrix with complex entries.*

1. *$A^* = [b_{ij}]$ has entries $b_{ij} = \bar{a}_{ji}$, where $\bar{a}_{ji}$ represents the complex conjugate of a_{ji}. In other words A^* is the **conjugate transpose** of A.*

2. *A is **Hermitian** if $A^* = A$.*

3. *A is **skew-Hermitian** if $A^* = -A$.*

4. *A is **unitary** if $A^* A = I$.*

Note that if A has all real number entries, then Hermitian is equivalent to symmetric, skew-Hermitian is equivalent to skew-symmetric and unitary is equivalent to orthogonal.

The definition of a **complex** inner product space, where the scalars are now complex numbers, differs from a **real** inner product space in one respect. The *symmetric property* of a real inner product is replaced by the following property:

$$\text{For all } v, w \in V, \ \overline{(v|w)} = (w|v).$$

Notice that this property reduces to the symmetric property of a real inner product space when the scalars are all real numbers.

Definition 5.9 *Consider the vector space $\mathbb{C}^n$. We define **dot product** on $\mathbb{C}^n$ as follows: For $v, w \in \mathbb{C}^n$, $v \cdot w = \bar{v}^T w$.*

We leave it as an exercise to show that $\mathbb{C}^n$ together with this dot product forms a complex inner product space. We point out that the Gram-Schmidt Process continues to function for a complex inner product space (see exercises). We also remark that one can also define a complex inner product on complex entry square matrices and complex valued functions, but our goal is to present the minimal amount of new information in order to prove the main result in this section which we present now.

Theorem 5.6 *Every square matrix is triangularizable over the complex numbers by means of a unitary matrix, i.e. given a matrix $A \in M_{nn}(\mathbb{C})$ there exists a unitary matrix $P \in M_{nn}(\mathbb{C})$ and a upper triangular matrix $B \in M_{nn}(\mathbb{C})$ such that $P^*AP = B$ where $P^*P = I$.*

Proof 5.11 *The proof is demonstrated by induction on n. For $n = 1$, take unitary matrix $P = I_1$ with $B = A$. For $n > 1$, note that the characteristic polynomial $p_A(t)$ has a root in $\mathbb{C}$, say λ_1, an eigenvalue for A. Let $v_1, v_2, \ldots, v_k$ be a basis for E_{λ_1}. Extend $v_1, v_2, \ldots, v_k$ to a basis $v_1, v_2, \ldots, v_k, v_{k+1}, \ldots, v_n$ for $\mathbb{C}^n$. Viewing $v_1, v_2, \ldots, v_n$ as elements of the complex inner product space $\mathbb{C}^n$, apply the Gram-Schmidt Process to obtain from this basis an orthonormal basis for $\mathbb{C}^n$ and form the unitary matrix P whose columns are these basis vectors. Then, represented in blocks,*

$$P^*AP = \begin{bmatrix} \lambda_1 I_k & B \\ 0_{n-k} & C \end{bmatrix},$$

where $B \in M_{k,n-k}$ and $C \in M_{n-k,n-k}$. By induction, there exists a unitary matrix $\hat{Q}$ such that $\hat{Q}^*C\hat{Q}$ is upper triangular. Define, in blocks, the unitary matrix

$$Q = \begin{bmatrix} I_k & 0_{n-k} \\ 0_{n-k} & \hat{Q} \end{bmatrix}.$$

Then $(PQ)^*A(PQ)$ is upper triangular via the unitary matrix PQ. □

A couple of remarks are in order.

1. If $A \in M_{nn}(\mathbb{R})$ and all of its eigenvectors are real numbers, then the matrix P which makes A similar to an upper triangular matrix will be real orthogonal.

2. In the proof, notice that $p_A(t) = (t - \lambda_1)^k p_C(t)$.

There are several methods for triangularizing a matrix. These methods include Householder matrices, Givens rotations, and the Gram-Schmidt Process. Indeed, our proof of Schur's Theorem gives us the algorithm for triangularization by means of the Gram-Schmidt Process.

Example 5.21 *In this example we will illustrate the Gram-Schmidt Process for triangularization. Consider the matrix*

$$A = \begin{bmatrix} -1 & -1 & -2 \\ 8 & -11 & -8 \\ -10 & 11 & 7 \end{bmatrix}.$$

One can check that A has real eigenvalues 1, -3 and -3 and the corresponding eigenspaces are

$$E_1 = \{ \ [a, 2a, -2a] \ : \ a \in \mathbb{R} \ \} \quad and \quad E_{-3} = \{ \ [a, 0, a] \ : \ a \in \mathbb{R} \ \}.$$

Notice that $\dim(E_1) + \dim(E_{-3}) = 1 + 1 = 2 < 3$, so A is not diagonalizable. However, by Schur's Theorem on triangularization we know that A is triangularizable. Randomly choose one of the eigenvalues, say $\lambda = 1$ and find a basis for E_1 which in our case is $v_1 = [1, 2, -2]$. Now extend to a basis for $\mathbb{R}^3$ by row reducing a matrix whose columns are v_1, e_1, e_2, e_3 to row-echelon form and selecting the vectors corresponding to the pivots in the row echelon form. In our case our extended basis is v_1, e_1, e_2. Now apply the Gram-Schmidt Process to this basis in order to create an orthogonal basis, namely

$$[1, 2, -2], \quad [8/9, -2/9, 2/9] \quad and \quad [0, 1/2, 1/2].$$

If we normalize these vectors we obtain the orthonormal basis

$$u_1 = [1/3, 2/3, -2/3], \quad u_2 = [\sqrt{8}/3, -\sqrt{8}/12, \sqrt{8}/12] \quad and \quad u_3 = [0, \sqrt{2}/2, \sqrt{2}/2].$$

Form the orthogonal matrix P whose columns are u_1, u_2 and u_3. Then

$$P^T A P = \begin{bmatrix} 1 & \frac{17\sqrt{2}}{2} & -\frac{77\sqrt{2}}{6} \\ 0 & -\frac{11}{2} & \frac{25}{6} \\ 0 & -\frac{3}{2} & -\frac{1}{2} \end{bmatrix}.$$

We now repeat the process on the 2×2 block in the bottom right corner, namely

$$C = \begin{bmatrix} -\frac{11}{2} & \frac{25}{6} \\ -\frac{3}{2} & -\frac{1}{2} \end{bmatrix}.$$

The eigenvalues for C are -3 and -3 with corresponding eigenspace

$$E_{-3} = \{ \, [5a, 3a] \, : \, a \in \mathbb{R} \, \} \quad \text{with basis} \ \ v_1 = [5, 3].$$

Once again we cannot diagonalize C so we extend v_1 to a basis for $\mathbb{R}^2$, namely v_1, e_1. Again we use Gram-Schmidt to create an orthogonal basis, namely

$$[5, 3] \quad \text{and} \quad [9/34, -15/34].$$

If we normalize these vectors we obtain the orthonormal basis

$$[5\sqrt{34}/34, 3\sqrt{34}/34] \quad \text{and} \quad [3\sqrt{34}/34, -5\sqrt{34}/34].$$

Form the orthogonal matrix $\hat{Q}$ whose columns are u_1 and u_2. Then define the orthogonal matrix in blocks

$$Q = \begin{bmatrix} 1 & 0 \\ 0 & \hat{Q} \end{bmatrix} = \begin{bmatrix} 1 & 0 & 0 \\ 0 & \frac{5\sqrt{34}}{34} & \frac{3\sqrt{34}}{34} \\ 0 & \frac{3\sqrt{34}}{34} & -\frac{5\sqrt{34}}{34} \end{bmatrix}.$$

Therefore,

$$(PQ)^T A(PQ) = \begin{bmatrix} 1 & \frac{4\sqrt{17}}{17} & \frac{269\sqrt{17}}{17} \\ 0 & -3 & -\frac{17}{3} \\ 0 & 0 & -3 \end{bmatrix},$$

which is upper triangular. Hence, A is similar to this upper triangular matrix via an orthogonal matrix PQ.

We mention some consequences of the Schur Triangularization Theorem.

1. The triangularization of a matrix has the roots of the characteristic polynomial (including multiplicity) on the diagonal.

2. The trace of a matrix is the sum of the roots of the characteristic polynomial (including multiplicity) for the matrix.

3. The determinant of a matrix is the product of the roots of the characteristic polynomial (including multiplicity) for the matrix.

Example 5.22 *Just as we did in Section 4.6, Example 4.32 we simplify the computation of performing matrix exponentiation. Consider the matrix A in the previous example. Suppose we wanted to compute A raised to a high power. Since we cannot diagonalize A, we cannot use the algorithm employed in Section 4.6, Example 4.32, however, we can use a similar technique with the triangularization of A. Just as an example, suppose we wish to compute A^{20}. From the work above*

$$A^{20} = \left((PQ) \begin{bmatrix} 1 & \frac{4\sqrt{17}}{17} & \frac{269\sqrt{17}}{17} \\ 0 & -3 & -\frac{17}{3} \\ 0 & 0 & -3 \end{bmatrix} (PQ)^T \right)^{20} = (PQ) \begin{bmatrix} 1 & \frac{4\sqrt{17}}{17} & \frac{269\sqrt{17}}{17} \\ 0 & -3 & -\frac{17}{3} \\ 0 & 0 & -3 \end{bmatrix}^{20} (PQ)^T.$$

Using the binomial theorem, notice that

$$\begin{bmatrix} 1 & \frac{4\sqrt{17}}{17} & \frac{269\sqrt{17}}{17} \\ 0 & -3 & -\frac{17}{3} \\ 0 & 0 & -3 \end{bmatrix}^{20} = \left(\begin{bmatrix} 1 & 0 & 0 \\ 0 & -3 & 0 \\ 0 & 0 & -3 \end{bmatrix} + \begin{bmatrix} 0 & \frac{4\sqrt{17}}{17} & \frac{269\sqrt{17}}{17} \\ 0 & 0 & -\frac{17}{3} \\ 0 & 0 & 0 \end{bmatrix} \right)^{20} =$$

$$\sum_{k=0}^{20} \binom{20}{k} \begin{bmatrix} 1 & 0 & 0 \\ 0 & -3 & 0 \\ 0 & 0 & -3 \end{bmatrix}^{20-k} \begin{bmatrix} 0 & \frac{4\sqrt{17}}{17} & \frac{269\sqrt{17}}{17} \\ 0 & 0 & -\frac{17}{3} \\ 0 & 0 & 0 \end{bmatrix}^{k}.$$

Note that one can verify that

$$\begin{bmatrix} 0 & \frac{4\sqrt{17}}{17} & \frac{269\sqrt{17}}{17} \\ 0 & 0 & -\frac{17}{3} \\ 0 & 0 & 0 \end{bmatrix}^{2} = \begin{bmatrix} 0 & 0 & -\frac{4\sqrt{17}}{3} \\ 0 & 0 & 0 \\ 0 & 0 & 0 \end{bmatrix},$$

and all higher powers of this matrix equals the zero matrix. Therefore,

$$\begin{bmatrix} 1 & \frac{4\sqrt{17}}{17} & \frac{269\sqrt{17}}{17} \\ 0 & -3 & -\frac{17}{3} \\ 0 & 0 & -3 \end{bmatrix}^{20}$$

$$
= \begin{bmatrix} 1 & 0 & 0 \\ 0 & -3 & 0 \\ 0 & 0 & -3 \end{bmatrix}^{20} + 20 \begin{bmatrix} 1 & 0 & 0 \\ 0 & -3 & 0 \\ 0 & 0 & -3 \end{bmatrix}^{19} \begin{bmatrix} 0 & \frac{4\sqrt{17}}{17} & \frac{269\sqrt{17}}{17} \\ 0 & 0 & -\frac{17}{3} \\ 0 & 0 & 0 \end{bmatrix}
$$

$$
+190 \begin{bmatrix} 1 & 0 & 0 \\ 0 & -3 & 0 \\ 0 & 0 & -3 \end{bmatrix}^{18} \begin{bmatrix} 0 & 0 & -\frac{4\sqrt{17}}{3} \\ 0 & 0 & 0 \\ 0 & 0 & 0 \end{bmatrix} =
$$

$$
= \begin{bmatrix} 1^{20} & 0 & 0 \\ 0 & (-3)^{20} & 0 \\ 0 & 0 & (-3)^{20} \end{bmatrix} + 20 \begin{bmatrix} 1^{19} & 0 & 0 \\ 0 & (-3)^{19} & 0 \\ 0 & 0 & (-3)^{19} \end{bmatrix} \begin{bmatrix} 0 & \frac{4\sqrt{17}}{17} & \frac{269\sqrt{17}}{17} \\ 0 & 0 & -\frac{17}{3} \\ 0 & 0 & 0 \end{bmatrix}
$$

$$
+190 \begin{bmatrix} 1^{18} & 0 & 0 \\ 0 & (-3)^{18} & 0 \\ 0 & 0 & (-3)^{18} \end{bmatrix} \begin{bmatrix} 0 & 0 & -\frac{4\sqrt{17}}{3} \\ 0 & 0 & 0 \\ 0 & 0 & 0 \end{bmatrix}.
$$

Thus, instead of performing twenty matrix multiplications, we perform much less. Furthermore, this minimal number of multiplications holds in our example no matter how high the exponent, an amazing computational savings. Keep in mind that this technique can be used for any square matrix.

EXERCISES

1. Consider the following matrix:

$$
A = \begin{bmatrix} 13 & 8 & 8 \\ -1 & 7 & -2 \\ -1 & -2 & 7 \end{bmatrix}.
$$

   a. Determine by investigating the eigenvalues of A whether or not it can be diagonalized.

   b. Use the Gram-Schmidt method to illustrate how A is similar to an upper triangular matrix.

   c. Compute A^{20} as we did in this section.

2. Prove the following facts about conjugate transpose for $A, B \in M_{nn}(\mathbb{C})$ and $a \in \mathbb{C}$:

   a. $(A^*)^* = A$.

   b. $(A + B)^* = A^* + B^*$.

   c. $(aA)^* = \bar{a}A^*$.

   d. $(AB)^* = B^*A^*$.

   e. $|A^*| = \overline{|A|}$.

3. Prove that the set of all Hermitian matrices are a subspace of $M_{nn}(\mathbb{C})$.

4. Prove that the set of all Skew-Hermitian matrices are a subspace of $M_{nn}(\mathbb{C})$.

5. What can we say about the diagonal entries of a Skew-Hermitian matrix? Prove your statement.

6. Prove that if P and Q are unitary matrices, then so is PQ.

7. Prove that a unitary matrix has determinant equal to ± 1.

8. Prove that the following are complex inner product spaces:

   a. Consider the complex vector space $\mathbb{C}^n$. For $v, w \in \mathbb{C}^n$, define the complex inner product $(v|w) = \bar{v}^T w$.

   b. Consider the complex vector space $M_{nn}(\mathbb{C})$. For $A, B \in M_{nn}(\mathbb{C})$, define the complex inner product $(A|B) = tr(B^*A)$.

9. Prove the following consequences for a complex inner product space:

   a. $(w|u + v) = (w|u) + (w|v)$, for all $u, v, w \in V$.

   b. $(v|aw) = \bar{a}(v|w)$, for all $v, w \in V$ and $a \in \mathbb{C}$.

10. Verify that the Gram-Schmidt Process continues to function for a complex inner product space.

5.6 ORTHOGONAL PROJECTIONS AND BEST APPROXIMATION

In this section, our ultimate goal is to define what it means to be a *best* approximation and how to find it. This approximation will then be applied in Section 5.9. Recall the following definitions for the inner product space $\mathbb{R}^n$ from Section 1.2:

$$\text{comp}_u v = \frac{u \cdot v}{|u|} \qquad\qquad \text{proj}_u v = \frac{u \cdot v}{u \cdot u}u.$$

We now formally extend these definitions to arbitrary inner product spaces. Although we will lose the geometric interpretation of these definitions in this general setting, it doesn't hurt to use the geometric interpretation to maintain an intuition of what is going on.

Definition 5.10 *Let V be an inner product space with inner product $(*|*)$ and $u, v \in V$. Then*

1. *The* **component** *of v along u, written $\mathrm{comp}_u v = \frac{(u|v)}{|u|}$.*

2. *The* **projection** *of v along u, written $\mathrm{proj}_u v = \frac{(u|v)}{(u|u)} u$.*

As before $\mathrm{comp}_u v$ is a scalar, while $\mathrm{proj}_u v$ is a vector. It's easy to check that the magnitude of $\mathrm{proj}_u v$ equals $|\mathrm{comp}_u v|$.

Example 5.23 *Let $V = \mathbb{R}^4$ with the standard inner product and take $u = [1, 0, -1, 2]$ and $v = [-2, 1, 1, -1]$. Then*

$$\mathrm{comp}_u v = \frac{([1, 0, -1, 2] | [-2, 1, 1, -1])}{|[1, 0, -1, 2]|} = \frac{-5}{\sqrt{6}} \quad and$$

$$\mathrm{proj}_u v = \frac{([1, 0, -1, 2] | [-2, 1, 1, -1])}{([1, 0, -1, 2] | [1, 0, -1, 2])} [1, 0, -1, 2]$$

$$= \frac{5}{6} [1, 0, -1, 2] = [5/6, 0, -(5/6), 5/3].$$

Example 5.24 *Let $V = M_{22}$ with the inner product defined in Section 5.1 and take $u = E_{12}$ and $v = E_{22}$. Then*

$$\mathrm{comp}_u v = \frac{(E_{12} | E_{22})}{|E_{12}|} = \frac{tr(E_{22}^T E_{12})}{\sqrt{tr(E_{12}^T E_{12})}} = \frac{tr(E_{22})}{\sqrt{tr(E_{21} E_{12})}}$$

$$= \frac{tr(E_{22})}{\sqrt{tr(E_{22})}} = \frac{1}{1} = 1 \quad and$$

$$\mathrm{proj}_u v = \frac{(E_{12} | E_{22})}{(E_{12} | E_{12})} E_{12} = 1 \cdot E_{12} = E_{12}.$$

Example 5.25 *Let $V = \mathcal{C}([0, \pi/2])$ with the inner product defined in Section 5.1 and take $u = \cos x$ and $v = \sin x$. Then*

$$\mathrm{comp}_u v = \frac{(\cos x | \sin x)}{|\cos x|} = \frac{\int_0^{\pi/2} \cos x \sin x \, dx}{\sqrt{\int_0^{\pi/2} \cos^2 x \, dx}} = \frac{1/2}{\sqrt{\pi/4}} = \frac{1}{\sqrt{\pi}}.$$

$$\mathrm{proj}_u v = \left(\frac{(\cos x | \sin x)}{(\cos x | \cos x)} \right) \cos(x) = \left(\frac{\int_0^{\pi/2} \cos x \sin x \, dx}{\int_0^{\pi/2} \cos^2 x \, dx} \right) \cos x = \frac{2}{\pi} \cos x.$$

Definition 5.11 *If U is a subspace of an inner product space V, then the annihilator of U, $U^\perp$, is called the* **orthogonal complement** *of U in V*

We point out that Lemma 5.4.i and v give us the fact that $U \cap U^\perp = \{0\}$. The next result is fundamental (refer to $U \oplus U^\perp$ at end of Section 3.2).

Theorem 5.7 *If U is a subspace of an inner product space V with $\dim(U) < \infty$, then $V = U \oplus U^\perp$.*

Proof 5.12 *We will prove that for any $v \in V$ there exist $u \in U$ and $u' \in U^\perp$ such that $v = u + u'$. The uniqueness of this representation follows from the fact that $U \cap U^\perp = \{\,0\,\}$ (see Lemma 3.2).*

We shall explicitly construct the appropriate $u \in U$ as follows: By the Gram-Schmidt Process, we first produce an orthonormal basis $u_1, \ldots, u_k$ for U. Now set

$$u = (v|u_1)u_1 + \cdots + (v|u_k)u_k \quad and \quad u' = v - u$$

Certainly $v = u + u'$ and $u \in U$, so we need only show that $u' \in U^\perp$. Take any $w \in U$ and we show that $(u'|w) = 0$. First write $w = a_1u_1 + \cdots + a_ku_k$ for some $a_1, \ldots, a_k \in \mathbb{R}$. Notice first that

$$(v|w) = (v|a_1u_1 + \cdots + a_ku_k) = a_1(v|u_1) + \cdots + a_k(v|u_k).$$

Secondly,

$$(u|w) = ((v|u_1)u_1 + \cdots + (v|u_k)u_k \mid a_1u_1 + \cdots + a_ku_k)$$

$$= a_1(v|u_1) + \cdots + a_k(v|u_k), \ by \ Lemma \ 5.3$$

Hence, $(v|w) = (u|w)$, and so $(u'|w) = (v - u|w) = (v|w) - (u|w) = 0$. ☐

The following is an immediate consequence of this theorem.

Corollary 5.2 *If U is a subspace of a Euclidean space V, then*

    i. $\dim V = \dim U + \dim U^\perp$.

    ii. $U^{\perp\perp} = U$.

Proof 5.13 *Part i follows immediately from Lemma 3.5. For part ii, note that since $U^\perp$ is a subspace of V we can use part i to conclude that $\dim V = \dim U^\perp + \dim U^{\perp\perp}$. Equating yields $\dim U + \dim U^\perp = \dim U^\perp + \dim U^{\perp\perp}$ and so $\dim U = \dim U^{\perp\perp}$. Now, by Lemma 5.4.ii, U is a subspace of $U^{\perp\perp}$, and so $U = U^{\perp\perp}$.* ☐

Definition 5.12 *Let U be a subspace of an inner product space V with $dim(U) < \infty$ and $u_1, \ldots, u_k$ an orthonormal basis for U. In the proof of Theorem 5.7, for any $v \in V$ we constructed a $u \in U$ with the property that $v - u \in U^\perp$, namely,*

$$u = (v|u_1)u_1 + \cdots + (v|u_k)u_k.$$

*We shall call this u the **orthogonal projection** of v onto U and denote it by* $\text{proj}_U v$.

Before we explore some examples, a few of observations are in order which we collect in the following proposition:

Proposition 5.1 *Let V be a Euclidean space with $v \in V$.*

1. *The definition of $\text{proj}_U v$ is unique and not dependent on our choice of orthonormal basis.*

2. *$v - \text{proj}_U v \in U^\perp$.*

3. *If $w \in V$ and $U = span(w)$ then $\text{proj}_w v = \text{proj}_U v$.*

Proof 5.14 *The first item follows from the fact that we have unique representation as $V = U \oplus U^\perp$. The second and third items are left as exercises.* □

Example 5.26 *As usual we will start with a more intuitive example for computing $\text{proj}_U v$. Take $V = \mathbb{R}^3$ with the standard dot product and U equal to the xy-plane. An orthonormal basis for U is $\hat{\imath}, \hat{\jmath}$. Set $v = [1, 2, 3]$ and we will compute $\text{proj}_U v$. By definition,*

$$\text{proj}_U v = (v \cdot \hat{\imath})\hat{\imath} + (v \cdot \hat{\jmath})\hat{\jmath} = 1\hat{\imath} + 2\hat{\jmath} = [1, 2, 0].$$

This makes intuitive geometric sense, since the point in the xy-plane closest to $[1, 2, 3]$ is indeed $[1, 2, 0]$.

Hence, $v - \text{proj}_U v = 3\hat{k}$ and is a vector perpendicular to the xy-plane U (as anticipated by Proposition 5.1) (see Figure 5.4).

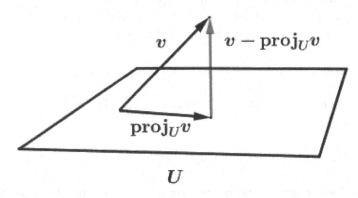

Figure 5.4 The projection of vector v onto the U.

Example 5.27 *Here is a less intuitive example. Take $V = P_2$ with the inner product defined in Section 5.1, i.e. $(p|q) = \int_0^1 p(x)q(x) \, dx$. Set $v = x^2$ and*

$$U = \{ \, a + bx \, : \, a, b \in \mathbb{R} \, \}.$$

Notice that a basis for U is 1 and x. Again, we will compute $\operatorname{proj}_U v$. This time we first need to obtain an orthogonal basis for U. Using Gram-Schmidt we will obtain an orthogonal basis: We set $w_1 = 1$ and

$$w_2 = x - \frac{(x|1)}{(1|1)}(1) = x - \frac{\int_0^1 x \, dx}{\int_0^1 dx} = -\frac{1}{2} + x.$$

Then we normalize the orthogonal vectors to get an orthonormal basis for U.

$$u_1 = \frac{w_1}{|w_1|} = \frac{1}{\sqrt{\int_0^1 dx}} = 1.$$

$$u_2 = \frac{w_2}{|w_2|} = \frac{-(1/2) + x}{\sqrt{\int_0^1 (-(1/2) + x)^2 \, dx}} = -\sqrt{3} + 2\sqrt{3}x.$$

Then, using the formula for $\operatorname{proj}_U v$ in terms of an orthogonal basis, we have that $\operatorname{proj}_U v$ equals

$$(v|u_1)u_1 + (v|u_2)u_2 = (x^2|1)(1) + (x^2| - \sqrt{3} + 2\sqrt{3}x)(-\sqrt{3} + 2\sqrt{3}x)$$

$$= \left(\int_0^1 x^2 \, dx \right) + \left(\int_0^1 -\sqrt{3}x^2 + 2\sqrt{3}x^3 \, dx \right) (-\sqrt{3} + 2\sqrt{3}x) = -\frac{1}{6} + x.$$

Proposition 5.1.2 gives us an alternative way to compute the orthogonal projection of a vector v onto a subspace U. Let $U = \operatorname{span}(v_1, v_2, \ldots, v_k)$ a subspace in an inner product space V. Since $v - \operatorname{proj}_U v \in U^\perp$ this implies that $v - \operatorname{proj}_U v \perp v_i$ for $i = 1, 2, \ldots, k$, or equivalently $(v - \operatorname{proj}_U v|v_i) = 0$ for $i = 1, 2, \ldots, k$. Rewriting, we have $(\operatorname{proj}_U v|v_i) = (v|v_i)$ for $i = 1, 2, \ldots, k$. Since $\operatorname{proj}_U v \in U$ we can write $\operatorname{proj}_U v = a_1 v_1 + a_2 v_2 + \cdots + a_k v_k$, for some scalars $a_1, a_2, \ldots, a_k \in \mathbb{R}$. Therefore,

$$(a_1 v_1 + a_2 v_2 + \cdots + a_k v_k|v_i) = (v|v_i) \quad \text{for} \quad i = 1, 2, \ldots, k.$$

Equivalently, we need to solve the following $k \times k$ linear system of equations in unknowns $a_1, a_2, \ldots, a_k$:

$$a_1(v_1|v_i) + a_2(v_2|v_i) + \cdots + a_k(v_k|v_i) = (v|v_i) \quad \text{for} \quad i = 1, 2, \ldots, k.$$

Hence, we have come up with an alternative algorithm for computing $\operatorname{proj}_U v$ which reduces to solving a linear system instead of having to use the Gram-Schmidt process to find an orthonormal basis for the subspace U.

Example 5.28 *Let's redo Example 5.27 using this alternative algorithm for comput-ing* $\text{proj}_U v$ *where* $v = x^2$, $U = \{ a + bx : a, b \in \mathbb{R} \}$ *and the inner product is defined by* $(p|q) = \int_0^1 p(x)q(x) \, dx$. *Since* $U = \text{span}(1, x)$, *we seek an* $a + bx \in U$ *such that*

$$(a + bx|1) = (x^2|1) \quad and \quad (a + bx|x) = (x^2|x).$$

$$\Leftrightarrow \int_0^1 a + bx \, dx = \int_0^1 x^2 \, dx \quad and \quad \int_0^1 ax + bx^2 \, dx = \int_0^1 x^3 \, dx.$$

$$\Leftrightarrow a + \frac{1}{2}b = \frac{1}{3} \quad and \quad \frac{1}{2}a + \frac{1}{3}b = \frac{1}{4}.$$

Solving this system yields $a = \frac{1}{6}$ *and* $b = 1$ *and so* $\text{proj}_U v = \frac{1}{6} + x$.

With the definition of $\text{proj}_U v$ we can acquire some intuition as to why the Gram-Schmidt process produces an orthogonal basis. As in Theorem 5.1, let $v_1, \ldots, v_n$ be any basis for a Euclidean space V and $w_1, \ldots, w_n$ be the orthogonal basis produced by the Gram-Schmidt process. Set $U_k = span(w_1, \ldots, w_k)$ for $k = 1, \ldots, n$. Then $w_1 = v_1$, $w_2 = v_2 - \text{proj}_{U_1} v_2$ and in general

$$w_{k+1} = v_{k+1} - \text{proj}_{U_k} v_{k+1} \quad \text{for } k = 1, \ldots, n - 1.$$

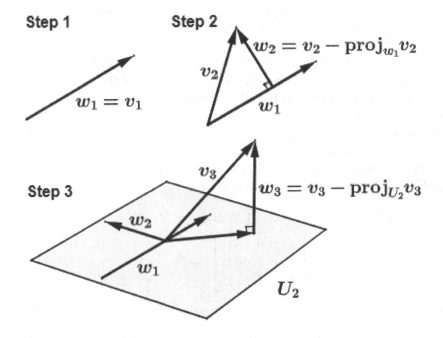

Figure 5.5 The Gram-Schmidt Process.

At each step of the process, we add a new vector w_{k+1} which is orthogonal to all the previous $w_1, \ldots, w_k$ (see Figure 5.5).

The next theorem brings us closer to our goal, which is to define a notion of *best* approximation. This theorem states that the closest vector in a given subspace U to a vector v is the vector $\text{proj}_U v$. This makes intuitive sense if we consider our previous example of $U = \text{span}(\hat{\imath}, \hat{\jmath})$. If we wish to minimize the distance $|v - u|$ for $u \in U$, since $v - \text{proj}_U v$ is perpendicular to U, we should take $u = \text{proj}_U v$.

Theorem 5.8 *Let U be a subspace of an inner product space V with $\dim U < \infty$ and take $v \in V$. Then for all $u \in U$ but $u \neq \text{proj}_U v$ we have $|v - \text{proj}_U v| < |v - u|$.*

Proof 5.15 *First, take any $u \in U$ but $u \neq \text{proj}_U v$ and write $v - u = (v - \text{proj}_U v) + (\text{proj}_U v - u)$ (note that $\dim U < \infty$ ensures that $\text{proj}_U v$ exists). Since $\text{proj}_U v - u \in U$ and we have seen that $v - \text{proj}_U v \in U^\perp$, we have $(v - \text{proj}_U v \mid \text{proj}_U v - u) = 0$. Now, by Lemma 5.1.g,*

$$|v - u|^2 = |v - \text{proj}_U v|^2 + |\text{proj}_U v - u|^2 > |v - \text{proj}_U v|^2 \quad (\text{since } u \neq \text{proj}_U v).$$

It follows that $|v - u| > |v - \text{proj}_U v|$. □

In the case of the inner product space $\mathbb{R}^n$ with the standard inner product, this approximation is called the **least squares** approximation, since if $v = [a_1, \ldots, a_n]$ and $u = [b_1, \ldots, b_n]$, then

$$|v - u| = \sqrt{(a_1 - b_1)^2 + \cdots + (a_n - b_n)^2}.$$

Hence, for a fixed v, if we minimize the sum of the squares $(a_1 - b_1)^2 + \cdots + (a_n - b_n)^2$, then we obtain the best approximation. We can view the $a_i - b_i$ for $i = 1, \ldots, n$ as the error between u and v in the ith coordinate.

EXERCISES

1. Compute $\text{comp} v$ and $\text{proj}_u v$ for each of the following inner product spaces V with inner product as defined in Section 5.1.

   a. $V = \mathbb{R}^5$ and $u = [1, 0, -2, -1, 2]$ and $v = [2, -1, -2, 0, 1]$.

   b. $V = M_{33}$ and $u = \begin{bmatrix} 1 & 0 & -1 \\ 2 & 1 & -1 \\ 0 & -1 & 2 \end{bmatrix}$ and $v = \begin{bmatrix} 2 & 1 & 0 \\ -1 & 1 & -2 \\ 2 & 0 & -1 \end{bmatrix}$

   c. $V = P_2$ and $u = 1 + 2x - x^2$ and $v = x - 2x^2$.

2. For each inner product space V given below with the inner product defined in Section 5.1 and subspace U and vector v, compute $\text{proj}_U v$ using the Gram-Schmidt process as was done in Example 5.27.

   a. $V = \mathbb{R}^3$ and $U = \{ [a, 2a + b, -a - 2b] \; : \; a, b \in \mathbb{R} \}$ and $v = [1, 1, 1]$.

b. $V = \mathbb{R}^4$ and $U = \{ [a+b, a, a-b, b] \ : \ a, b \in \mathbb{R} \}$ and $v = [1, 2, 0, -1]$.

c. $V = \mathbb{R}^4$ and $U = \{ [a, b, 2a-b, c] \ | \ a, b, c \in \mathbb{R} \}$ and $v = [1, -2, 1, -1]$.

d. $V = P_2$ and $U = \{ (a+b) + ax - bx^2 \ : \ a, b \in \mathbb{R} \}$ and $v = 1 - x - x^2$.

e. $V = M_{22}$ and $U = \left\{ \begin{bmatrix} a & b \\ 0 & 0 \end{bmatrix} \ : \ a, b \in \mathbb{R} \right\}$ and $v = I$.

f. $V = P_1$ and $U = \{a - 2ax \ | \ a \in F \}$ and $v = 1 - x$.

g. $V = M_{22}$ and U is the matrices in V where the sum of the non-diagonal entries is zero, and $v = \begin{bmatrix} 1 & 1 \\ 1 & 1 \end{bmatrix}$.

3. Redo do Exercise 2 by solving the appropriate linear system as was done in Example 5.28.

4. Prove that the magnitude of $\mathrm{proj}_u v$ equals $|\mathrm{comp}_u v|$.

5. Prove that if U is a finite dimensional subspace of an inner product space V, then $V/U \simeq U^{\perp}$.

6. Let V be the inner product space of continuous, real valued functions with inner product

$$(f|g) = \int_{-1}^{1} f(x)g(x) \ dx.$$

Set U equal to the subspace of all odd functions, i.e. $f(-x) = -f(x)$ for all $x \in \mathbb{R}$. Show that $U^{\perp}$ is the subspace of all even functions, i.e. $f(-x) = f(x)$ for all $x \in \mathbb{R}$.

Hints: First show that (1) Every even function is in $U^{\perp}$, then (2) Every function can be written as an even plus an odd function (very similar to how we proved every square matrix is a sum of a symmetric plus a skew-symmetric matrix).

7. For those who studied Section 4.9 and referring to Exercise 8 of Section 5.1, prove for U is a subspace of an inner product space V that $v \in U^{\perp}$ iff $v^* \in U^{\circ}$ (justifying in a sense the use of the name <u>annihilator</u> in two different settings).

8. Prove Proposition 5.1 parts b and c.

5.7 REAL SYMMETRIC MATRICES

In this section, we prove some facts about real symmetric matrices which will be used in a big way in later applied sections of this text. In this section A is an $n \times n$ symmetric matrix whose entries are real numbers. We now focus our attention on their eigenvalues and eigenvectors.

Theorem 5.9

a. *All the eigenvalues of a real symmetric matrix are real numbers with corresponding real eigenvectors. More specifically, if λ is a complex eigenvalue for a symmetric matrix with corresponding complex eigenvector $w = u + vi$, then $\lambda \in \mathbb{R}$ and either u or v is the eigenvector corresponding to λ.*

    b. *Any real symmetric matrix has a real eigenvalue with corresponding real eigenvector.*

Proof 5.16 *For part a, let λ be a (potentially) complex eigenvalue for a real symmetric matrix A with corresponding complex eigenvector $w = u + vi$. Since $Aw = \lambda w$, applying the conjugate to this equation we have $\overline{Aw} = \overline{\lambda w}$ or equivalently, $\overline{A}\,\overline{w} = \overline{\lambda}\,\overline{w}$. Since A has real number entries, we get $A\overline{w} = \overline{\lambda}\,\overline{w}$. Notice that*

$$\lambda \overline{w}^T w = \overline{w}^T \lambda w = \overline{w}^T A w = \overline{w}^T A^T w = (A\overline{w})^T w = (\overline{A}\overline{w})^T w = (\overline{\lambda}\overline{w})^T w = \overline{\lambda}\,\overline{w}^T w.$$

    This implies that $(\lambda - \overline{\lambda})\overline{w}^T w = 0$. Since $w \neq 0$, it must be that $\overline{\lambda} = \lambda$ and so $\lambda \in \mathbb{R}$.

    Now we show that w can be chosen in $\mathbb{R}^n$. Since $Aw = \lambda w$, this implies that $A(u + vi) = \lambda(u + vi) = \lambda u + \lambda vi$. Equating yields $Au = \lambda u$ and $Av = \lambda v$. Since $w \neq 0$ either $u \neq 0$ or $v \neq 0$, so either u or v is a real eigenvector for A corresponding to real eigenvalue λ.

    For part b, consider the characteristic polynomial $p_A(t)$. By the Fundamental Theorem of Algebra, $p_A(t)$ has a complex root and so A has an eigenvalue. By part a, that eigenvalue must be a real number with a corresponding real eigenvector. □

Theorem 5.10 *Eigenvectors from distinct eigenvalues of a real symmetric matrix are orthogonal.*

Proof 5.17 *Let $Av_1 = \lambda_1 v_1$ and $Av_2 = \lambda_2 v_2$ for $v_1, v_2 \neq 0$ and $\lambda_1 \neq \lambda_2$. Notice that*

$$\lambda_1 (v_1 \cdot v_2) = (\lambda_1 v_1) \cdot v_2 = (Av_1) \cdot v_2 = (Av_1)^T v_2$$

$$= v_1^T A^T v_2 = v_1^T A v_2 = v_1^T \lambda_2 v_2 = \lambda_2 v_1^T v_2 = \lambda_2 (v_1 \cdot v_2).$$

    This implies that $(\lambda_1 - \lambda_2)(v_1 \cdot v_2) = 0$. Since $\lambda_1 \neq \lambda_2$ this implies that $v_1 \cdot v_2 = 0$.
□

Theorem 5.11 *A real symmetric $n \times n$ matrix has n orthogonal real eigenvectors corresponding to real eigenvalues.*

Proof 5.18 *Let $V = \mathbb{R}^n$ and A be a real symmetric $n \times n$ matrix and let $T_A \in L(V)$ be defined as $T_A(v) = Av$. Note that the eigenvalues and eigenvectors for T_A and A are identical. We will prove this result by induction on the dimension of V.*

For $n = 1$, by Theorem 5.9.b, A has a real eigenvalue with corresponding real eigenvector v. This single vector serves as the required 1 orthogonal real eigenvector.

Now assume the statement of the theorem is true for vector spaces of dimension less than $n > 1$. Once again, by Theorem 5.9.b, A has a real eigenvalue λ with corresponding real eigenvectors $u_1, \ldots, u_k$ forming a basis for the eigenspace E_λ. Apply the Gram-Schmidt process to $u_1, \ldots, u_k$ to obtain an orthogonal basis for E_λ, say $v_1, \ldots, v_k$. If $k = n$, then we are done, otherwise set $U = E_\lambda$ and by Theorem 5.7, write $V = U \oplus U^\perp$.

We claim that if $w \in U^\perp$, then $Aw \in U^\perp$. To verify this it's enough to show that $Aw \cdot v_i = 0$ for $i = 1, \ldots, k$. Indeed,

$$Aw \cdot v_i = (Aw)^T v_i = w^T A^T v_i = w^T A v_i = w^T \lambda v_i = \lambda(w^T v_i) = \lambda(w \cdot v_i) = 0.$$

Therefore T_A is a linear operator on $U^\perp$. Now $\dim U^\perp < n$, so by induction T_A and therefore A has orthogonal real eigenvectors with real eigenvalues, say $v_{k+1}, \ldots, v_n$. Now since $v_{k+1}, \ldots, v_n \in U^\perp$ this implies that $v_1, v_2, \ldots, v_n$ are orthogonal. □

In lieu of Theorem 5.11 we have the following Corollary:

Corollary 5.3 *A real symmetric matrix is similar to a real diagonal matrix via an orthogonal matrix. In other words, if A is real symmetric, then there exists a diagonal matrix D and a matrix P such that $P^T P = I$ such that $P^T A P = D$.*

Proof 5.19 *Let A be a real symmetric $n \times n$ matrix. By Theorem 5.11 there exist real eigenvectors for A with respect to real eigenvalues which form an orthogonal basis for $\mathbb{R}^n$. Normalize these vectors and let them be the columns of an orthogonal matrix P. The rest of the proof relies on the fact that matrix representations of a linear operator are similar.* □

Example 5.29 *In this example we illustrate Corollary 5.3. Consider the 3×3 symmetric matrix*

$$A = \begin{bmatrix} 1 & 1 & 3 \\ 1 & 3 & 1 \\ 3 & 1 & 1 \end{bmatrix}.$$

One can compute the eigenvalues of A to be -2, 2 and 5 with corresponding eigenvectors (normalized) as $[-1/\sqrt{2}, 0, 1/\sqrt{2}]$, $[1/\sqrt{6}, -2/\sqrt{6}, 1/\sqrt{6}]$ and $[1/\sqrt{3}, 1/\sqrt{3}, 1/\sqrt{3}]$. Drop these vectors in columns to form the orthogonal matrix

$$P = \begin{bmatrix} -1/\sqrt{2} & 1/\sqrt{6} & 1/\sqrt{3} \\ 0 & -2/\sqrt{6} & 1/\sqrt{3} \\ 1/\sqrt{2} & 1/\sqrt{6} & 1/\sqrt{3} \end{bmatrix}.$$

The reader can check that $P^T P = I$ and that $P^T A P$ produces a diagonal matrix with the eigenvalues -2, 2 and 5 on the diagonal.

EXERCISES

1. Illustrate Corollary 5.3 as we did in Example 5.29 with each of the following real symmetric matrices:

   a. $A = \begin{bmatrix} 3 & 1 \\ 1 & 3 \end{bmatrix}$.

   b. $B = \begin{bmatrix} 3 & 1 & -1 \\ 1 & 3 & -1 \\ -1 & -1 & 5 \end{bmatrix}$.

   c. $C = \begin{bmatrix} 3 & -2 & 4 \\ -2 & 6 & 2 \\ 4 & 2 & 3 \end{bmatrix}$.

   d. $D = \begin{bmatrix} 6 & -2 & -1 \\ -2 & 6 & -1 \\ -1 & -1 & 5 \end{bmatrix}$.

2. Prove that a symmetric matrix is orthogonal iff all its eigenvalues are either 1 or -1.

5.8   SINGULAR VALUE DECOMPOSITION

As we have seen, not every square matrix is diagonalizable (although they are traingularizable). And certainly, not every square matrix is diagonalizable in the way symmetric matrices are, i.e. via an orthogonal matrix. The Singular Value Decomposition (SVD) of a matrix is perhaps the next best thing. It says any (not necessarily square) matrix $A = QSP$ where $Q^T Q = I = P^T P$ and S has zeros off the principal diagonal. In this section, we will derive this decomposition and look at some examples. There are many uses for this decomposition. In this section, we will see an application in data compression while in the next chapter we will apply this decomposition to a technique in data analytics called **dimensionality reduction**.

For $A \in M_{mn}$ consider the symmetric matrix $A^T A$. By Theorem 5.11, $A^T A$ has an orthonormal basis consisting of real eigenvectors $v_1, v_2, \ldots, v_n$ with corresponding real eigenvalues $\lambda_1 \geq \lambda_2 \geq \cdots \geq \lambda_n$ (put in descending order). We now show the eigenvalues of $A^T A$ are non-negative real numbers. Notice that for $i = 1, 2, \ldots, n$,

$$|Av_i|^2 = (Av_i)^T (Av_i) = v_i^T A^T A v_i = v_i^T (\lambda_i v_i) = \lambda_i v_i^T v_i = \lambda_i |v_i|^2.$$

Since $|Av_i|^2$ and $|v_i|^2$ are non-negative, then so is λ_i. Furthermore, a similar argument (left as an exercise) can be given to show that $Av_1, Av_2, \ldots, Av_n$ form an orthogonal set of vectors (not necessarily a basis, since some of the Av_i may be the zero vector when the eigenvector v_i corresponds to a zero eigenvalue).

Note that if $v_1, v_2, \ldots, v_n$ are normalized to form an orthonormal basis for $\mathbb{R}^n$ they are still eigenvectors of $A^T A$ with the same eigenvalues. Indeed, for $i = 1, 2, \ldots, n$,

$$A^T A u_i = A^T A \left(\frac{v_i}{|v_i|} \right) = \frac{1}{|v_i|} A^T A v_i = \frac{1}{|v_i|} \lambda_i v_i = \lambda_i \frac{v_i}{|v_i|} = \lambda_i u_i.$$

In this setting, $\lambda_i = |Au_i|^2$ and $Au_i = 0$ iff $\lambda_i = 0$. Putting this all together, we see that if $\lambda_1 \geq \lambda_2 \geq \cdots \geq \lambda_r > 0$ are the non-zero eigenvalues of $A^T A$, then $Av_1, Av_2, \ldots, Av_r$ is an orthogonal and linearly independent set of vectors. In addition, $Av_1, Av_2, \ldots, Av_r$ forms a basis for the rowspace of A and thus the rank of A equals r. In other words the rank of A is the number of non-zero eigenvalues of $A^T A$.

Theorem 5.12 *If $A \in M_{mn}$, then there exist $U \in M_{mm}$, $S \in M_{mn}$, $V \in M_{nn}$ such that $A = USV^T$ with $U^T U = I = V^T V$ and S is a matrix with zeros off the diagonal.*

Proof 5.20 *As detailed in the above remarks, we construct $v_1, v_2, \ldots, v_n$ an orthonormal basis for $\mathbb{R}^n$ consisting eigenvectors of $A^T A$ with corresponding eigenvalues $\lambda_1 \geq \lambda_2 \geq \cdots \geq \lambda_n \geq 0$. Let r equal the number of non-zero eigenvalues. As we saw above, $Av_1, Av_2, \ldots, Av_r$ is an orthogonal basis for the rowspace of A. Normalize this set of vectors, i.e. for $i = 1, 2, \ldots, r$ set*

$$u_i = \frac{Av_i}{|Av_i|} = \frac{1}{|Av_i|} Av_i = \frac{1}{\sqrt{\lambda_i}} Av_i.$$

Therefore, $Av_i = \sqrt{\lambda_i} u_i$ for $i = 1, 2, \ldots, r$. Extend via Gram-Schmidt, $u_1, u_2, \ldots, u_r$ to $u_1, u_2, \ldots, u_m$ an orthonormal basis for $\mathbb{R}^m$. Consider the matrices $V = [v_1 \ v_2 \ \cdots \ v_n]$ (in columns) and $U = [u_1 \ u_2 \ \cdots \ u_m]$ (in columns). As per our construction, we have $U^T U = I = V^T V$. Note that

$$AV = [Av_1 \ Av_2 \ \cdots \ Av_r \ 0 \ \cdots \ 0] = [\sqrt{\lambda_1} u_1 \ \sqrt{\lambda_2} u_2 \ \cdots \ \sqrt{\lambda_r} u_r \ 0 \ \cdots \ 0].$$

Set

$$D = diag(\sqrt{\lambda_1}, \sqrt{\lambda_2}, \ldots, \sqrt{\lambda_r}) \quad and \quad S = \left[\begin{array}{c|c} D & 0_{r,n-r} \\ \hline 0_{n-r,r} & 0_{n-r,n-r} \end{array} \right].$$

Then it follows that $AV = US$ and so $A = USV^T$. □

Definition 5.13 *The values on the principal diagonal the matrix S in Theorem 5.12, i.e. $\sqrt{\lambda_1}, \sqrt{\lambda_2}, \ldots, \sqrt{\lambda_n}$ are called the* **singular values** *of A.*

The standard notation for singular values is $\sigma_i = \sqrt{\lambda_i}$ for $i = 1, 2, \ldots, n$. We remind the reader that the number of non-zero singular values of A equals the rank of A.

Example 5.30 *We carefully compute a small example so that the reader can with understanding follow the steps of the proof which led to the singular value decomposition. Indeed, the proof of Theorem 5.12 is constructive and outlines an algorithm for this decomposition.*

$$Set \quad A = \begin{bmatrix} 3\sqrt{3} & -3\sqrt{3} & \sqrt{3} & \sqrt{3} \\ 0 & 0 & 2\sqrt{6} & 2\sqrt{6} \\ 3\sqrt{3} & -3\sqrt{3} & \sqrt{3} & \sqrt{3} \end{bmatrix}. \quad Then \quad A^T A = \begin{bmatrix} 45 & -45 & 27 & 27 \\ -45 & 45 & -27 & -27 \\ 27 & -27 & 45 & 45 \\ 27 & -27 & 45 & 45 \end{bmatrix}.$$

The eigenvalues are $\lambda_1 = 144$, $\lambda_2 = 36$, $\lambda_3 = 0$, $\lambda_4 = 0$ *with corresponding unit eigenvectors*

$$v_1 = [1/2, -1/2, 1/2, 1/2], \quad v_2 = [-1/2, 1/2, 1/2, 1/2],$$
$$v_3 = [0, 0, -1/\sqrt{2}, 1/\sqrt{2}], \quad v_4 = [1/\sqrt{2}, 1/\sqrt{2}, 0, 0].$$

Following the construction in the theorem,

$$S = \begin{bmatrix} 12 & 0 & 0 & 0 \\ 0 & 6 & 0 & 0 \\ 0 & 0 & 0 & 0 \end{bmatrix} \quad and$$

$$V = [v_1\ v_2\ v_3\ v_4] \ \ (in\ columns) \ \ = \begin{bmatrix} 1/2 & -1/2 & 0 & 1/\sqrt{2} \\ -1/2 & 1/2 & 0 & 1/\sqrt{2} \\ 1/2 & 1/2 & -1/\sqrt{2} & 0 \\ 1/2 & 1/2 & 1/\sqrt{2} & 0 \end{bmatrix}.$$

$$u_1 = \frac{1}{12} A v_1 = [\sqrt{3}/3, \sqrt{6}/6, \sqrt{2}/2] \quad and \quad u_2 = \frac{1}{6} A v_2 = [-\sqrt{3}/3, \sqrt{6}/6, 0].$$

We extend u_1, u_2 *to a basis for* $\mathbb{R}^3$ *by dropping* u_1, u_2 *together with* e_1, e_2, e_3 *into the columns of a matrix and row reduce*

$$\begin{bmatrix} \sqrt{3}/3 & -\sqrt{3}/3 & 1 & 0 & 0 \\ \sqrt{6}/6 & \sqrt{6}/6 & 0 & 1 & 0 \\ \sqrt{2}/2 & 0 & 0 & 0 & 1 \end{bmatrix} \quad to \quad \begin{bmatrix} 1 & 0 & 0 & 0 & \sqrt{2} \\ 0 & 1 & 0 & \sqrt{6}/2 & -\sqrt{2}/2 \\ 0 & 0 & 1 & \sqrt{2}/2 & -\sqrt{6}/2 \end{bmatrix}.$$

Looking at the pivots of the reduced row-echelon we form the basis is u_1, u_2, e_1. *Using Gram-Schmidt we create* u_1, u_2, u_3 *an orthonormal basis for* $\mathbb{R}^3$. *Since* u_1, u_2 *are already orthonormal we start with*

$$w_3 = e_1 - \left(\frac{e_1^T u_1}{|u_1|^2} u_1 + \frac{e_1^T u_2}{|u_2|^2} u_2 \right) = e_1 - \frac{\sqrt{3}}{3} u_1 - \frac{-\sqrt{3}}{3} u_2 = [1/3, \sqrt{2}/6, -\sqrt{6}/6]$$

Now, we normalize w_3 to get

$$u_3 = \frac{w_3}{|w_3|} = [\sqrt{3}/3, \sqrt{6}/6, -\sqrt{2}/2] \quad and$$

$$U = [u_1 \; u_2 \; u_3] \;\; (\textit{in columns}) \;\; = \begin{bmatrix} \sqrt{3}/3 & -\sqrt{3}/3 & \sqrt{3}/3 \\ \sqrt{6}/6 & \sqrt{6}/3 & \sqrt{6}/6 \\ \sqrt{2}/2 & 0 & -\sqrt{2}/2 \end{bmatrix}.$$

Hence, the decomposition is complete and one can check that $USV^T = A$.

Example 5.31 *This examples illustrates how SVD can be used as a compression algorithm. We start with a 129 by 154 grayscale image of my pug Penelope. It is a 129×154 matrix A filled with numbers from 0 (black) to 255 (white). We decompose $A = USV^T$ via SVD. We now create what are called **truncated** SVD decompositions which are "best" approximations to A. We do this by replacing with zero a certain number of the smallest singular values as well as dropping the corresponding columns of U and rows of V. Figure 5.6 illustrates this technique.*

Figure 5.6 Truncated SVD on a black and white image

From left to right, Photo 1 is the original 129 by 154 photo with 129 singular values. The singular values are 77.9271, 25.0281, 14.0747, 12.2264 down to the smallest value 0.0063. Photo 2 includes the top 29 singular values (which are all > 1). Photo 3 includes the top 15 singular values (> 2). Photo 4 includes only the top 4 singular values (> 10).

In terms of data compression even Photo 2 makes an incredible space saving. The original photo contains $129 \times 154 = 19866$ values from 0 to 255. Photo 2 has three matrices whose combined number of values is $(129 \times 29) + 29 + (29 \times 154) = 8236$, more than a 50% reduction in size without losing that much clarity in the image. Photo 3 has $(129 \times 15) + 15 + (15 \times 154) = 4260$ and Photo 4 has only 1136 values. Depending on the application, even the last photo may be sufficient (more to come in the Data Analytics section).

We will now carefully define what it means to be a "best" approximation of a matrix. It's pretty much the same as our previous least squared definition for n-tuples. Indeed, as we have seen, a matrix in M_{mn} can be viewed as an mn-tuple in $\mathbb{R}^{mn}$.

Definition 5.14 *Let $A, B \in M_{mn}$.*

1. *The distance between A and B, written*

$$d(A, B) = \sqrt{\sum_{i=1}^{m}\sum_{j=1}^{n}(a_{ij} - b_{ij})^2} \quad \text{where} \quad A = [a_{ij}], \ B = [b_{ij}].$$

2. *The norm of a matrix $A \in M_{mn}$, written $|A| = \sqrt{d(A, 0_{mn})}$, i.e. the square root of the sum of the squares of the entries in A.*

3. *Let $A \in M_{mn}$ and $X \subseteq M_{mn}$. We say $B \in X$ is the **best approximation** of A if B is the element in X closest to A (with respect to the distance defined above).*

As we have seen, the number of non-zero singular values of a matrix A equals the rank of A. When we consider the truncated SVD of A which uses s non-zero singular values we have created an approximation of A with rank s. We shall show now the following result:

Theorem 5.13 *Let $A \in M_{mn}$ with rank r. Out of all the $m \times n$ matrices of rank $s \leq r$, the truncated SVD of A with s non-zero singular values is the best approximation to A.*

Proof 5.21 *Let $\sigma_1, \sigma_2, \ldots, \sigma_n$ be the singular values of A. We first show for any $B \in M_{mn}$ that $|B|^2$ equals the sum of the squares of its singular values. To see this, using the SVD decomposition write $B = USV^T$ and set $B = [b_{ij}]$, $U = [u_{ij}]$, $V = [v_{ij}]$ and $S = [s_{ij}]$. Note that*

$$b_{ij} = \sum_{k=1}^{m}\sum_{l=1}^{n} = u_{ik}s_{kl}v_{jl} \quad \text{and so}$$

$$b_{ij}^2 = \sum_{k=1}^{m}\sum_{l=1}^{n}\sum_{r=1}^{m}\sum_{s=1}^{n} u_{ik}s_{kl}v_{jl}u_{ir}s_{rs}v_{js} = \sum_{k=1}^{m}\sum_{l=1}^{n}\sum_{r=1}^{m}\sum_{s=1}^{n} u_{ik}u_{ir}s_{kl}s_{rs}v_{jl}v_{js}.$$

Examining the terms of b_{ij}^2 and using the fact that $U^T U = I = V^T V$ and that S has zero entries off the diagonal,

$$b_{ij}^2 = \sum_{k=1}^{m}\sum_{s=1}^{n} u_{ik}u_{is}s_{kk}s_{ss}v_{jk}v_{js} \quad \text{and so}$$

$$|B|^2 = \sum_{i=1}^{m}\sum_{j=1}^{n}\sum_{k=1}^{m}\sum_{s=1}^{n} u_{ik}u_{is}s_{kk}s_{ss}v_{jk}v_{js} = \sum_{j=1}^{n}\sum_{k=1}^{m} s_{kk}s_{ss}v_{jk}v_{js} = \sum_{k=1}^{m} s_{kk}^2 = \sum_{k=1}^{m} s_{kk}^2.$$

Now assume B has rank s so that $|B|^2 = \sum_{k=1}^{s} s_{kk}^2$ and

$$d(A, B) = \sum_{k=1}^{s}(\sigma_k - s_{kk})^2 + \sum_{k=s+1}^{n} \sigma_k^2.$$

Therefore, the rank s matrix which minimizes $d(A, B)$ is the truncated SVD of A with only the first s singular values of A. ☐

EXERCISES

1. Compute the singular value decomposition for each of the following matrices:

   a. $A = \begin{bmatrix} 1 & 2 \\ 1 & 1 \\ 2 & 1 \end{bmatrix}.$

   b. $B = \begin{bmatrix} 1 & 1 & 2 \\ 3 & 1 & 1 \end{bmatrix}.$

   c. $C = \begin{bmatrix} 1 & -1 & 2 \\ 3 & 1 & 1 \\ 2 & 2 & -1 \\ 1 & 3 & -1 \end{bmatrix}.$

   d. $D = \begin{bmatrix} 1 & -1 & 2 \\ 1 & 1 & 1 \\ 2 & 2 & -1 \\ 1 & 0 & -1 \end{bmatrix}.$

   e. $E = \begin{bmatrix} 3 & 1 & -1 \\ 1 & 3 & -1 \\ -1 & -1 & 5 \end{bmatrix}.$

   f. $F = \begin{bmatrix} 1 & -1 & 1 & 1 \\ 1 & -1 & 1 & 1 \end{bmatrix}.$

2. Let $A \in M_{mn}$ and $v_1, v_2, \ldots, v_n$ be the orthogonal eigenvectors for the symmetric matrix $A^T A$. Prove that $Av_1, Av_2, \ldots, Av_n$ is an orthogonal set of vectors.

3. Prove that if A is symmetric, then SVD produces the same decomposition as diagonalizing the matrix does (up to an ordering of the columns).

4. Select a small black and white photo and explore different truncated SVD decompositions as we did with the photo of my pug Penelope.

5.9 APPLICATION: LEAST SQUARES OPTIMIZATION

In this section, we look at several applications of the use of *best* approximation developed in Section 5.6. We restrict ourselves to the standard inner product space, i.e. $\mathbb{R}^n$ with dot product for all but the last subsection.

5.9.1   Overdetermined Systems

For our first application, consider a system of linear equations $AX = B$, where $A \in M_{mn}$, X is a column of unknowns and $B \in \mathbb{R}^m$. In general, $AX = B$ may not have a solution. In this case we will say that $X = X_0$ is the **best approximation** to a solution to $AX = B$ if X_0 minimizes the value of $|AX_0 - B|$. In other words, the distance between AX_0 and B is minimized. This application is useful in the case of **overdetermined** in which the number of equations exceeds the number of unknowns and in the case of inconsistent systems. Overdetermined systems are at times inconsistent.

Let's put this notion into the context of the discussion in the previous section. Set $U = \{\, Av \mid v \in \mathbb{R}^n \,\}$. We are looking for an $AX_0 \in U$ such that $|AX_0 - B|$ is minimal. By Theorem 5.8, to find the best approximation to a solution to $AX = B$, we should choose $AX_0 = \mathrm{proj}_U B$. For the remainder of this section we seek a simple and direct way to compute X_0 and the conditions for when X_0 is unique, for although $AX_0 = \mathrm{proj}_U B$ is unique, it doesn't necessarily follow that X_0 is unique.

Recall that $B - \mathrm{proj}_U B \in U^\perp$ and as we have seen in Section 3.6, $U = \mathrm{colsp}(A)$. Therefore, by Lemma 5.5.ii, $U^\perp = \mathrm{colsp}(A)^\perp = \mathrm{nullsp}(A^T)$. Hence, $A^T(B - \mathrm{proj}_U B) = 0$ which implies that $A^T B = A^T \mathrm{proj}_U B = A^T A X_0$. Thus, X_0 is a solution to the $n \times n$ linear system $(A^T A)X = A^T B$. This equation is sometimes called the **normal equation** (basically because $AX_0 - B$ is normal to the range of $A^T A$). Furthermore, if $A^T A$ is invertible, then X_0 is uniquely determined as $X_0 = (A^T A)^{-1} A^T B$. Standard notation for $(A^T A)^{-1} A^T$ is $A^\dagger$ and is called the **pseudo-inverse** or **generalized inverse** of A, since it generalizes the method of solving a system $AX = B$ having a unique solution and because when A is invertible, $A^\dagger = A^{-1}$. Let's summarize our results in a theorem.

Theorem 5.14 *The best approximation to a solution to the $m \times n$ linear system $AX = B$ is precisely a solution to the $n \times n$ linear system $(A^T A)X = A^T B$. In the case when $A^T A$ is invertible, the unique solution is $X_0 = (A^T A)^{-1} A^T B$.*

Now, we give an equivalent and more straightforward condition for determining when $A^T A$ is invertible.

Lemma 5.7 *For $A \in M_{mn}(F)$, $A^T A$ is invertible iff the columns of A are linearly independent.*

Proof 5.22 *Assuming that $A^T A$ is invertible, we show that $AX = 0$ has only the trivial solution (and hence, by Theorem 3.4, we are done). If $Au = 0$ for some $u \in \mathbb{R}^n$, then $A^T A u = A^T 0 = 0$ and so u is a solution to $(A^T A)X = 0$. Now, by Theorem 3.11, $(A^T A)X = 0$ has only the trivial solution and so $u = 0$.*

Assuming now that the columns of A are linearly independent, we show that $(A^T A)X = 0$ has only the trivial solution (and so, by Theorem 3.11, we are done). Suppose that $(A^T A)u = 0$ for some $u \in \mathbb{R}^n$. Then Au is a solution to $A^T X = 0$, i.e. $Au \in \mathrm{nullsp}(A^T) = \mathrm{colsp}(A)^\perp$, by Lemma 5.5. By Lemma 3.8.ii, $Au \in \mathrm{colsp}(A)$ as well. Therefore, $Au \in \mathrm{colsp}(A)^\perp \cap \mathrm{colsp}(A) = \{\, 0 \,\}$. Thus, $Au = 0$ which means u is *a solution to $AX = 0$. Now, by Theorem 3.4 and our assumption, $AX = 0$ has only the trivial solution, i.e. $u = 0$.* □

Example 5.32 *Consider the following 3×2 linear system:*

$$
\begin{aligned}
x - y &= -1 \\
2x + y &= 3 \\
-x - 2y &= 2
\end{aligned}
$$

Notice that this system is inconsistent, since the corresponding augmented matrix

$$
\begin{bmatrix} 1 & -1 & -1 \\ 2 & 1 & 3 \\ -1 & -2 & 2 \end{bmatrix} \quad \text{reduces to} \quad \begin{bmatrix} 1 & -1 & -1 \\ 0 & 3 & 5 \\ 0 & 0 & 6 \end{bmatrix},
$$

and the last line of the matrix says that $0 = 6$, a contradiction. Notice also that the columns of the coefficient matrix $A = \begin{bmatrix} 1 & -1 \\ 2 & 1 \\ -1 & -2 \end{bmatrix}$ are linearly independent, since neither column is a scalar multiple of the other. Hence, by Theorems 5.14 and 5.7, the linear system has a unique best approximation, namely

$$
\begin{bmatrix} x \\ y \end{bmatrix} = \left(\begin{bmatrix} 1 & -1 \\ 2 & 1 \\ -1 & -2 \end{bmatrix}^T \begin{bmatrix} 1 & -1 \\ 2 & 1 \\ -1 & -2 \end{bmatrix} \right)^{-1} \begin{bmatrix} 1 & -1 \\ 2 & 1 \\ -1 & -2 \end{bmatrix}^T \begin{bmatrix} -1 \\ 3 \\ 2 \end{bmatrix}
$$

$$
= \left(\begin{bmatrix} 1 & 2 & -1 \\ -1 & 1 & -2 \end{bmatrix} \begin{bmatrix} 1 & -1 \\ 2 & 1 \\ -1 & -2 \end{bmatrix} \right)^{-1} \begin{bmatrix} 1 & 2 & -1 \\ -1 & 1 & -2 \end{bmatrix} \begin{bmatrix} -1 \\ 3 \\ 2 \end{bmatrix}
$$

$$
= \begin{bmatrix} 6 & 3 \\ 3 & 6 \end{bmatrix}^{-1} \begin{bmatrix} 3 \\ 0 \end{bmatrix} = \begin{bmatrix} 2/9 & -1/9 \\ -1/9 & 2/9 \end{bmatrix} \begin{bmatrix} 3 \\ 0 \end{bmatrix} = \begin{bmatrix} 2/3 \\ -1/3 \end{bmatrix}.
$$

In Figure 5.7, we plot the three lines in the linear system together with the best approximation to a solution, X_0.

Set $X_0 = \begin{bmatrix} 2/3 \\ -1/3 \end{bmatrix}$, our best approximation to the inconsistent system $AX = B$. Let's describe in another way what is happening geometrically. First we compute $\mathrm{colsp}(A)$. Since

$$
A^T = \begin{bmatrix} 1 & 2 & -1 \\ -1 & 1 & -2 \end{bmatrix} \quad \text{reduces to} \quad \begin{bmatrix} 1 & 0 & 1 \\ 0 & 1 & -1 \end{bmatrix},
$$

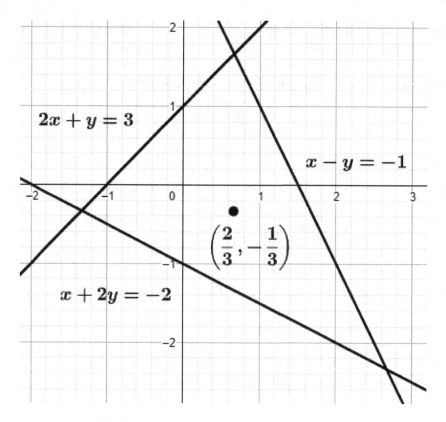

Figure 5.7 The plot of a linear system with its best solution in the xy-plane.

we have that

$$\text{colsp}(A) = \text{span}([1,0,1],[0,1,-1]) = \{ [a,b,a-b] \mid a,b \in \mathbb{R} \}.$$

This corresponds to the plane $z = x - y$ or $x - y - z = 0$. Second,

$$AX_0 = \begin{bmatrix} 1 \\ 1 \\ 0 \end{bmatrix}.$$

Hence, $(1,1,0)$ is the point in the plane $x - y - z = 0$ closest to $B = (-1,3,2)$ (see Figure 5.8).

Just to reiterate, X_0 is closest to a solution to $AX = B$, since it minimizes the value of

$$|AX_0 - B| = |[1,1,0] - [-1,3,2]| = |[2,-2,-2]| = \sqrt{4+4+4} = 2\sqrt{3}.$$

5.9.2 Best Fitting Polynomial

In our next application of best approximation, we put forth a method for fitting a collection of points in $\mathbb{R}^2$ with a polynomial. In general, given a collection of points

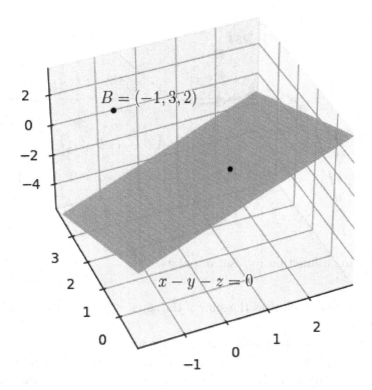

Figure 5.8 The plot a point in plane closest to a given point off the plane.

$(x_1, y_1), \ldots, (x_n, y_n)$, one hopes for a mathematical relationship between the variables x and y derived from these points. In other words, we want a function f with the property that $f(x_i) = y_i$ for all $i = 1, \ldots, n$. Examples of such functions were computed in the exercises in Section 2.2, Problem 3, Section 2.6, Problems 7 and 8, and Section 2.10, Problems 2, 7 and 9.

However, it is not always possible to obtain an exact fit for a given set of points. One can imagine an inconsistent system of equations arising in certain instances when we attempt to fit the points with a curve. In such a case we attempt to find a *best* fit to the points. In other words, we would like a *best approximation* to a function which fits the points. We will be relying heavily on the results derived in Section 5.6 to obtain this best approximation.

Consider Figure 5.9 depicting our set of points and a curve $y = f(x)$ approximating a fit to the points:

In the diagram, $e_i = f(x_i) - y_i$ for $i = 1, \ldots, n$ represents the error is approximating the y-coordinate of the ith point by the curve at $x = x_i$. Minimizing the sum of all these errors might seem a natural way to obtain a best fitting curve. These errors, however, are signed and therefore subject to cancellation when added. For this reason, we minimize the sum of the squares of the errors,

$$e_1^2 + e_2^2 + \cdots + e_n^2 = (f(x_1) - y_1)^2 + (f(x_2) - y_2)^2 + \cdots + (f(x_n) - y_n)^2.$$

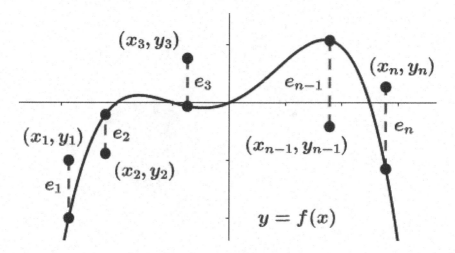

Figure 5.9 Fitting a function to a set of points.

Such a curve which minimizes the sum of the squares of the errors is called the **least squares** curve.

We now put this discussion in terms of what was done in the previous section. Set $Y = [y_1, y_2, \ldots, y_n]$ and $\hat{Y} = [f(x_1), f(x_2), \ldots, f(x_n)]$. We seek to minimize $|\hat{Y} - Y|^2$ or equivalently to minimize $|\hat{Y} - Y|$. Let's narrow our scope a bit and look for a best fitting polynomial function which minimizes $|\hat{Y} - Y|$. Set $f(x) = a_0 + a_1 x + \cdots + a_m x^m$. Notice that

$$\hat{Y} = [a_0 + a_1 x_1 + \cdots + a_m x_1^m, a_0 + a_1 x_2 + \cdots + a_m x_2^m, \ldots, a_0 + a_1 x_n + \cdots + a_m x_n^m]$$

$$= \begin{bmatrix} a_0 + a_1 x_1 + \cdots + a_m x_1^m \\ a_0 + a_1 x_2 + \cdots + a_m x_2^m \\ \vdots \\ a_0 + a_1 x_n + \cdots + a_m x_n^m \end{bmatrix} = \begin{bmatrix} 1 & x_1 & \cdots & x_1^m \\ 1 & x_2 & \cdots & x_2^m \\ \vdots & \vdots & \ddots & \vdots \\ 1 & x_n & \cdots & x_n^m \end{bmatrix} \begin{bmatrix} a_0 \\ a_1 \\ \vdots \\ a_m \end{bmatrix}.$$

Now we set

$$A = \begin{bmatrix} 1 & x_1 & \cdots & x_1^m \\ 1 & x_2 & \cdots & x_2^m \\ \vdots & \vdots & \ddots & \vdots \\ 1 & x_n & \cdots & x_n^m \end{bmatrix} \quad \text{and} \quad X_0 = \begin{bmatrix} a_0 \\ a_1 \\ \vdots \\ a_m \end{bmatrix}.$$

Hence, our goal is to find a X_0 which minimizes $|AX_0 - Y|$. But this is exactly the topic of discussion in the previous application where we found a best solution to an overdetermined system.

Theorem 5.15 *Let $(x_1, y_1), \ldots, (x_n, y_n)$ be a collection of points in $\mathbb{R}^2$ and $m < n$ a* *positive integer. If at least $m + 1$ of the x-coordinates of the points are distinct, then there exists a unique polynomial of degree m, $a_0 + a_1 x + \cdots + a_m x^m$ which best fits the points. The coefficients of this polynomial are given by*

$$\begin{bmatrix} a_0 \\ a_1 \\ \vdots \\ a_m \end{bmatrix} = (A^T A)^{-1} A^T Y,$$

where

$$A = \begin{bmatrix} 1 & x_1 & \cdots & x_1^m \\ 1 & x_2 & \cdots & x_2^m \\ \vdots & \vdots & \ddots & \vdots \\ 1 & x_n & \cdots & x_n^m \end{bmatrix} \quad and \quad Y = \begin{bmatrix} y_0 \\ y_1 \\ \vdots \\ y_m \end{bmatrix}.$$

Proof 5.23 *Theorem 5.14 assures us that such a polynomial exists, so the proof entails showing the uniqueness of this polynomial under the condition that at least $m + 1$ of the x-coordinates of the points are distinct. To show uniqueness, by Theorem 5.7, it is enough to show that A has linearly independent columns. Set $A = [c_0 \ c_1 \ \cdots \ c_m]$ as columns. Therefore we need to prove*

Claim: *If at least $m + 1$ of the x-coordinates of the points are distinct, then $c_0, c_1, \ldots, c_m$ are linearly independent.*

We will prove the contrapositive statement, namely if $c_0, c_1, \ldots, c_m$ are linearly dependent, then at most m of the x-coordinates of the points are distinct. So suppose that $c_0, c_1, \ldots, c_m$ are linearly dependent. Then there exist $b_0, b_1, \ldots, b_m \in \mathbb{R}$ not all zero such that $b_0 c_0 + b_1 c_1 + \cdots + b_m c_m = 0$. In expanded notation we have

$$\begin{bmatrix} b_0 + b_1 x_1 + \cdots + b_m x_1^m \\ b_0 + b_1 x_2 + \cdots + b_m x_2^m \\ \vdots \\ b_0 + b_1 x_n + \cdots + b_m x_n^m \end{bmatrix} = \begin{bmatrix} 0 \\ 0 \\ \vdots \\ 0 \end{bmatrix}.$$

Equating the entries gives

$$b_0 + b_1 x_1 + \cdots + b_m x_1^m = 0$$
$$b_0 + b_1 x_2 + \cdots + b_m x_2^m = 0$$
$$\vdots$$
$$b_0 + b_1 x_n + \cdots + b_m x_n^m = 0$$

which means that $x_1, x_2, \ldots, x_n$ are roots of the polynomial $g(x) = b_0 + b_1 x + \cdots + b_m x^m$. Now since the degree of g is at most m, this implies (by a basic fact about polynomials) that g has at most m distinct roots. Hence, at most m of the x-coordinates of the points are distinct. □

Example 5.33 *Let's first take the simplest setup and best fit three points with a line. Note that the theorem requires that the points not lie on the same vertical line (i.e. at least two of the x-coordinates are distinct). Consider the points* $(0,2), (1,0)$ *and* $(2,3)$*. Then*

$$A = \begin{bmatrix} 1 & 0 \\ 1 & 1 \\ 1 & 2 \end{bmatrix} \quad and$$

$$\begin{bmatrix} a_0 \\ a_1 \end{bmatrix} = \left(\begin{bmatrix} 1 & 0 \\ 1 & 1 \\ 1 & 2 \end{bmatrix}^T \begin{bmatrix} 1 & 0 \\ 1 & 1 \\ 1 & 2 \end{bmatrix} \right)^{-1} \begin{bmatrix} 1 & 0 \\ 1 & 1 \\ 1 & 2 \end{bmatrix}^T \begin{bmatrix} 2 \\ 0 \\ 3 \end{bmatrix}$$

$$= \begin{bmatrix} 3 & 3 \\ 3 & 5 \end{bmatrix}^{-1} \begin{bmatrix} 2 \\ 0 \\ 3 \end{bmatrix} = \begin{bmatrix} 5/6 & -1/2 \\ -1/2 & 1/2 \end{bmatrix} \begin{bmatrix} 2 \\ 0 \\ 3 \end{bmatrix} = \begin{bmatrix} 7/6 \\ 1/2 \end{bmatrix}.$$

Hence, the best fitting line is $y = \frac{7}{6} + \frac{1}{2}x$ *(see Figure 5.10).*

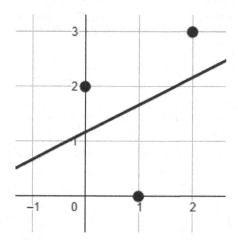

Figure 5.10 Fitting a line to a set of points.

Example 5.34 *Now let's fit five points with a parabola. Consider the five points* $(-1,-1), (-1,0), (0,-1), (1,0)$ *and* $(2,1)$*. Notice that four of the five x-coordinates are distinct and we need at least three being distinct to apply the theorem.*

$$A = \begin{bmatrix} 1 & -1 & 1 \\ 1 & -1 & 1 \\ 1 & 0 & 0 \\ 1 & 1 & 1 \\ 1 & 2 & 4 \end{bmatrix} \quad and$$

$$\begin{bmatrix} a_0 \\ a_1 \\ a_2 \end{bmatrix} = \left(\begin{bmatrix} 1 & -1 & 1 \\ 1 & -1 & 1 \\ 1 & 0 & 0 \\ 1 & 1 & 1 \\ 1 & 2 & 4 \end{bmatrix}^T \begin{bmatrix} 1 & -1 & 1 \\ 1 & -1 & 1 \\ 1 & 0 & 0 \\ 1 & 1 & 1 \\ 1 & 2 & 4 \end{bmatrix} \right)^{-1} \begin{bmatrix} 1 & -1 & 1 \\ 1 & -1 & 1 \\ 1 & 0 & 0 \\ 1 & 1 & 1 \\ 1 & 2 & 4 \end{bmatrix}^T \begin{bmatrix} -1 \\ 0 \\ -1 \\ 0 \\ 1 \end{bmatrix}$$

$$= \begin{bmatrix} 5 & 1 & 7 \\ 1 & 7 & 7 \\ 7 & 7 & 19 \end{bmatrix}^{-1} \begin{bmatrix} -1 \\ 3 \\ 3 \end{bmatrix} = \begin{bmatrix} 7/13 & 5/26 & -7/26 \\ 5/26 & 23/78 & -7/39 \\ -7/26 & -7/39 & 17/78 \end{bmatrix} \begin{bmatrix} -1 \\ 3 \\ 3 \end{bmatrix} = \begin{bmatrix} -10/13 \\ 2/13 \\ 5/13 \end{bmatrix}.$$

Therefore, the parabola we seek is $y = -\frac{10}{13} + \frac{2}{13}x + \frac{5}{13}x^2$ *(see Figure 5.11).*

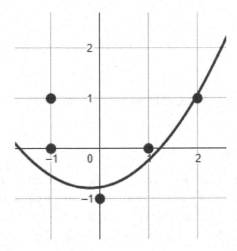

Figure 5.11 Fitting a parabola to a set of points.

Example 5.35 *This example illustrates the fact that Theorem 5.15 can be used to find an exact fit to a set of points (when it exists), thus giving an alternate method to Gaussian Elimination used in Chapter 2.*

We find a parabola which passes through the points $(0, -1), (1, 2)$ *and* $(-1, 0)$. *In this case,*

$$A = \begin{bmatrix} 1 & 0 & 0 \\ 1 & 1 & 1 \\ 1 & -1 & 1 \end{bmatrix} \quad and$$

$$\begin{bmatrix} a_0 \\ a_1 \\ a_2 \end{bmatrix} = \left(\begin{bmatrix} 1 & 0 & 0 \\ 1 & 1 & 1 \\ 1 & -1 & 1 \end{bmatrix}^T \begin{bmatrix} 1 & 0 & 0 \\ 1 & 1 & 1 \\ 1 & -1 & 1 \end{bmatrix} \right)^{-1} \begin{bmatrix} 1 & 0 & 0 \\ 1 & 1 & 1 \\ 1 & -1 & 1 \end{bmatrix}^T \begin{bmatrix} -1 \\ 2 \\ 0 \end{bmatrix} =$$

$$\begin{bmatrix} 3 & 0 & 2 \\ 0 & 2 & 0 \\ 2 & 0 & 2 \end{bmatrix}^{-1} \begin{bmatrix} 1 \\ 2 \\ 2 \end{bmatrix} = \begin{bmatrix} 1 & 0 & -1 \\ 0 & 1/2 & 0 \\ -1 & 0 & 3/2 \end{bmatrix} \begin{bmatrix} 1 \\ 2 \\ 2 \end{bmatrix} = \begin{bmatrix} -1 \\ 1 \\ 2 \end{bmatrix}.$$

Therefore, the parabola we seek is $y = -1 + x + 2x^2$ (see Figure 5.12).

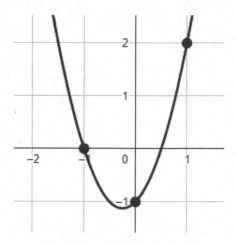

Figure 5.12 Fitting a parabola exactly to a set of points.

5.9.3 Linear Regression

In this subsection, our goal is to obtain the "best fit" of a set of data points in $\mathbb{R}^n$ to a hyperplane. First, let's establish our notation. Our data will be of the form $(x_1^{(i)}, x_2^{(i)}, \ldots, x_{d-1}^{(i)}, y_i)$ for $i = 1, 2, \ldots, n$ and our hyperplane will be $y = f(x_1, x_2, \ldots, x_{d-1})$. Since f represents a hyperplane, we can be more specific, namely

$$y = a_0 + a_1 x_1 + a_2 x_2 + \cdots + a_{d-1} x_{d-1}.$$

For each data point $(x_1^{(i)}, x_2^{(i)}, \ldots, x_{d-1}^{(i)}, y_i)$ as before define the error

$$e_i = f(x_1^{(i)}, x_2^{(i)}, \ldots, x_{d-1}^{(i)}) - y_i.$$

In order to find the best fit, we wish to minimize the sum of the squares of these errors, i.e.

$$\sum_{i=1}^{n} e_i^2 = \sum_{i=1}^{n} [f(x_1^{(i)}, x_2^{(i)}, \ldots, x_{d-1}^{(i)}) - y_i]^2 = \sum_{i=1}^{n} [a_0 + a_1 x_1^{(i)} + a_2 x_2^{(i)} + \cdots + a_{d-1} x_{d-1}^{(i)} - y_i]^2.$$

In other words, just as before, we are looking to minimize $|\hat{Y} - Y|^2$ or equivalently $|\hat{Y} - Y|$, where

$$\hat{Y} = [a_0 + a_1 x_1^{(1)} + a_2 x_2^{(1)} + \cdots + a_{d-1} x_{d-1}^{(1)}, \ldots, a_0 + a_1 x_1^{(n)} + a_2 x_2^{(n)} + \cdots + a_{d-1} x_{d-1}^{(n)}]$$

$$\text{and} \quad Y = [y_1, y_2, \ldots, y_n].$$

One can see that $\hat{Y} = A X_0$, where

$$A = \begin{bmatrix} 1 & x_1^{(1)} & x_2^{(1)} & \cdots & x_{d-1}^{(1)} \\ 1 & x_1^{(2)} & x_2^{(2)} & \cdots & x_{d-1}^{(2)} \\ \vdots & \vdots & \vdots & \ddots & \vdots \\ 1 & x_1^{(n)} & x_2^{(n)} & \cdots & x_{d-1}^{(n)} \end{bmatrix} \quad \text{and} \quad X_0 = \begin{bmatrix} a_0 \\ a_1 \\ a_2 \\ \vdots \\ a_{d-1} \end{bmatrix}.$$

Hence, we are looking to find an X_0 which minimizes the quantity $|A X_0 - Y|$. But just as before, we are looking to find a best solution to an overdetermined system which we already know how to find, namely $X_0 = (A^T A)^{-1} A^T Y$.

Example 5.36 *Consider the following set of data points:*

$$(1, 2, 1), (2, -1, 1), (-2, 1, 1), (-1, 1, -2), (1, 1, -2), (-2, -1, 1), (2, -1, -1),$$
$$(-2, -1, -2).$$

We seek a best fitting plane to this set of data. To this end we construct

$$A = \begin{bmatrix} 1 & 1 & 2 \\ 1 & 2 & -1 \\ 1 & -2 & 1 \\ 1 & -1 & 1 \\ 1 & 1 & 1 \\ 1 & -2 & -1 \\ 1 & 2 & -1 \\ 1 & -2 & -1 \end{bmatrix} \quad \text{and} \quad Y = \begin{bmatrix} 1 \\ 1 \\ 1 \\ -2 \\ -2 \\ 1 \\ -1 \\ -2 \end{bmatrix}.$$

Therefore,

$$\begin{bmatrix} a_0 \\ a_1 \\ a_2 \end{bmatrix} = (A^T A)^{-1} A^T \approx \begin{bmatrix} -0.3759 \\ 0.0271 \\ 0.0342 \end{bmatrix},$$

so that the equation of our plane is $z = -0.3759 + 0.0271x + 0.0342y$. *Figure 6.3 depicts the resulting best fitting plane to the data.*

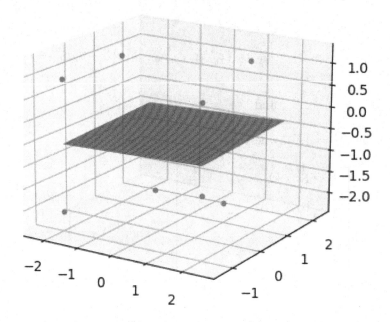

Figure 5.13 Fitting a plane to a set of points.

5.9.4 Underdetermined Systems

The next application deals with the case of **underdetermined** linear systems in which the number of unknowns exceeds the number of equations. Undetermined systems often have infinite solutions. Given an linear system $AX = B$ with infinite solutions, we wish to find the solution of smallest magnitude called the **minimal norm solution**.

Before addressing how to find such a minimal norm solution we present a more efficient way to compute $\text{proj}_U v$ in the special case that U is a subspace of $\mathbb{R}^n$ with the standard dot product. This algorithm will be ultimately used in finding the minimal norm solution.

Theorem 5.16 *Given a subspace U of $\mathbb{R}^n$ and $v \in \mathbb{R}^n$, then $\text{proj}_U v = (AA^\dagger)v$ where A is a matrix whose columns consist of any basis for U.*

Proof 5.24 *It was shown in Exercise 20 of Section 3.5 that $U = \{ Aw \; : \; w \in \mathbb{R}^m \}$. Therefore, there exists an $w^* \in \mathbb{R}^m$ such that $\text{proj}_U v = Aw^*$. Recall that $\text{proj}_U v$ minimizes the distance between v and U and from above*

$$|v - Aw^*| \leq |v - Aw| \quad \text{for all} \quad w \in \mathbb{R}^m$$

But this implies that w^* is a best approximation to a solution to $Aw = v$. Therefore, using the work in the first application of this section, $w^* = A^\dagger v$ and so $\text{proj}_U v = AA^\dagger v$ □

Example 5.37 *Let's redo the simple Example 5.26 we gave in the Section 5.6 for*
$V = \mathbb{R}^3$, $U = \text{span}(\hat{\imath}, \hat{\jmath})$ *and* $v = [1, 2, 3]$. *Then the matrix* $A = \begin{bmatrix} 1 & 0 \\ 0 & 1 \\ 0 & 0 \end{bmatrix}$ *and*

$$
\text{proj}_U v = \begin{bmatrix} 1 & 0 \\ 0 & 1 \\ 0 & 0 \end{bmatrix} \left(\begin{bmatrix} 1 & 0 \\ 0 & 1 \\ 0 & 0 \end{bmatrix}^T \begin{bmatrix} 1 & 0 \\ 0 & 1 \\ 0 & 0 \end{bmatrix} \right)^{-1} \begin{bmatrix} 1 & 0 \\ 0 & 1 \\ 0 & 0 \end{bmatrix}^T \begin{bmatrix} 1 \\ 2 \\ 3 \end{bmatrix}
$$

$$
= \begin{bmatrix} 1 & 0 & 0 \\ 0 & 1 & 0 \\ 0 & 0 & 0 \end{bmatrix} \begin{bmatrix} 1 \\ 2 \\ 3 \end{bmatrix} = \begin{bmatrix} 1 \\ 2 \\ 0 \end{bmatrix}.
$$

Theorem 5.17 *The minimal norm solution to an underdetermined linear system* $AX = B$ *is* $X^* = A^T(AA^T)^{-1}B$.

Proof 5.25 *Let* $AX = B$ *be an* $m \times n$ *undetermined linear system and set* $U = \text{nullsp}(A) = \{ X_h : AX_h = 0 \}$. *We've seen that the solution set to* $AX = B$ *can be expressed as* $\{ X_p + X_h : X_h \in U \}$ *for a fixed particular solution* X_p *to* $AX = B$. *Equivalently, we can write this solution set as* $\{ X_p - X_h : X_h \in U \}$. *Set* X^* *to be the minimal norm solution to* $AX = B$ *and let* X_h^* *be such that* $X^* = X_p - X_h^*$. *Since* X_h^* *minimizes* $|X_p - X_h|$ *this implies that* $X_h^* = \text{proj}_U X_p$ *and so, by Proposition 5.1,* $X^* = X_p - \text{proj}_U X_p \in U^\perp$. *By Lemma 5.5 and Corollary 5.2,* $U^\perp = \text{rowsp}(A^T)$. *Therefore,* $X^* \in \text{rowsp}(A^T)$ *and so* $X^* = A^TY$ *for some* $Y \in \mathbb{R}^m$. *Since* X^* *is a solution to* $AX = B$ *we have* $AA^TY = B$ *and so* $Y = (AA^T)^{-1}B$, *but then* $X^* = A^T(AA^T)^{-1}B$. □

Example 5.38 *When we find the point on the intersection of the two planes* $x + y + z = 1$ *and* $-x + y + z = 1$ *closest to the origin we are solving a minimal norm solution problem, namely we are finding the solution to*

$$
\begin{cases} x + y + z &= 1 \\ -x + y + z &= 1 \end{cases} \quad \text{of smallest magnitude.}
$$

Following Theorem 5.17, we set $A = \begin{bmatrix} 1 & 1 & 1 \\ -1 & 1 & 1 \end{bmatrix}$ *and* $B = \begin{bmatrix} 1 \\ 1 \end{bmatrix}$ *and compute the minimal norm solution to be*

$$
X^* = \begin{bmatrix} 1 & 1 & 1 \\ -1 & 1 & 1 \end{bmatrix}^T \left(\begin{bmatrix} 1 & 1 & 1 \\ -1 & 1 & 1 \end{bmatrix} \begin{bmatrix} 1 & 1 & 1 \\ -1 & 1 & 1 \end{bmatrix}^T \right)^{-1} \begin{bmatrix} 1 \\ 1 \end{bmatrix} = \begin{bmatrix} 0 \\ 1/2 \\ 1/2 \end{bmatrix}.
$$

Hence, the solution we seek is $(0, 1/2, 1/2)$. *The solution is illustrated in Figure 5.14.*

We make one last aesthetic observation about best solutions to overdetermined systems and minimal norm solutions to underdetermined systems: Notice the beautiful similarity between the two solutions, namely $X_0 = (A^TA)^{-1}A^TB$ and $X^* = A^T(AA^T)^{-1}B$.

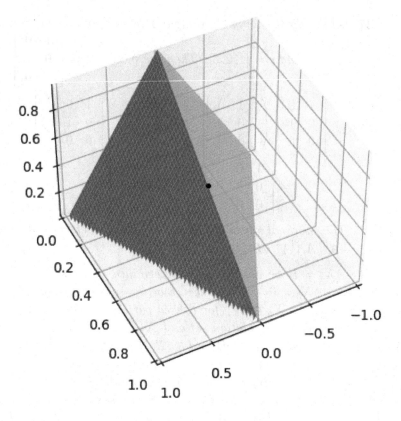

Figure 5.14 Minimal norm solution to a 2×3 linear system.

5.9.5 Approximating Functions

Our goal is to make a best approximation of a function by a linear combination of another collection of functions. Perhaps the reader has run across this in an integral calculus course, in particular, approximating a function by a Taylor polynomial, which is a linear combination of powers of x. Our best approximation will be in the sense of the inner product $(f(x)|g(x)) = \int_a^b f(x)g(x) \, dx$. For now we focus on approximating a function by a polynomial, say of degree k. Therefore, our vector space will be continuous functions on the interval $[a, b]$ and we consider the subspace U of polynomials of degree k or less. Now $U = \text{span}(1, x, \ldots, x^k)$ and if $f(x)$ is the function we wish to best approximate by an element in U, we therefore wish to compute $\text{proj}_U f(x)$. In other words we seek a polynomial $p(x)$ closest to $f(x)$ by minimizing

$$|p(x) - f(x)|^2 = \int_a^b (p(x) - f(x))^2 \, dx.$$

This we already know how to do as we saw in Section 5.6. We will take the second approach to solving this problem in which we solve a linear system as opposed to using the Gram-Schmidt process.

Example 5.39 *We will find the best fitting quadratic polynomial to* $\cos x$ *on the interval* $[-1, 1]$. *As before, the quadratic* $a + bx + cx^2$ *will satisfy the following equations:*

$$(a+bx+cx^2|1) = (\cos x|1), \quad (a+bx+cx^2|x) = (\cos x|x), \quad (a+bx+cx^2|x^2) = (\cos x|x^2).$$

Equivalently,

$$\int_{-1}^{1} (a + bx + cx^2) \, dx = \int_{-1}^{1} \cos x \, dx,$$

$$\int_{-1}^{1} (ax + bx^2 + cx^3) \, dx = \int_{-1}^{1} x \cos x \, dx,$$

$$\int_{-1}^{1} (ax^2 + bx^3 + cx^4) \, dx = \int_{-1}^{1} x^2 \cos x \, dx.$$

Computing these intergrals results in the following linear system:

$$\begin{cases} 2a + \frac{2}{3}c &= 2\sin 1 \\[2mm] \frac{2}{3}b &= 0 \\[2mm] \frac{2}{3}a + \frac{2}{5}c &= 4\cos 1 - 2\sin 1 \end{cases}.$$

Solving this 3×3 *linear system yields* $a \approx 0.9966$, $b = 0$ *and* $c \approx 0.4653$. *Therefore, our best approximating quadratic of* $\cos x$ *is* $0.9966 - 0.4653x^2$. *It's interesting to note that if one were to compute the 2nd degree Taylor polynomial centered at zero, the result would be* $1 - \frac{1}{2}x^2$ *which is a slightly different result from our best approximation. Recall that Taylor polynomials focus on approximating well a function close to its center, while the best approximation we computed is attempting to approximate well the function on the entire interval* $[-1, 1]$. *Figure 5.15 illustrates* $\cos x$ *as well as the two approximations just mentioned.*

At this point we could continue the discussion and present Fourier approximations and the Fourier series which would naturally follow from our discussion of best approximating functions and is a very important field of applied mathematics having many uses. However, we begin to go too far afield should we enter into that discourse. At the very least the reader should be reassured with the knowledge that they are now fully prepared for such a topic after having read this subsection.

One of the motivations for this particular subsection was to show a very concrete use for an inner product besides the usual dot product. We hope the reader agrees that the ability to approximate functions on intervals using an inner product besides the dot product is impressive.

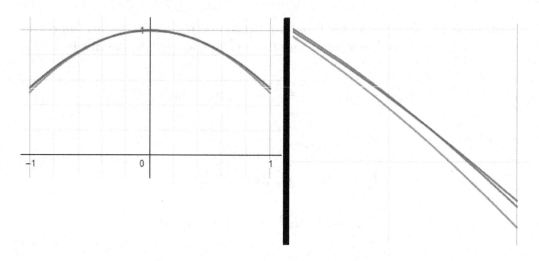

Figure 5.15 On the left is the graph of $\cos x$ together with the 2nd degree Taylor polynomial centered at zero and the best approximating quadratic. On the right is a close up of the right hand tails of same three graphs. In the close up, the best approximation is above, the Taylor polynomial is below and $\cos x$ is in between.

EXERCISES

1. Consider the following linear system:

$$
\begin{aligned}
y &= x + 1 \\
x + y &= 1 \\
2x - y &= 4
\end{aligned}
$$

a. Verify that the linear system is inconsistent.

b. Find the best approximation to a solution to the system.

c. Plot the lines in the system as well as the best approximation.

2. Repeat the previous exercise with the following linear system:

$$
\begin{aligned}
x &= 2 - 2y \\
x &= 3y + 3 \\
y &= -2x \\
y &= x
\end{aligned}
$$

3. Repeat the previous exercise with the following linear system:

$$
\begin{aligned}
4x - 2y + 2z &= -3 \\
2x + 5y + z &= 1 \\
-2x + y - z &= 2
\end{aligned}
$$

4. Consider the following points: $(1, 2), (0, 1), (0, -1)$

   a. Find the best fitting line

   b. Explain why the theorem doesn't apply for a best fitting parabola.

5. Consider the following points: $(1, -1), (2, 2), (-1, 2), (0, 1)$.

   a. Find the best fitting line.

   b. Find the best fitting parabola.

   c. Find the exact fitting cubic (third degree polynomial).

   d. Plot the points with your results from parts a,b and c.

6. Consider the following points: $(-2, -2), (-1, 0), (0, 2), (1, -1), (2, -2)$.

   a. Find the best fitting parabola.

   b. Find the best fitting cubic.

   c. Plot the points with your results from parts a and b.

7. Find an exact fit for the points $(1, 0), (2, 2), (-2, -1)$ with a parabola and plot the points with your result.

8. Recompute $\text{proj}_U v$ in Exercise 5.6 for Problem 2.a,b,c using the more efficient method discussed in this section.

9. Find the point on the intersection of the planes $x + 2y - z = 1$ and $2x - y + z = -1$ closest to the origin.

10. Find the minimal norm solution to the following underdetermined system:

$$\begin{cases} w - y &= 0 \\ w - x &= -1 \\ z - w &= 2 \end{cases}$$

11. Find the best approximating linear polynomial to the function e^x on the interval $[0, 1]$, then plot the two functions on the same graph.

12. Find the best approximating cubic polynomial to the function $\sin x$ on the interval $[-1, 1]$, then plot the two functions on the same graph.

CHAPTER 6

Applications in Data Analytics

IN THIS CHAPTER, some examples of linear algebra applied to the field of data analytics and machine learning are given. Topics in the field of data analytics and machine learning were sought which applied linear algebra techniques and were easily within grasp without too much background development. Section 6.1 is a general introduction to the topics presented. Section 6.2 determines in what direction a data set is most spread. Section 6.3 presents the multi-use topic of principal component analysis. Section 6.4 introduces an integral technique in data analytics called dimension reduction. A distance more statistically based is presented in Section 6.5 called Mahalanobis distance. A useful tool called data sphering is developed in Section 6.6. The remaining three sections deal with linear discriminant functions. In Section 6.7, the Fisher linear discriminant function is presented while the minimal square error linear discriminant function is presented in Section 6.9. The general notion of a linear discriminant function is discussed in Section 6.8.

6.1 INTRODUCTION

The general topics to be discussed in this chapter fall under three categories: Classification, feature selection/reduction and data preprocessing.

In Machine Learning, in the subdiscipline of classification, a set of data might represent the *features* of objects, called **classes** of data. When we speak of **feature data** we are referring to aspects of an object under consideration. It may be **numerical** or **categorical**. For instance, consider a glass of wine. A numerical feature would be the percentage of sugar in the wine, while a categorical feature might be the grape used to make the wine.

Oftentimes we wish to identify/classify/distinguish different classes. For instance, maybe we wish to distinguish between two types of handwritten digits [8]: Written zeros and written ones (see Figure 6.1).

Perhaps we collect two features on the handwritten digits: Number of pixels in the digit and Minimal distance from the pixels in the digit to its mean (disregarding

DOI: 10.1201/9781003217794-6

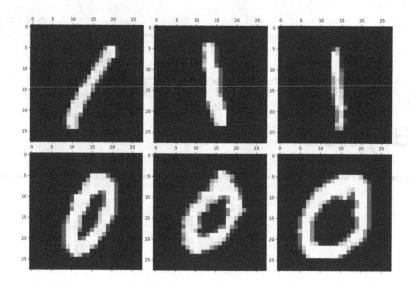

Figure 6.1 28×28 Handwritten zeros and ones.

the intensity of the pixels in the digit). We then plot the two classes in what we call **feature space**. We can distinguish the two classes by creating a linear demarkation between the two data sets called a **linear discriminant function** (LDF). Figure 6.2 depicts an example of an LDF dividing two classes.

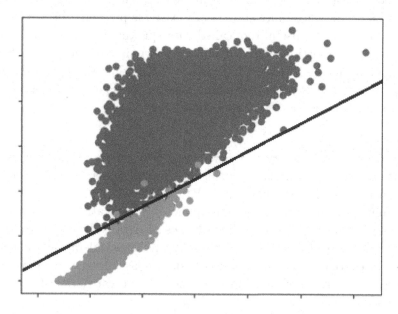

Figure 6.2 Two classes divided by an LDF in 2D feature space.

Feature selection and reduction aims to reduce the dimension of a data set while minimizing the reduction in information lost and predictive power. This is done so that algorithms which use the data will be more computationally efficient. Oftentimes a data set can be unruly due to the fact that the data points are of high dimension. For instance if we need to run an algorithm on this data the high dimension can cause the algorithm to run very slow to the point where we are unable to perform our intended data analysis. In addition, high dimensional data can exist in a sparse distribution across this high dimensional space and this can also cause issues. For instance, the distance between points in the data set tends to lose significance if it is needed in our data analysis. These issues are often referred to as the **curse of dimensionality**. For this reason it is useful to reduce the dimension of our data set in such a way as to minimize the loss of information. This is aptly called **dimensionality reduction** and there are numerous ways to do this.

Data preprocessing refers to preliminary manipulation of data sets in order to make them more amenable to data analytic applications. One such technique which will be presented in this chapter is called data sphering.

As we shall see many of the first applications in this chapter center around a matrix we shall define and which will be called the *scatter matrix* of a data set. Finally, the reader should take note that only in this chapter will vectors have the conventional arrow notation primarily to conform better to existing literature in this field.

6.2 DIRECTION OF MAXIMAL SPREAD

The term *projection pursuit* was coined by Friedman and Tukey. It is a means of analyzing data sets and finding interesting structural properties. This is done by projecting multi-dimensional data onto lines in all possible directions through the mean and finding the direction which maximizes or minimizes a particular property of interest.

Consider a collection of data points $\vec{x}_1, \vec{x}_2, \ldots, \vec{x}_n \in \mathbb{R}^d$ with mean $\vec{\mu}$. Let ℓ be a line passing through $\vec{\mu}$ and $\vec{a}$ be a unit vector parallel to ℓ. For a given data point $\vec{x}_k$, let x_k be the projection of $\vec{x}_k$ on the line ℓ.

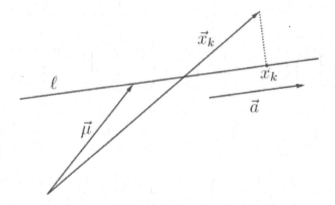

Figure 6.3 Projection of a point onto a line through the mean of a data set.

Recall that x_k is the component of $\vec{x}_k - \vec{\mu}$ along $\vec{a}$, i.e.

$$x_k = \frac{(\vec{x}_k - \vec{\mu}) \cdot \vec{a}}{||\vec{a}||} = (\vec{x}_k - \vec{\mu}) \cdot \vec{a} = \vec{a}^T(\vec{x}_k - \vec{\mu}).$$

The variance of all the projections $x_1, x_2, \ldots, x_m$ is then

$$\frac{1}{n}\sum_{k=1}^{n}\left[\vec{a}^T(\vec{x}_k - \vec{\mu})\right]^2$$

Note, we are viewing vectors as $d \times 1$ column matrices. We seek to find the line through the mean of the data which maximizes the variance. We can dispense with the fraction $\frac{1}{n}$ and simply maximize

$$\sum_{k=1}^{n}\left[\vec{a}^T(\vec{x}_k - \vec{\mu})\right]^2 = \sum_{k=1}^{n}\vec{a}^T(\vec{x}_k - \vec{\mu})\vec{a}^T(\vec{x}_k - \vec{\mu})$$

$$= \sum_{k=1}^{n}\vec{a}^T(\vec{x}_k - \vec{\mu})(\vec{x}_k - \vec{\mu})^T\vec{a} = \vec{a}^T\left(\sum_{k=1}^{n}(\vec{x}_k - \vec{\mu})(\vec{x}_k - \vec{\mu})^T\right)\vec{a}$$

Set $S = \sum_{k=1}^{n}(\vec{x}_k - \vec{\mu})(\vec{x}_k - \vec{\mu})^T$ called the $d \times d$ **scatter matrix** of the data.

Example 6.1 *Consider the four data points* $(0, 1)$, $(0, -1)$, $(1, 2)$, $(3, 2)$. *The scatter matrix is computed as follows:*

$$\vec{\mu} = \frac{1}{4}\left(\begin{bmatrix} 0 \\ 1 \end{bmatrix} + \begin{bmatrix} 0 \\ -1 \end{bmatrix} + \begin{bmatrix} 1 \\ 2 \end{bmatrix} + \begin{bmatrix} 3 \\ 2 \end{bmatrix}\right) = \begin{bmatrix} 1 \\ 1 \end{bmatrix}.$$

$$S = \left(\begin{bmatrix} 0 \\ 1 \end{bmatrix} - \begin{bmatrix} 1 \\ 1 \end{bmatrix}\right)\left(\begin{bmatrix} 0 \\ 1 \end{bmatrix} - \begin{bmatrix} 1 \\ 1 \end{bmatrix}\right)^T + \left(\begin{bmatrix} 0 \\ -1 \end{bmatrix} - \begin{bmatrix} 1 \\ 1 \end{bmatrix}\right)\left(\begin{bmatrix} 0 \\ -1 \end{bmatrix} - \begin{bmatrix} 1 \\ 1 \end{bmatrix}\right)^T +$$

$$\left(\begin{bmatrix} 1 \\ 2 \end{bmatrix} - \begin{bmatrix} 1 \\ 1 \end{bmatrix}\right)\left(\begin{bmatrix} 1 \\ 2 \end{bmatrix} - \begin{bmatrix} 1 \\ 1 \end{bmatrix}\right)^T + \left(\begin{bmatrix} 3 \\ 2 \end{bmatrix} - \begin{bmatrix} 1 \\ 1 \end{bmatrix}\right)\left(\begin{bmatrix} 3 \\ 2 \end{bmatrix} - \begin{bmatrix} 1 \\ 1 \end{bmatrix}\right)^T$$

$$= \begin{bmatrix} -1 \\ 0 \end{bmatrix}\begin{bmatrix} -1 & 0 \end{bmatrix} + \begin{bmatrix} -1 \\ -2 \end{bmatrix}\begin{bmatrix} -1 & -2 \end{bmatrix} + \begin{bmatrix} 0 \\ 1 \end{bmatrix}\begin{bmatrix} 0 & 1 \end{bmatrix} + \begin{bmatrix} 2 \\ 1 \end{bmatrix}\begin{bmatrix} 2 & 1 \end{bmatrix}$$

$$= \begin{bmatrix} 1 & 0 \\ 0 & 0 \end{bmatrix} + \begin{bmatrix} 1 & 2 \\ 2 & 4 \end{bmatrix} + \begin{bmatrix} 0 & 0 \\ 0 & 1 \end{bmatrix} + \begin{bmatrix} 4 & 2 \\ 2 & 1 \end{bmatrix} = \begin{bmatrix} 6 & 4 \\ 4 & 6 \end{bmatrix}.$$

Note that this is not the most efficient method for finding the scatter matrix, but we did it this way to illustrate its definition. An easier way would be to subtract the mean from each point and put them in the rows of a matrix A and then compute A^TA. Indeed,

$$A^T A = \begin{bmatrix} -1 & -1 & 0 & 2 \\ 0 & -2 & 1 & 1 \end{bmatrix} \begin{bmatrix} -1 & 0 \\ -1 & -2 \\ 0 & 1 \\ 2 & 1 \end{bmatrix} = \begin{bmatrix} 6 & 4 \\ 4 & 6 \end{bmatrix}.$$

In order to find the direction of maximal variance we need to solve the following constrained optimization problem: Maximize $\vec{a}^T S \vec{a}$ subject to $\vec{a}^T \vec{a} = 1$. We use Lagrange multipliers to do this and solve the vector equation

$$\nabla \left[\vec{a}^T S \vec{a} \right] = \lambda \nabla \left[\vec{a}^T \vec{a} \right]$$

Equivalently,

$$2S\vec{a} = \lambda(2\vec{a}) \quad \text{or} \quad S\vec{a} = \lambda \vec{a}.$$

In other words, the direction $\vec{a}$ which maximizes variance is an eigenvector of the scatter matrix. The question now is which of the eigenvectors is the direction of maximal variance. Notice, if $\vec{a}^*$ is an eigenvector of S with respect to the eigenvalue λ^*, then the variance in that direction equals

$$\frac{1}{n} [\vec{a}^*]^T S \vec{a}^* = \frac{1}{n} [\vec{a}^*]^T \lambda^* \vec{a}^* = \frac{1}{n} \lambda^* [\vec{a}^*]^T \vec{a}^* = \frac{1}{n} \lambda^*$$

Thus, the eigenvector corresponding to the largest eigenvalue is the direction of maximal variance.

Example 6.2 *(revisited) In Example 6.1, the eigenvalues of the scatter matrix are 2 and 10 with corresponding eigenvectors*

$$\begin{bmatrix} -\sqrt{2}/2 \\ \sqrt{2}/2 \end{bmatrix} \quad and \quad \begin{bmatrix} \sqrt{2}/2 \\ \sqrt{2}/2 \end{bmatrix}.$$

Therefore, the direction of maximal variance corresponds to the angle $\theta = \tan^{-1}(1) = \pi/4$.

Figure 6.4 shows the data plus their projections onto the line of maximal spread.

Now let's do an exhaustive search of projected variance for this data set. Let $\vec{a} = [\cos\theta, \sin\theta]$ where θ varies from 0 to 2π. We've seen that the variance of the projected data is then

$$\frac{1}{4} \vec{a}^T S \vec{a} = \frac{1}{4} \begin{bmatrix} \cos\theta & \sin\theta \end{bmatrix} \begin{bmatrix} 6 & 4 \\ 4 & 6 \end{bmatrix} \begin{bmatrix} \cos\theta \\ \sin\theta \end{bmatrix}$$

$$= \frac{1}{4} \begin{bmatrix} \cos\theta & \sin\theta \end{bmatrix} \begin{bmatrix} 6\cos\theta + 4\sin\theta \\ 4\cos\theta + 6\sin\theta \end{bmatrix} = \frac{3}{2}\cos^2\theta + 2\cos\theta\sin\theta + \frac{3}{2}\sin^2\theta.$$

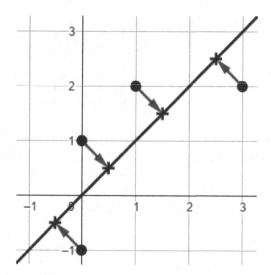

Figure 6.4 Projection of points on the line producing the maximal spread.

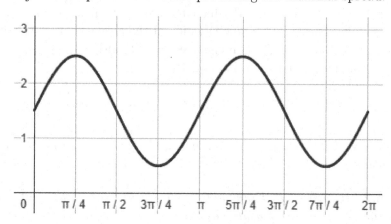

Figure 6.5 The graph of angle versus projected variance

Figure 6.5 shows the graph of this function.

As one can see, the angle where variance is maximal is roughly $\frac{\pi}{4}$. This confirms the exact value of θ computed using eigenvalues of S is $\frac{\pi}{4}$.

EXERCISES

1. Consider the following two dimensional data:

$$\{ (0,0),\ (0,1),\ (1,1),\ (1,2),\ (2,3) \}.$$

   a. Following the same steps as in Example 6.1 in this section, compute the scatter matrix for this data.

   b. Following the same steps as in Example 6.2 in this section, compute the direction of maximal variance and the corresponding angle θ.

c. Following the same steps as in Example 6.2 in this section, plot the graph of angle versus projected variance.

2. Consider the following three dimensional data:

$$\{ (0,0,0), (0,1,1), (1,1,1), (1,2,1), (1,2,3) \}.$$

a. Following the same steps as in Example 6.1 in this section, compute the scatter matrix for this data.

b. Following similar steps to Example 6.2 in this section, compute the direction of maximal variance and the corresponding angles ϕ and θ (spherical coordinates).

c. Plot a surface with axes θ, ϕ and the projected maximal variance for values $0 \leq \theta \leq 2\pi$ and $0 \leq \phi \leq \pi$.

6.3 PRINCIPAL COMPONENT ANALYSIS

In this section, we consider a way of "best" representing data by a lower dimensional set of data called Principal Component Analysis (PCA). Consider a collection of data points $X = \{ \vec{x}_1, \vec{x}_2, \ldots, \vec{x}_n \} \subseteq \mathbb{R}^d$ with mean $\vec{\mu}$. We will now ask a series of questions which will motivate this topic.

Zero Dimensional Question: What point $\vec{x}_0 \in \mathbb{R}^d$ "best represents" the data points X?

Our definition of "best represents" will be a least squares definition. In other words, we want a point $\vec{x}_0 \in \mathbb{R}^d$ which minimizes the quantity

$$\sum_{k=1}^{n} ||\vec{x}_k - \vec{x}_0||^2.$$

Claim: $\vec{x}_0 = \vec{\mu}$.

To see this, notice that

$$\sum_{k=1}^{n} ||\vec{x}_k - \vec{x}_0||^2 = \sum_{k=1}^{n} ||(\vec{x}_k - \vec{\mu}) - (\vec{x}_0 - \vec{\mu})||^2 = \sum_{k=1}^{n} [(\vec{x}_k - \vec{\mu}) - (\vec{x}_0 - \vec{\mu})] \cdot [(\vec{x}_k - \vec{\mu}) - (\vec{x}_0 - \vec{\mu})]$$

$$\sum_{k=1}^{n} ||\vec{x}_k - \vec{\mu}||^2 - 2\sum_{k=1}^{n} (\vec{x}_0 - \vec{\mu}) \cdot (\vec{x}_k - \vec{\mu}) + \sum_{k=1}^{n} ||\vec{x}_0 - \vec{\mu}||^2$$

Note that

$$\sum_{k=1}^{n} (\vec{x}_0 - \vec{\mu}) \cdot (\vec{x}_k - \vec{\mu}) = (\vec{x}_0 - \vec{\mu}) \cdot \sum_{k=1}^{n} (\vec{x}_k - \vec{\mu}) = (\vec{x}_0 - \vec{\mu}) \cdot \left(\sum_{k=1}^{n} \vec{x}_k - n\vec{\mu} \right) = (\vec{x}_0 - \vec{\mu}) \cdot \vec{0} = 0.$$

Therefore,

$$\sum_{k=1}^{n} ||\vec{x}_k - \vec{x}_0||^2 = \sum_{k=1}^{n} ||\vec{x}_k - \vec{\mu}||^2 + \sum_{k=1}^{n} ||\vec{x}_0 - \vec{\mu}||^2.$$

Now the first term in the sum above is a constant and the second can be made smallest (indeed, zero) when $\vec{x}_0 = \vec{\mu}$, which proves the Claim. Hence, $\vec{\mu}$ is the best zero dimensional representation of X.

One Dimensional Question: What line "best represents" the data points X?

Let's assume our line passes through $\vec{\mu}$, since $\vec{\mu}$ is the best zero dimensional representation of the data points. Let $\vec{a}$ be a unit vector parallel to the line through $\vec{\mu}$. Then the vector equation of the line is

$$\vec{x} = \vec{\mu} + t\vec{a} \quad \text{where} \quad t \in \mathbb{R}.$$

This One Dimensional Question involves one subordinate question, namely

One Dimensional Subquestion: How do we "best represent" a data point on a line?

Our definition of "best represents" will again be a least squares definition. In other words for all $\vec{x}_k$ we wish to find $t_k \in \mathbb{R}$ which minimizes the quantity

$$\sum_{k=1}^{n} ||(\vec{\mu} + t_k \vec{a}) - \vec{x}_k||^2.$$

Set $L(\vec{t}) = \sum_{k=1}^{n} ||(\vec{\mu} + t_k \vec{a}) - \vec{x}_k||^2$. Notice that

$$L(\vec{t}) = \sum_{k=1}^{n} ||t_k \vec{a} - (\vec{x}_k - \vec{\mu})||^2 = \sum_{k=1}^{n} [t_k \vec{a} - (\vec{x}_k - \vec{\mu})] \cdot [t_k \vec{a} - (\vec{x}_k - \vec{\mu})]$$

$$= \sum_{k=1}^{n} ||t_k \vec{a}||^2 - 2 \sum_{k=1}^{n} t_k \vec{a} \cdot (\vec{x}_k - \vec{\mu}) + \sum_{k=1}^{n} ||\vec{x}_k - \vec{\mu}||^2 = \sum_{k=1}^{n} t_k^2 - 2 \sum_{k=1}^{n} t_k \vec{a} \cdot (\vec{x}_k - \vec{\mu}) + \sum_{k=1}^{n} ||\vec{x}_k - \vec{\mu}||^2.$$

We wish to minimize L, so we first find critical points by solving $\nabla L = \vec{0}$. This leads to the equations

$$\frac{\partial L}{\partial t_k} = 2t_k - 2\vec{a} \cdot (\vec{x}_k - \vec{\mu}) = 0 \quad \text{or} \quad t_k = \vec{a} \cdot (\vec{x}_k - \vec{\mu}).$$

In other words, t_k is the component of $\vec{x}_k - \vec{\mu}$ along $\vec{a}$, which answers the subquestion. Set

$$\vec{t}^* = [\vec{a} \cdot (\vec{x}_1 - \vec{\mu}), \ \vec{a} \cdot (\vec{x}_2 - \vec{\mu}), \dots, \vec{a} \cdot (\vec{x}_n - \vec{\mu})].$$

If we evaluate L at $\vec{t}^*$, we have

$$L(\vec{t}^*) = \sum_{k=1}^{n} [\vec{a} \cdot (\vec{x}_k - \vec{\mu})]^2 - 2 \sum_{k=1}^{n} [\vec{a} \cdot (\vec{x}_k - \vec{\mu})][\vec{a} \cdot (\vec{x}_k - \vec{\mu})] + \sum_{k=1}^{n} ||\vec{x}_k - \vec{\mu}||^2$$

$$= -\sum_{k=1}^{n} [\vec{a} \cdot (\vec{x}_k - \vec{\mu})]^2 + \sum_{k=1}^{n} ||\vec{x}_k - \vec{\mu}||^2 = -\sum_{k=1}^{n} [\vec{a}^T (\vec{x}_k - \vec{\mu})]^2 + \sum_{k=1}^{n} ||\vec{x}_k - \vec{\mu}||^2.$$

Note that

$$[\vec{a}^T (\vec{x}_k - \vec{\mu})]^2 = \vec{a}^T (\vec{x}_k - \vec{\mu}) \vec{a}^T (\vec{x}_k - \vec{\mu})$$

$$= \vec{a}^T (\vec{x}_k - \vec{\mu}) \left(\vec{a}^T (\vec{x}_k - \vec{\mu}) \right)^T = \vec{a}^T (\vec{x}_k - \vec{\mu})(\vec{x}_k - \vec{\mu})^T \vec{a}.$$

Therefore,

$$L(\vec{t}^*) = -\vec{a}^T \left(\sum_{k=1}^{n} (\vec{x}_k - \vec{\mu})(\vec{x}_k - \vec{\mu})^T \right) \vec{a} + \sum_{k=1}^{n} ||\vec{x}_k - \vec{\mu}||^2.$$

Set $S = \sum_{k=1}^{n} (\vec{x}_k - \vec{\mu})(\vec{x}_k - \vec{\mu})^T$ (which we have seen already) called the scatter matrix for the data points. Thus, $L(\vec{t}^*) = -\vec{a}^T S \vec{a} + \sum_{k=1}^{n} ||\vec{x}_k - \vec{\mu}||^2$. Now view L as a function of $\vec{a}$ and find the minimum which will answer the one dimensional question. Since, $\sum_{k=1}^{n} ||\vec{x}_k - \vec{\mu}||^2$ is constant it's enough to minimize $-\vec{a}^T S \vec{a}$. This leads to the following constrained optimization problem:

$$\text{Maximize} \quad \vec{a}^T S \vec{a} \quad \text{subject to} \quad \vec{a}^T \vec{a} = 1.$$

This optimization problem we have already encountered in Section 6.2 and found that the answer is the eigenvector of S corresponding to the largest eigenvalue. Therefore, we have answered the One Dimensional Question.

We now list the general result which answers the m Dimensional Question (proof omitted).

Theorem 6.1 *The "best" m dimensional representation of a collection of data points (for $m = 1, 2, \ldots, d$) is the eigenvectors of the scatter matrix corresponding to the m largest eigenvalues.*

Note that since the scatter matrix is real symmetric we are guaranteed an orthonormal set of eigenvectors which diagonalise the scatter matrix. Therefore, all d eigenvectors can be used to create a change of coordinate system for the data points. This change of coordinate system transforms the data point about its mean. Here is the general algorithm for this transforming process:

1. Shift the data points to have $\vec{\mu} = \vec{0}$ by subtracting the mean from each data point.

2. Assuming $\vec{\mu} = \vec{0}$ we let A be an $n \times d$ matrix whose rows consist of the data points. Set $S = A^T A$ which is real symmetric. Therefore, there exists a $P \in M_{dd}$ such that $P^T P = I$ and $P^T S P = D$ is diagonal. Indeed, P is a matrix whose columns are composed of the orthonormal eigenvectors of S and D is a diagonal matrix with diagonal entries are composed of the corresponding eigenvalues. Notice that

$$D = P^T S P = P^T A^T A P = (AP)^T (AP).$$

If we set $B = AP$ we see that the scatter matrix of B is D which means the PCA coordinate system coincides with the principal axes. Therefore multiplying A on the right by P transforms the points into the PCA coordinate system.

3. Shift the points back to their original mean by adding the mean from each transformed data point.

Example 6.3 *We will compute the PCA coordinate system for the four data points* $(0, 1)$, $(0, -1)$, $(1, 2)$, $(3, 2)$ *in Example 6.1 and transform the data points about their mean into the PCA coordinate system.*

We have already computed the scatter matrix

$$S = \begin{bmatrix} 6 & 4 \\ 4 & 6 \end{bmatrix}.$$

If we put the unit eigenvectors in the columns of a matrix, we have

$$P = \begin{bmatrix} 1/\sqrt{2} & -1/\sqrt{2} \\ 1/\sqrt{2} & 1/\sqrt{2} \end{bmatrix}.$$

Notice that P is the orthogonal matrix corresponding to a counter-clockwise rotation through a 45° angle (see Section 5.3). We shift the points to have mean equaling zero, rotate the points via P, then shift the points back to their original mean. In this way we obtain points with the principal axes coinciding with the directions of the principal components. Figure 6.6 depicts the original points and the points rotated about the mean counter-clockwise by 45°.

Indeed, it is sometimes useful to switch to the PCA coordinate system. For instance, Mahalanobis distance utilizes this coordinate system to define the distance between a point and a collection of points in terms of the collection's mean and covariance matrix (hence, relying on first and second order statistics of the collection). We will define this distance in Section 6.5.

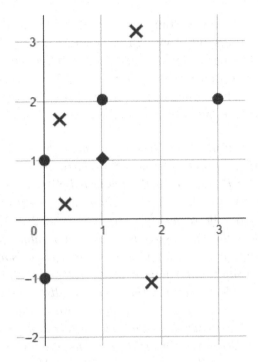

Figure 6.6 The original points are circles, while the rotated points are X's. The diamond is the mean of the points.

EXERCISES

1. Consider the following two dimensional data:

$$\{ (0,0),\ (0,1),\ (1,1),\ (1,2),\ (2,3) \}.$$

Following the same steps as in Example 6.3 in this section, find the PCA coordinate system, perform the change of basis and plot the original points together with the change of basis points. You can use the work done on the exercise in Section 6.2.

2. Consider the following three dimensional data:

$$\{ (0,0,0),\ (0,1,1),\ (1,1,1),\ (1,2,1),\ (1,2,3) \}.$$

Following the same steps as in Example 6.3 in this section, find the PCA coordinate system, perform the change of basis and plot the original points together with the change of basis points. You can use the work done on the exercise in Section 6.2.

6.4 DIMENSIONALITY REDUCTION

We will focus on two ways of performing feature reduction which involve techniques for which we already have the background: PCA and SVD. PCA can be used to retain only the highest principal components and thus reduce the dimension of the set. Truncated SVD retains only the highest singular values. When using truncated SVD the number of dimensions of the data set is not reduced, however it will exist in a lower dimensional subspace (since the rank will be reduced), i.e. every singular value replaced by zero reduces the rank by one and thus the rows of the matrix exists in a subspace of dimension one less than before.

Example 6.4 *Let's illustrate these techniques with three dimensional data. Let A be a matrix representing the data in rows. For simplicity, the mean of our data will be the zero vector. We will start with the truncated SVD method for dimensionality reduction. Let $A = USV^T$ be the singular value decomposition of A. Note that the singular values for A are approximately 55.6, 45.7 and 0.2. Set $\hat{S}$ to be the truncated SVD in which we replace the smallest non-zero singular value of S by 0. The intuition is that by dropping such a small singular value in comparison with the other two, we should not lose too much information about our data set. Now set $\hat{A} = U\hat{S}V^T$, the truncated singular value decomposition. The resulting data set will sit in a two dimensional, rank 2, subspace of $\mathbb{R}^3$, i.e. a plane through the origin.*

In Figure 6.7, on the left we have the original data in $\mathbb{R}^3$ and in the center we have its truncated SVD decomposition with two non-zero singular values. Notice how the truncated points sit in a two dimensional plane. Also included in the figure is the data transformed from the plane into the xy-plane using the orthogonal matrix V from the singular value decomposition (one can show that V is a rotoreflection—see the end of Section 5.3).

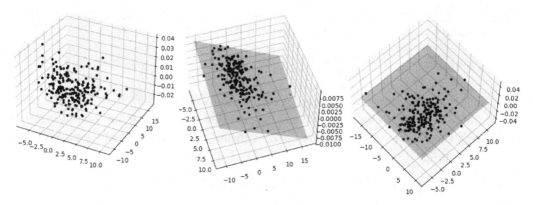

Figure 6.7 On the left are the original points. In the center is the truncated SVD points in a two dimensional subspace. On the right is the transformed data into the xy-plane.

Starting with the same original data we employ the PCA technique of dimensionality reduction by extracting the first two principal components of the data set. Let $P^T(A^T A)P = D$, where P is the orthogonal matrix whose columns consist of orthonormal eigenvectors. We do this by multiplying A by P and dropping the third component of each data point. In Figure 6.8, we plot these points.

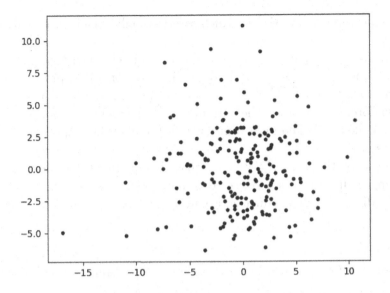

Figure 6.8 The result of keeping the first two principal components of the three di-mensional data.

If we plot both two-dimensional data points in Figure 6.7, we see that they correspond exactly (see Figure 6.9).

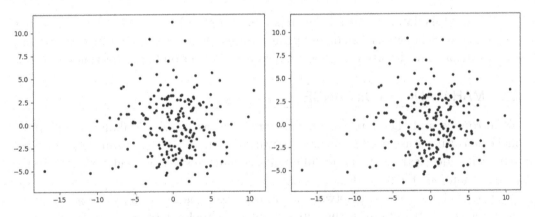

Figure 6.9 On the left is the truncated SVD points in a two dimensional subspace. On the right is the first two principal components of the three dimensional data.

Hence, we see that there really is no difference between the two versions of di-mension reduction. The only difference is that truncated SVD keeps the data set in its original dimensional space. We now give a formal proof of this statement which was illustrated in the example above.

Let $A \in M_{nd}$ be a data set in which each row is a data point and assume the mean of the data set is $\vec{0}$. Let $A = USV^T$ be its singular value decomposition. Set r equal to the number of non-zero singular values of A. Consider the truncated singular value decomposition $\hat{A} = U\hat{S}V^T$ in which $\hat{S}$ retains only the highest k singular values, $k < r$.

Recall that $V = [v_1 \ v_2 \ \cdots \ v_d]$ consists of columns which form an orthonormal basis for $\mathbb{R}^d$ and are the eigenvectors of $A^T A$. Therefore, V is precisely the orthogonal matrix such that $V^T(A^T A)V$ is a diagonal matrix of eigenvalues in descending order. Therefore, we transform the data into the principal coordinate system by multiplying by V, i.e. AV. The PCA dimensionality reduction simply retains the first k components of AV.

Multiply $\hat{A}$ by the same V. We show that these points correspond exactly to the PCA dimensionality reduction. To see this, notice that $AV = [Av_1 \ Av_2 \ \cdots \ Av_d]$ and so the PCA dimensionality reduction is simply $[Av_1 \ Av_2 \ \cdots \ Av_k \ 0 \ \cdots \ 0]$. Now look at

$$\hat{A}V = U\hat{S} = [\sqrt{\lambda_1}u_1 \ \sqrt{\lambda_2}u_2 \ \cdots \ \sqrt{\lambda_k}u_k \ 0 \ \cdots \ 0] = [Av_1 \ Av_2 \ \cdots \ Av_k \ 0 \ \cdots \ 0].$$

EXERCISES

1. Consider the following three dimensional data:

$$\{ \ (0,0,0), \ (0,1,1), \ (1,1,1), \ (1,2,1), \ (1,2,3) \ \}.$$

Using this data, repeat all the steps illustrated in Example 6.4. Be sure to plot your results just as was done in the example using a computer algebra system. You can use the work done on the exercise in Section 6.2 or Section 6.4.

6.5 MAHALANOBIS DISTANCE

To introduce and therefore get an understanding as to why Mahalanobis distance makes intuitive sense, we first look at univariate data. Consider two normal distributions with identical means μ yet different standard deviations, and a value x in the distribution (see Figure 6.10).

The Euclidean distance between x and μ is the same for both distributions. However, for the distribution on the top x is intuitively closer to μ, since its likelihood of occurring in that distribution is higher. The natural way to account for this is by looking at the standard deviation, σ, of the distribution, so that instead of the distance from x to μ being simply $|x - \mu|$ we define the distance to be

$$\frac{|x - \mu|}{\sigma}.$$

Let $\vec{\mu} = [\mu_1, \mu_2, \ldots, \mu_d]$ and $\vec{x} = [x_1, x_2, \ldots, x_d]$. The Euclidean distance is defined to be

$$d(\vec{x}, \vec{\mu}) = \sqrt{\sum_{k=1}^{d}(x_i - \mu_i)^2}.$$

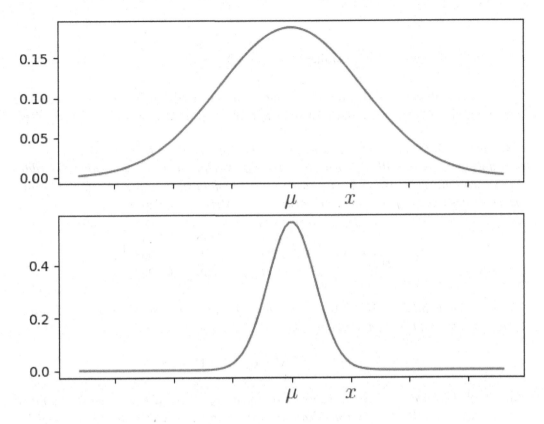

Figure 6.10 The result of keeping the first two principal components of the three dimensional data.

The distance we defined for univariate normal distributions generalizes to d-dimensional Gaussian distributions as

$$d(\vec{x}, \vec{\mu}) = \sqrt{\sum_{k=1}^{d} \frac{(x_i - \mu_i)^2}{\sigma_i^2}}.$$

This distance is sometimes called **standardized Euclidean** distance. Now consider a collection of data points $X = \{ \vec{x}_1, \vec{x}_2, \ldots, \vec{x}_n \} \subseteq \mathbb{R}^d$ with mean $\vec{\mu}$ and standard deviation $\vec{\sigma}$. The **covariance** matrix for X is the scatter matrix S divided by the number of data points n. Set $C = S/n$ to be the covariance matrix.

We define a particular distance from a point $\vec{x}$ to a set X, Mahal$(\vec{x}, X)$, called **Mahalanobis** distance. It is defined as follows:

$$\text{Mahal}(\vec{x}, X) = \sqrt{(\vec{x} - \vec{\mu})^T C^{-1} (\vec{x} - \vec{\mu})}.$$

Suppose $\vec{x}$ and X are such that $\vec{\mu} = \vec{0}$ with the PCA directions coinciding with the principal axes. In this case the inverse of the covariance matrix is diagonal with diagonal entries consisting of the reciprocal of the variance in each of the PCA directions. If we set $\vec{x} = [x_1, x_2, \ldots, x_d]$ and the entries in C^{-1} to be $1/\sigma_1^2$, $1/\sigma_2^2$, $\ldots$, $1/\sigma_d^2$ then

$$\text{Mahal}(\vec{x}, X) = \sqrt{\sum_{k=1}^{d} \frac{x_i^2}{\sigma_i^2}}.$$

Hence, Mahalanobis distance is in fact standardized Euclidian distance, since shifting and transforming a data set does not affect Mahalanobis distance (exercise).

Example 6.5 *Consider the 212 points of data we used in Example 6.4, which we will designate as X with mean $\mu = \vec{0}$. We can skip subtracting the mean of the data from each data point, since it is zero. We put the data in the rows of a matrix which we shall designate by A. Then the covariance matrix is approximately*

$$C = \frac{1}{212} A^T A = \begin{bmatrix} 10.2170 & 1.2838 & -0.0047 \\ 1.2838 & 14.2170 & -0.0068 \\ -0.0047 & -0.0068 & 0.0002 \end{bmatrix}.$$

Consider the point $\vec{x} = (15, 20, 0.1)$ (see Figure 6.11). Then the Mahalanobis distance from $\vec{x}$ to the data X is approximately

$$\text{Mahal}(P, X) = \sqrt{\vec{x}^T C^{-1} \vec{x}} = 11.4658.$$

Note that there was no need to subtract the mean of the data from $\vec{x}$ since the mean is zero. Compare this to the Euclidean distance from $\vec{x}$ to X (i.e. to the mean of the data which is zero) which is approximately $\sqrt{\vec{x}^T \vec{x}} = 25.0002$. One way to interpret the fact that the Mahalanobis distance is smaller is that the data is spread more so in the direction of the point $\vec{x}$, loosely speaking.

EXERCISES

1. Consider the following two dimensional data:

$$\{ (0,0), (0,1), (1,1), (1,2), (2,3) \}.$$

   As in Example 6.5,

   a. Compute the Mahalanobis distance from $(3, 2)$ to this data set.

   b. Compute the Euclidean distance from $(3, 2)$ to this data set.

2. Consider the following three dimensional data:

$$\{ (0,0,0), (0,1,1), (1,1,1), (1,2,1), (1,2,3) \}.$$

   As in Example 6.5,

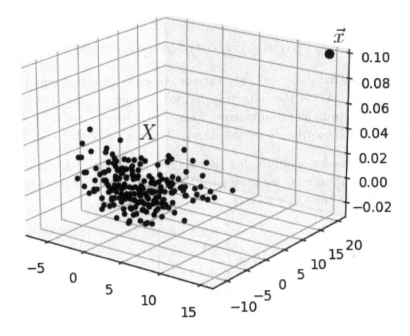

Figure 6.11 Plot of the data set X and the point $\vec{x}$.

    a. Compute the Mahalanobis distance from $(3, 2, 1)$ to this data set.

    b. Compute the Euclidean distance from $(3, 2, 1)$ to this data set.

3. Prove that rigidly shifting a point and a data set does not change the Mahalanobis distance from the point to the data set.

4. Prove that transforming a point and a data set via an orthogonal matrix does not change the Mahalanobis distance from the point to the data set.

6.6 DATA SPHERING

Data sphering is loosely speaking a way to normalize data so that different data can be compared and also so that one can compute invariants.

In the case of feature data, data sphering levels the playing field of the various features so that no feature plays a stronger role than the others. Mathematically, it results in data which has variance equal to one in any direction through the mean of the data. Equivalently, it transforms the data so that Mahalanobis distance has been reduced to Euclidean distance (exercise).

Let $A \in M_{nd}$ be a matrix whose rows consist of points in the data and let's assume the mean is $\vec{0}$. We must also assume that the points have distinct values in each of its given coordinates in order for $A^T A$ to be invertible (which is often the case with data sets). Since $A^T A$ is symmetric, it is diagonalizable by an orthogonal matrix, i.e. there exists $P \in M_{dd}$ with $P^T P = I = P P^T$ such that $P^T(A^T A)P = D$, where D is a diagonal matrix. If we set $S = APD^{-1/2}$, we have

$$S^T S = (APD^{-1/2})^T (APD^{-1/2}) = D^{-1/2} P^T A^T APD^{-1/2} = D^{-1/2} DD^{-1/2} = I.$$

Hence, S is sphered data, since the variance in every direction equals one. Recall in Section 6.3 that AP is the data transformed so that the its coordinates are in terms of the data's principal components. Therefore, when we multiply AP by $D^{-1/2}$ we are dividing each coordinate by the standard deviation of that coordinate.

Example 6.6 *Consider again the four data points* $(0,1)$, $(0,-1)$, $(1,2)$, $(3,2)$ *from Example 6.1. In that example we subtracted the mean of the data from each point and inserted them in the rows of a matrix*

$$B = \begin{bmatrix} -1 & 0 \\ -1 & -2 \\ 0 & 1 \\ 2 & 1 \end{bmatrix}.$$

In Example 6.3. we computed

$$P = \begin{bmatrix} 1/\sqrt{2} & -1/\sqrt{2} \\ 1/\sqrt{2} & 1/\sqrt{2} \end{bmatrix} \quad and \quad D = \begin{bmatrix} 10 & 0 \\ 0 & 2 \end{bmatrix}.$$

Therefore, the sphered data will be

$$S = BPD^{1/2} =$$

$$\begin{bmatrix} -1 & 0 \\ -1 & -2 \\ 0 & 1 \\ 2 & 1 \end{bmatrix} \begin{bmatrix} 1/\sqrt{2} & -1/\sqrt{2} \\ 1/\sqrt{2} & 1/\sqrt{2} \end{bmatrix} \begin{bmatrix} 10 & 0 \\ 0 & 2 \end{bmatrix}^{-1/2} = \begin{bmatrix} -\sqrt{5}/10 & 1/2 \\ -3\sqrt{5}/10 & -1/2 \\ \sqrt{5}/10 & 1/2 \\ 3\sqrt{5}/10 & -1/2 \end{bmatrix}.$$

One can easily check that $S^T S = I$.

Example 6.7 *Let's return to the data with mean* $\vec{0}$ *explored in Example 6.4. If A is a matrix whose rows consist of the points in the data, we computed for the scatter matrix* $A^T A$ *that (approximately)*

$$P = \begin{bmatrix} 0.0004 & -0.9596 & -0.2815 \\ 0.0004 & 0.2815 & -0.9596 \\ 1.0000 & 0.0003 & 0.0005 \end{bmatrix} \quad and \quad D = \begin{bmatrix} 0.03 & 0 & 0 \\ 0 & 2086.18 & 0 \\ 0 & 0 & 3093.84 \end{bmatrix}.$$

One can again check that $S^T S \approx I$. *In Figure 6.12. we plot the original data versus the sphered data. We set the axes to have the same aspect ratio. As you can see the original data is very flat in the z direction, however after sphering we removed the flatness of the data.*

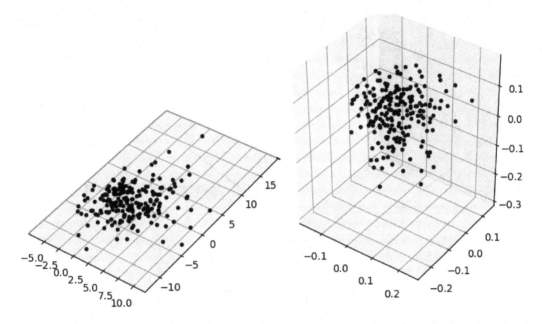

Figure 6.12 On the left is the original data while on the right is the resulting sphered data.

EXERCISES

1. Consider the following two dimensional data:

$$\{ (0,0),\ (0,1),\ (1,1),\ (1,2),\ (2,3) \}.$$

As was done in Example 6.6, sphere this data.

2. Consider the following three dimensional data:

$$\{ (0,0,0),\ (0,1,1),\ (1,1,1),\ (1,2,1),\ (1,2,3) \}.$$

As was done in Example 6.6, sphere this data.

3. Prove that for sphered data, Mahalanobis distance and Euclidean distance coincide.

6.7 FISHER LINEAR DISCRIMINANT FUNCTION

The Fisher LDF is one example of a linear discriminant function and has a very intuitive definition which we explain now. Consider two different classes of objects C_1 and C_2 each of which is represented by a data set of features. As a first step we shall project all the data in both classes onto a line thus creating univariate data P_1 and P_2. Figure 6.13 illustrates two classes with two possible lines (above) and their corresponding projections (below).

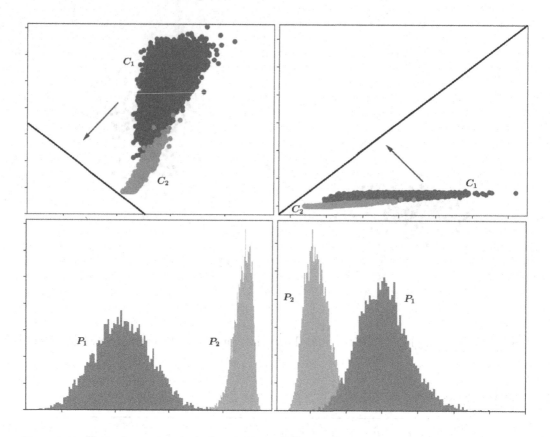

Figure 6.13 Two classes projected onto two different lines.

Notice that in the figure on the top left the line of projection better separates the projected data into two separate distributions as opposed to the projection on the right. Our goal is to find the line of projection that "best" separates the two classes of projected data. Once we find the "best" separation, a line perpendicular to that line of projection will be our Fisher LDF. All this will become clear in Example 6.8. Let $\vec{a}$ be a unit vector parallel to the line on which we project the data and $\vec{\mu}_i$ be the mean of class C_i for $i = 1, 2$ and μ_i be the mean of projection P_i for $i = 1, 2$. The projection x of any vector $\vec{x}$ onto a line parallel to $\vec{a}$ is defined by

$$x = comp_{\vec{a}}\vec{x} = \frac{\vec{a} \cdot \vec{x}}{|\vec{a}|} = \vec{a} \cdot \vec{x} = \vec{a}^T \vec{x}.$$

We now define two measures on the two classes. Recall, our goal is the find the line which "best" separates the two classes C_1 and C_2.

(1.) **Scatter between two classes:** This will simply be the Euclidian distance between the two projected means, i.e.

$$|\mu_1 - \mu_2| = |\vec{a}^T \vec{\mu}_1 - \vec{a}^T \vec{\mu}_2| = |\vec{a}^T (\vec{\mu}_1 - \vec{\mu}_2)|$$

A line which "best" separates the classes should maximize this measure, or equivalently maximize

$$(\mu_1 - \mu_2)^2 = [\vec{a}^T(\vec{\mu}_1 - \vec{\mu}_2)]^2 = \vec{a}^T(\vec{\mu}_1 - \vec{\mu}_2)\vec{a}^T(\vec{\mu}_1 - \vec{\mu}_2) = \vec{a}^T(\vec{\mu}_1 - \vec{\mu}_2)(\vec{\mu}_1 - \vec{\mu}_2)^T\vec{a}.$$

Define $S_B = (\vec{\mu}_1 - \vec{\mu}_2)(\vec{\mu}_1 - \vec{\mu}_2)^T$ called the **scatter between matrix**.

(2.) **Scatter within a class:** This is nearly the variance of the class C_i and will be denoted by $\tilde{\sigma}_i^2$ for $i = 1, 2$ and is defined by

$$\tilde{\sigma}_i^2 = \sum_{x_k \in P_i} (x_k - \mu_i)^2 = \sum_{\vec{x}_k \in C_i} [\vec{a}^T\vec{x}_k - \vec{a}^T\vec{\mu}_i]^2 = \sum_{\vec{x}_k \in C_i} [\vec{a}^T(\vec{x}_k - \vec{\mu}_i)]^2$$

$$= \sum_{\vec{x}_k \in C_i} \vec{a}^T(\vec{x}_k - \vec{\mu}_i)\vec{a}^T(\vec{x}_k - \vec{\mu}_i) = \sum_{\vec{x}_k \in C_i} \vec{a}^T(\vec{x}_k - \vec{\mu}_i)(\vec{x}_k - \vec{\mu}_i)\vec{a}$$

$$= \vec{a}^T \left[\sum_{\vec{x}_k \in C_i} (\vec{x}_k - \vec{\mu}_i)(\vec{x}_k - \vec{\mu}_i) \right] \vec{a}.$$

Define $S_i = \sum_{\vec{x}_k \in C_i}(\vec{x}_k - \vec{\mu}_i)(\vec{x}_k - \vec{\mu}_i)$, which we've seen is the scatter matrix for class C_i. Then the **scatter within matrix**, $S_W = S_1 + S_2$. A line which "best" separates the classes should minimize

$$\tilde{\sigma}_i^2 + \tilde{\sigma}_i^2 = \vec{a}^T S_1 \vec{a} + \vec{a}^T S_2 \vec{a} = \vec{a}^T S_W \vec{a}.$$

We need to construct a function which takes into account both these measures, a function which at the same time maximizes $\vec{a}^T S_B \vec{a}$ while minimizing $\vec{a}^T S_W \vec{a}$. One such function is

$$f(\vec{a}) = \frac{\vec{a}^T S_B \vec{a}}{\vec{a}^T S_W \vec{a}}.$$

If we maximize $f(\vec{a})$, we will be maximizing $\vec{a}^T S_B \vec{a}$ while minimizing $\vec{a}^T S_W \vec{a}$. The following theorem tells us how to find this maximum.

Theorem 6.2 *A unit vector parallel to the line of projection, which maximizes $f(\vec{a})$ is parallel to a vector which is a solution to the linear system $S_W X = \vec{\mu}_1 - \vec{\mu}_2$.*

Proof 6.1 *In order to find the maximum, we need to find the critical points of f by solving $\nabla f = \vec{0}$, i.e. solve*

$$\nabla f(\vec{a}) = \frac{(2S_B\vec{a})(\vec{a}^T S_W \vec{a}) - (\vec{a}^T S_B \vec{a})(2S_W \vec{a})}{(\vec{a}^T S_B \vec{a})^2} = \vec{0}.$$

This equation reduces to

$$(2S_B\vec{a})(\vec{a}^T S_W \vec{a}) = (\vec{a}^T S_B \vec{a})(2S_W \vec{a}),$$

or equivalently,

$$S_B \vec{a} = \lambda S_W \vec{a} \quad \text{where} \quad \lambda = \frac{\vec{a}^T S_B \vec{a}}{\vec{a}^T S_W \vec{a}}, \quad \text{a scalar.}$$

The solution to this equation involves generalized eigenvectors, something we have not covered in this text, however we can further simplify this problem. Note that $S_B \vec{a}$ is parallel to $\vec{\mu}_1 - \vec{\mu}_2$. Indeed, since

$$S_B = (\vec{\mu}_1 - \vec{\mu}_2)(\vec{\mu}_1 - \vec{\mu}_2)^T,$$

which implies

$$S_B \vec{a} = (\vec{\mu}_1 - \vec{\mu}_2)(\vec{\mu}_1 - \vec{\mu}_2)^T \vec{a} = \alpha(\vec{\mu}_1 - \vec{\mu}_2) \quad \text{where} \quad \alpha = (\vec{\mu}_1 - \vec{\mu}_2)^T \vec{a}, \quad \text{a scalar.}$$

Hence, $\alpha(\vec{\mu}_1 - \vec{\mu}_2) = \lambda S_W \vec{a}$ or equivalently $S_W \vec{a} = \beta(\vec{\mu}_1 - \vec{\mu}_2)$ where $\beta = \lambda/\alpha$. Now β can be dropped, since we are simply looking for a vector parallel to the optimal line. To obtain the unit vector $\vec{a}$ we simply normalize the solution to the system $S_W X = \vec{\mu}_1 - \vec{\mu}_2$. □

Example 6.8 *We find a unit vector $\vec{a}$ parallel to the line of projection which maximizes $f(\vec{a})$ for the classes $C_1 = \{ (-2,0), (-2,-1) \}$ and $C_2 = \{ (0,2), (1,2) \}$.*

$$\vec{\mu}_1 = \frac{1}{2}[(-2,0) + (-2,-1)] = (-2,-1/2) \quad \text{and} \quad \vec{\mu}_2 = \frac{1}{2}[(0,2) + (1,2)] = (1/2,2).$$

To form the scatter matrix for C_1, first subtract $\vec{\mu}_1$ from each point in C_1 and place the results in rows of a matrix, say

$$A_1 = \begin{bmatrix} 0 & 1/2 \\ 0 & -1/2 \end{bmatrix}.$$

Then the scatter matrix for C_1,

$$S_1 = A_1^T A_1 = \begin{bmatrix} 0 & 0 \\ 1/2 & -1/2 \end{bmatrix} \begin{bmatrix} 0 & 1/2 \\ 0 & -1/2 \end{bmatrix} = \begin{bmatrix} 0 & 0 \\ 0 & 1/2 \end{bmatrix}.$$

To form the scatter matrix for C_2, first subtract $\vec{\mu}_2$ from each point in C_2 and place the results in rows of a matrix, say

$$A_2 = \begin{bmatrix} -1/2 & 0 \\ 1/2 & 0 \end{bmatrix}.$$

Then the scatter matrix for C_2,

$$S_2 = A_2^T A_2 = \begin{bmatrix} -1/2 & 1/2 \\ 0 & 0 \end{bmatrix} \begin{bmatrix} -1/2 & 0 \\ 1/2 & 0 \end{bmatrix} = \begin{bmatrix} 1/2 & 0 \\ 0 & 0 \end{bmatrix}.$$

Therefore,

$$S_W = S_1 + S_2 = \begin{bmatrix} 1/2 & 0 \\ 0 & 1/2 \end{bmatrix} \text{ and } S_W^{-1} = \begin{bmatrix} 2 & 0 \\ 0 & 2 \end{bmatrix}.$$

Thus, we can solve $X = S_W^{-1}(\vec{\mu}_1 - \vec{\mu}_2) =$

$$\begin{bmatrix} 2 & 0 \\ 0 & 2 \end{bmatrix} \left(\begin{bmatrix} 0 \\ -1/2 \end{bmatrix} - \begin{bmatrix} 3/2 \\ 1 \end{bmatrix} \right) = \begin{bmatrix} 2 & 0 \\ 0 & 2 \end{bmatrix} \begin{bmatrix} -3/2 \\ -3/2 \end{bmatrix} = \begin{bmatrix} -3 \\ -3 \end{bmatrix}.$$

Hence, the unit vector we seek is $\vec{a} = [\sqrt{2}/2, \sqrt{2}/2]$.

Note that the Fisher LDF is a line perpendicular to the line of projection. There are several possibilities for which line to choose. One option is to choose the line passing through the midpoint of $\vec{\mu}_1$ and $\vec{\mu}_2$. Another option is to choose a line which minimizes the number of data points on the wrong side of the line, called the **minimal total error** (MTE) line.

Example 6.9 *Returning to Example 6.8, let's illustrate each of the two lines mentioned above and graph them both together with the points in each class. However, in this example the two lines coincide, since the first line perfectly separates the two classes and thus also minimizes the total error (which equals zero, since nothing has been misclassified). In Figure 6.14, the dotted line represents the optimal line on which to project the data and the solid line is the Fisher LDF.*

Example 6.10 *We will repeat the algorithm for a real world 3-dimensional data set and determine the Fisher LDF. One place to find real world data is at the UC, Irvine Machine Learning Repository [2]. We chose the first, second and fifth features of a data set entitled "Breast Cancer Wisconsin (Diagnostic)". In this case the Fisher LDF will be a plane dividing the two classes, but since the two classes overlap it will not be a perfect separation as is typical for real world data (see Figure 6.15).*

We begin by subtracting from each class its mean and place the results in the rows of a matrix, say A_1 and A_2 for classes C_1 and C_2, respectively. Then we compute the scatter matrix for each class, $S_1 = A_1^T A_1$ and $S_2 = A_2^T A_2$ so that the scatter within matrix is $S_W = S_1 + S_2$, a 3×3 matrix with very large entries not worth listing here. Then the vector parallel to the Fisher optimal line of projection is obtained by computing $S_W^{-1}(\vec{\mu}_1 - \vec{\mu}_2)$ where

$$\vec{\mu}_1 - \vec{\mu}_2 \approx [5.3163, 3.6901, 0.0104].$$

If we normalize $S_W^{-1}(\vec{\mu}_1 - \vec{\mu}_2)$ *we get*

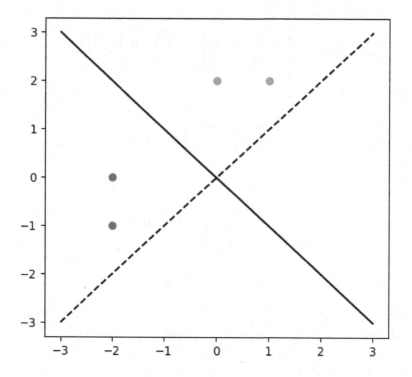

Figure 6.14 Two classes projected on the dotted line and divided by the solid Fisher LDF line.

$$\vec{a} \approx [0.8041, -0.2408, -0.5436].$$

Now $\vec{a}$ is normal to the plane of separation and we want the midpoint of the two class means to be a point on the plane, i.e. the point

$$\frac{1}{2}(\vec{\mu_1} + \vec{\mu_2}) \approx [14.8047, 19.7598, 0.0977].$$

Therefore, our plane has the equation

$$\begin{bmatrix} 0.8041 & -0.2408 & -0.5436 \end{bmatrix} \begin{bmatrix} x \\ y \\ z \end{bmatrix} = \begin{bmatrix} 0.8041 & -0.2408 & -0.5436 \end{bmatrix} \begin{bmatrix} 14.8047 \\ 19.7598 \\ 0.0977 \end{bmatrix}$$

or equivalently, $\quad z = 1.4793x - 0.4431y - 13.0476.$

In Figure 6.16, we include the Fisher LDF plane between the two classes of data.

As stated earlier, we cannot expect to have perfect separation when it comes to real world data. In fact, around 16.98% of the first class is on the wrong side of the plane and 14.57% of the second class. The Fisher LDF plane becomes the classifier in the following sense: Given a new object for which we do not know to which class

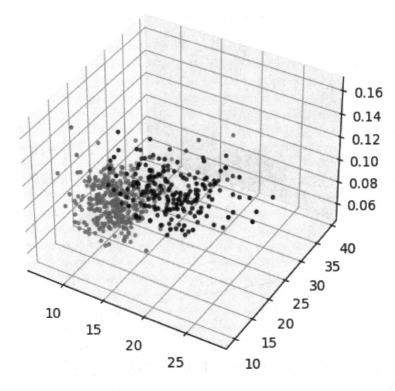

Figure 6.15 Two classes, black and grey, plotted in 3-dimensional space.

it belongs, we collect the same three features for that object and map it in feature space. Our prediction as to which class the unknown object belongs depends on which side of the Fisher LDF plane the point lies. In Section 6.8, we will explain in more generality this process.

EXERCISES

1. Consider the following two classes of data:

$$C_1 = \{ (0,0),\ (0,1) \} \quad \text{and} \quad C_2 = \{ (1,1),\ (1,2),\ (2,3) \}.$$

   a. Following the same steps as in Example 6.8 in this section, find a unit vector corresponding to the line of optimal projection.

   b. Following the same steps as in Example 6.9 in this section, plot the classes, the line of projection and a Fisher LDF though the midpoint of the two class means.

2. Consider the following two classes of data:

$$C_1 = \{ (0,0,-1),\ (1,0,1) \} \quad \text{and} \quad C_2 = \{ (1,1,0),\ (1,1,2),\ (2,3,1) \}.$$

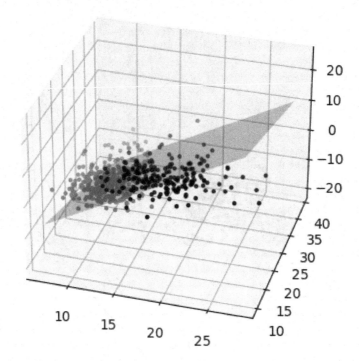

Figure 6.16 Two classes, black and grey, plotted in 3-dimensional space with the Fisher LDF plane.

   a. Following the same steps as in Example 6.10 in this section, find a unit vector corresponding to the line of optimal projection.

   b. Following the same steps as in Example 6.10 in this section, plot the classes and the Fisher LDF (which is a plane) though the midpoint of the two class means.

3. Consider the following two classes of data:

$$C_1 = \{\ (6,0),\ (0,6)\ \} \quad \text{and} \quad C_2 = \{\ (-1,-1),\ (0,-5),\ (-5,0)\ \}.$$

Find the equation of the Fisher LDF (in the form $y = mx + b$) which passes through the midpoint of the means of the two classes.

6.8   LINEAR DISCRIMINANT FUNCTIONS IN FEATURE SPACE

In Section 6.7, we discussed a specific example of a linear discriminant function (LDF) called the Fisher LDF. In this section, we discuss LDFs in general as well as modifications of feature space. Recall that feature space is $\mathbb{R}^d$ for some dimension d and points in feature space consist of features collected from objects, and that features come in two varieties: Numerical and categorical. We will be focusing on numerical features. Of course, you can make categorical features numerical simply by replacing words by numbers, but in many cases this is a weak labeling when the categorical features do not have an ordering of any sort (e.g. the type of grape used to make a wine).

We begin with a discussion of the general idea of an LDF, but at first we shall make two assumptions. First, that the two classes of data are linearly separable. Second, that our line of separation passes through the origin. Later in the section we will remove these assumptions.

Definition 6.1 *A* **linear discriminant function** *is a map* $g : \mathbb{R}^d \to \mathbb{R}$ *having the form*

$$g(x_1, x_2, \ldots, x_d) = a_1 x_1 + a_2 x_2 + \cdots + a_d x_d, \quad \text{where } a_1, a_2, \ldots, a_d \in \mathbb{R}.$$

We can denote the LDF using shorthand vector notation as follows. Set $\vec{x} = [x_1, x_2, \ldots, x_d]$ and $\vec{a} = [a_1, a_2, \ldots, a_d]$, then

$$g(\vec{x}) = \vec{a} \cdot \vec{x} = \vec{a}^T \vec{x}, \quad \text{viewing vectors as column matrices.}$$

In Section 6.7, we equated the LDF with the separating line/plane/hyperplane, but in fact this is not so. The **hyperplane of separation** is in fact the equation $g(\vec{x}) = 0$. Let's focus on two and three dimensions so that we have a mental picture of what things look like.

Example 6.11 *In* $\mathbb{R}^2$, $g(x, y) = ax + by$ *and so the line of separation is* $ax + by = 0$ *or* $y = -\frac{a}{b}x$ *(see Figure 6.17).*

Example 6.12 *In* $\mathbb{R}^3$, $g(x, y, z) = ax + by + cz$ *and so the plane of separation is* $ax + by + cz = 0$ *or* $z = -\frac{a}{c}x - \frac{b}{c}y$ *(see Figure 6.18).*

Observe that the vector $\vec{a}$ is always normal to the separating line/plane/hyperplane, since $\vec{a} \cdot \vec{x} = 0$. Let's assume that $\vec{a}$ is pointing at the class C_1. Then any point in C_1 (viewed as a vector) makes an acute angle with $\vec{a}$, while any vector in C_2 makes an obtuse angle with $\vec{a}$. Indeed, since $\vec{a} \cdot \vec{x} = |\vec{a}||\vec{x}| \cos \theta$ this is equivalent to saying for any $\vec{x}^* \in C_1 \cup C_2$,

$$\text{If } g(\vec{x}^*) > 0 \text{ then } \vec{x}^* \in C_1, \quad \text{and}$$

$$\text{If } g(\vec{x}^*) < 0 \text{ then } \vec{x}^* \in C_2.$$

Figures 6.17 and 6.18 illustrate these comments as well. Now we would like a statement about our separating LDF which does not depend on what class we are in as

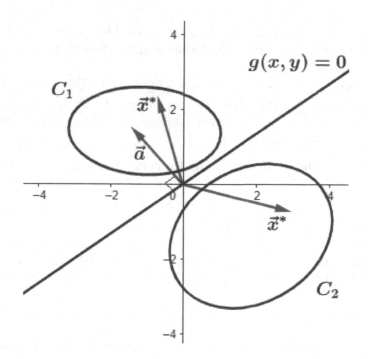

Figure 6.17 Two classes, C_1 and C_2, together with a line of separation.

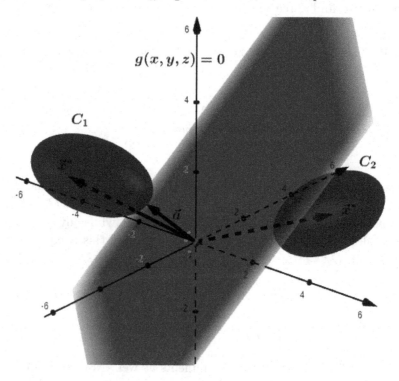

Figure 6.18 Two classes, C_1 and C_2, together with a plane of separation.

is the case for the two conditional statements above. To resolve this we look at what is called **normalized feature space** which is very easy to explain. Assuming again that $\vec{a}$ points at C_1, we first define

$$-C_2 = \{ -\vec{x} \mid \vec{x} \in C_2 \}.$$

Set $\hat{C} = C_1 \cup -C_2$. Now the separating LDF we seek simply must satisfy the condition $g(\vec{x}) > 0$ for all $\vec{x} \in \hat{C}$ (see Figure 6.19).

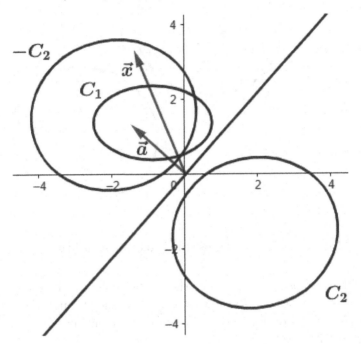

Figure 6.19 Two classes, C_1 and C_2, together with $-C_2$.

In general, $\vec{a}$ will not be uniquely determined by this condition. For instance, in $\mathbb{R}^2$, choices of $\vec{a}$ which produce separable LDFs lie within a sector of 2-space (see Figure 6.20).

If we hope to produce an algorithm for determining $\vec{a}$, then we need to add an additional condition in order to make $\vec{a}$ unique. There are a number of different conditions one can add in order to make $\vec{a}$ unique, and different conditions will lead to different $\vec{a}$. Some examples of conditions one can add include

1. Maximize the minimal distance between points in $C_1 \cup C_2$ and the separating LDF. This leads to what is called a *linear support vector machine* (which is beyond the scope of this text).

2. Minimize the sum of the squares of the distances between points in $C_1 \cup C_2$ and the separating LDF. This leads to a *minimal square error LDF* which will be discussed in Section 6.9.

There is one last change we need to make to feature space before we proceed to the next section. This alteration will allow us to remove the assumption that our

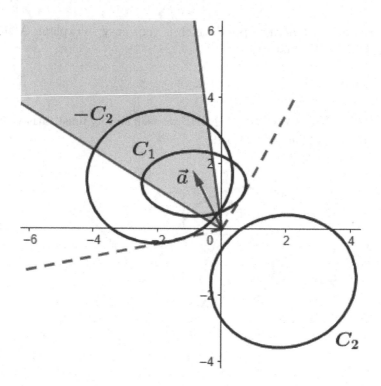

Figure 6.20 Shaded region indicates possible choice of separating $\vec{a}$.

separating LDF must pass through the origin. The basic idea is to lift the classes one dimension higher, divide the results with a hyper-plane through the origin and then project the hyper-plane back down to the original dimension of the classes to obtain the separating LDF.

Definition 6.2 *Given two classes C_1 and C_2 in feature space and $\vec{x} \in C_1 \cup C_2$, define $\vec{y} = [\vec{x}, 1]$ and $\hat{C}_i = \{\, \vec{y} = [\vec{x}, 1] \mid \vec{x} \in C_i \,\}$ for $i = 1, 2$. We say that $\hat{C}_1$ and $\hat{C}_2$ exist in* **augmented feature space.**

We now give an algorithm for finding a separating LDF for classes C_1 and C_2 which does not have to pass through the origin (refer to Figure 6.21).

1. Given two classes C_1 and C_2, form the sets $\hat{C}_1$ and $\hat{C}_2$.

2. Find a separating LDF for $\hat{C}_1$ and $\hat{C}_2$, say $g(\vec{x}, z) = 0$.

3. The separating LDF for C_1 and C_2 is then $g(\vec{x}, 1) = 0$, i.e. replace z by 1 in $g(\vec{x}, z) = 0$.

In Section 6.9, we will illustrate this algorithm using concrete data and a specific kind of LDF called the *minimal square error* LDF.

EXERCISES

1. Prove that if $g(\vec{x}, z) = 0$ separates $\hat{C}_1$ and $\hat{C}_2$, then $g(\vec{x}, 1) = 0$ separates C_1 and C_2. Hint: use the normalized feature space definition of a separating LDF.

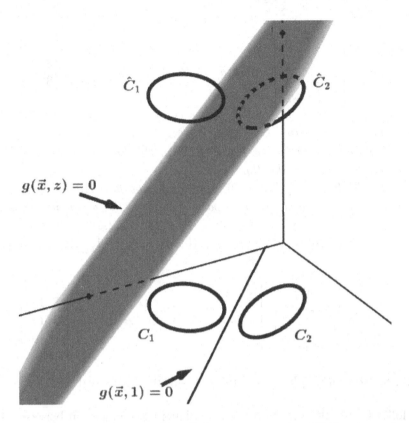

$g(\vec{x}, z) = 0$

$g(\vec{x}, 1) = 0$

Figure 6.21 Illustration of algorithm for finding a separating LDF not passing through the origin.

6.9 MINIMAL SQUARE ERROR LINEAR DISCRIMINANT FUNCTION

The Minimal Square Error Linear Discriminant Function (MSE LDF) will give us another application of best approximating an overdetermined system. Recall that a separating LDF with nomal vector $\vec{a}$ will have the property that $\vec{a} \cdot \vec{x} > 0$ for all $\vec{x} \in C_1 \cup -C_2$ in normalized feature space and such an $\vec{a}$ is not unique in general. We can pin down $\vec{a}$ to be unique if we add the following conditions: Suppose $C_1 \cup -C_2 = \{ \vec{x}_1, \vec{x}_2, \ldots, \vec{x}_n \}$. The conditions we will add are

$$\vec{a} \cdot \vec{x}_i = b_i, \quad \text{a fixed positive number, } i = 1, 2, \ldots, n.$$

If we let A be the matrix whose rows are $\vec{x}_1, \vec{x}_2, \ldots, \vec{x}_n$ and B a column vector with entries $b_1, b_2, \ldots, b_n$, then $\vec{a}$ is a solution to $AX = B$. In general this linear system will be overdetermined, so we will choose $\vec{a}$ to be the best approximation to a solution to $AX = B$, i.e. $\vec{a} = (A^T A)^{-1} A^T B$.

Example 6.13 *Consider again the classes $C_1 = \{ (-2, 0), (-2, -1) \}$ and $C_2 = \{ (0, 2), (1, 2) \}$ from Example 6.8. First, we will find an MSE LDF through the origin. Let's choose $b_1 = b_2 = b_3 = b_4 = 1$ for our MSE LDF. Now $-C_2 = \{ (0, -2), (-1, -2) \}$, so we form the matrix*

$$A = \begin{bmatrix} -2 & 0 \\ -2 & -1 \\ 0 & -2 \\ -1 & -2 \end{bmatrix} \quad \text{and} \quad \vec{a} = (A^T A)^{-1} A^T B = \begin{bmatrix} -5/13 \\ -5/13 \end{bmatrix}.$$

Therefore, the separating line is $[-5/13, -5/13] \cdot [x, y] = 0$ or $y = -x$, which leads to the same solution found for the Fisher LDF in Example 6.9. There is a reason for this which will be explained shortly.

Next, we will find an MSE LDF that does not have to go through the origin by looking in augmented feature space. We form the augmented classes $\hat{C}_1 = \{ (-2, 0, 1), (-2, -1, 1) \}$ and $\hat{C}_2 = \{ (0, 2, 1), (1, 2, 1) \}$. again, we will set $b_1 = b_2 = b_3 = b_4 = 1$. Now $-\hat{C}_2 = \{ (0, -2, -1), (-1, -2, -1) \}$, so we form the matrix

$$A = \begin{bmatrix} -2 & 0 & 1 \\ -2 & -1 & 1 \\ 0 & -2 & -1 \\ -1 & -2 & -1 \end{bmatrix} \quad \text{and} \quad \vec{a} = (A^T A)^{-1} A^T B = \begin{bmatrix} -5/13 \\ -5/13 \\ 0 \end{bmatrix}.$$

Therefore, the separating plane (and line) is $[-5/13, -5/13, 0] \cdot [x, y, z] = 0$ or $y = -x$

The following result (proof omitted) explains the connection between the Fisher LDF and MSE LDF.

Theorem 6.3 *Let C_1 and C_2 be two classes of size n_1 and n_2, respectively. Set $n = n_1 + n_2$. The Fisher LDF is a special case of an MSE LDF with the following values of $b_1, b_2, \ldots, b_n$:*

$$\underbrace{\frac{n_1}{n}, \ldots, \frac{n_1}{n}}_{n_1 \ times}, \underbrace{\frac{n_2}{n}, \ldots, \frac{n_2}{n}}_{n_2 \ times}.$$

This explains why in Example 6.13 we obtained the same results as the Fisher LDF, since when the classes are equal in size, the values of b_i will all be equal.

Example 6.14 *Let's consider an example with the classes not of the same size. Let $C_1 = \{ (-2, 0), (-2, -1) \}$ and $C_2 = \{ (0, 2), (1, 2), (2, 2) \}$. Then $\hat{C}_1 = \{ (-2, 0, 1), (-2, -1, 1) \}$, $C_2 = \{ (0, 2, 1), (1, 2, 1), (2, 2, 1) \}$ and $-C_2 = \{ (0, -2, -1), (-1, -2, -1), (-2, -2, -1) \}$. First, we will compute the MSE LDF with all the $b_i = 1$, $(i = 1, 2, 3, 4, 5)$. We form the matrix*

$$A = \begin{bmatrix} -2 & 0 & 1 \\ -2 & -1 & 1 \\ 0 & -2 & -1 \\ -1 & -2 & -1 \\ -2 & -2 & -1 \end{bmatrix} \quad \text{and} \quad \vec{a} = (A^T A)^{-1} A^T B = \begin{bmatrix} -18/107 \\ -60/107 \\ 35/107 \end{bmatrix}.$$

Therefore, the separating plane for $\hat{C}_1$ and $\hat{C}_2$ is

$$[-18/107, -60/107, 35/107] \cdot [x, y, z] = 0 \quad or \quad 18x + 60y - 35z = 0.$$

To obtain the separating line for C_1 and C_2 we set $z = 1$ and get $18x + 60y = 35$. Figure 6.22 is a concrete example of the general illustration presented in Figure 6.21.

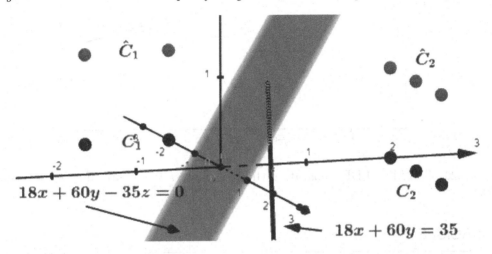

Figure 6.22 Concrete example of algorithm for finding a separating LDF not passing through the origin.

Second, we produce the Fisher LDF by assigning the values $b_1 = 2/5$, $b_2 = 2/5$, $b_3 = 3/5$, $b_4 = 3/5$ and $b_5 = 3/5$. In this case,

$$\vec{a} = (A^T A)^{-1} A^T B = \begin{bmatrix} -9/107 \\ -30/107 \\ 34/535 \end{bmatrix}.$$

Therefore, the separating plane for $\hat{C}_1$ and $\hat{C}_2$ is

$$[-9/107, -30/107, 34/535] \cdot [x, y, z] = 0 \quad or \quad 45x + 150y - 34z = 0.$$

To obtain the separating line for C_1 and C_2 we set $z = 1$ and get $45x + 150y = 34$. In Figure 6.23, we plot both MSE separating lines together with the two classes.

We make a couple of remarks before ending this section. First, as we saw with the Fisher LDF, the MSE LDF still functions even when the two classes are not linearly separable, however there are instances of linearly separable classes for which the line corresponding to an MSE LDF does not succeed in separating these two classes. A solution to this problem is the Ho-Kashyap LDF which is a variation of the MSE LDF which guarantees to separate classes which are linearly separable, however the details of this LDF is beyond the scope of this text.

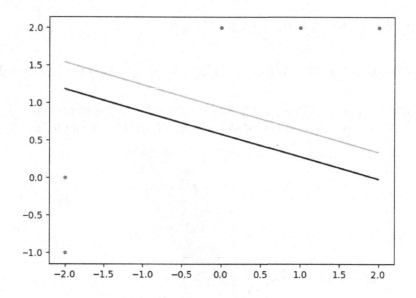

Figure 6.23 Two MSE LDFs separating the two classes. The lighter line is the Fisher LDF

EXERCISES

1. Consider the following two classes of data:

$$C_1 = \{ (0,0), (0,1) \} \quad \text{and} \quad C_2 = \{ (1,1), (1,2), (2,3) \}.$$

  a. Using augmented feature space, find the MSE LDF for C_1 and C_2 having all b_i's equal to zero.

  b. Using augmented feature space, find the Fisher LDF for C_1 and C_2.

  c. Graph both lines and the classes as was done in Example 6.14.

2. Consider the following two classes of data:

$$C_1 = \{ (0,0,-1), (1,0,1) \} \quad \text{and} \quad C_2 = \{ (1,1,0), (1,1,2), (2,3,1) \}.$$

  a. Using augmented feature space, find the MSE LDF for C_1 and C_2 having all b_i's equal to zero.

  b. Using augmented feature space, find the Fisher LDF for C_1 and C_2.

CHAPTER 7

Quadratic Forms

I N THIS CHAPTER, we introduce the notion of a quadratic form. Quadratic forms are prevalent in linear algebra and have many applications. In Section 7.1, we introduce the notion of a quadratic form and the associated definitions of positive and negative definite and semi-definite and indefinite. In Section 7.2, we derive the first test for determining if a quadratic form is positive or negative, definite or semi-definite, or indefinite called the Principal Minor Criterion. In Section 7.3, we derive the second test for determining if a quadratic form is positive or negative, definite or semi-definite, or indefinite called the Eigenvalue Criterion. In Section 7.4, we apply the criteria developed in Sections 7.2 and 7.3 to analyze critical points to determine if they are extrema for a multivariate function. In Section 7.5 we generalize the notion of a quadratic form.

7.1 INTRODUCTION TO QUADRATIC FORMS

In this section, we introduce the notion of a quadratic form and the associated definitions of positive and negative definite and semi-definite.

Definition 7.1 *Let A be a symmetric $n \times n$ matrix. The* **quadratic form** *associated with A, written Q_A, is a map from $\mathbb{R}^n$ to $\mathbb{R}$ defined by $Q_A(x) = x^T A x$ for $x \in \mathbb{R}^n$.*

Example 7.1 *Let $A = \begin{bmatrix} 1 & -2 \\ -2 & 5 \end{bmatrix}$, then*

$$Q_A(x_1, x_2) = \begin{bmatrix} x_1 & x_2 \end{bmatrix} \begin{bmatrix} 1 & -2 \\ -2 & 5 \end{bmatrix} \begin{bmatrix} x_1 \\ x_2 \end{bmatrix} = \begin{bmatrix} x_1 & x_2 \end{bmatrix} \begin{bmatrix} x_1 - 2x_2 \\ -2x_1 + 5x_2 \end{bmatrix}$$

$$= x_1(x_1 - 2x_2) + x_2(-2x_1 + 5x_2) = x_1^2 - 4x_1x_2 + 5x_2^2$$

A couple of things to notice about Q_A: First, the formula consists of a sum of terms each of which is degree two. Secondly the coefficients of the terms can quickly be obtained from the entries in the matrix A.

DOI: 10.1201/9781003217794-7

Example 7.2 *To illustrate the last comment about quadratic forms, let's recover the matrix A associated with the quadratic form* $2x_1^2 - x_2^2 + 4x_1x_2 - 2x_2x_3$. *The diagonal entry* a_{ii} *of A will consist of the coefficient of the pure square terms* x_i^2, *while the entries off the diagonal,* a_{ij} *and* a_{ji} *(i ≠ j), will each consist of half the coefficient of the mixed term* x_ix_j, *so that*

$$A = \begin{bmatrix} 2 & 2 & 0 \\ 2 & -1 & -1 \\ 0 & -1 & 0 \end{bmatrix}$$

We now introduce some fundamental definitions associated with quadratic forms.

Definition 7.2 *The symmetric* $n \times n$ *matrix A and its associated quadratic form* Q_A *is called*

1. **positive semi-definite** *if* $Q_A(x) \geq 0$ *for all* $x \in \mathbb{R}^n$.

2. **positive definite** *if* $Q_A(x) > 0$ *for all non-zero* $x \in \mathbb{R}^n$.

3. **negative semi-definite** *if* $Q_A(x) \leq 0$ *for all* $x \in \mathbb{R}^n$.

4. **negative definite** *if* $Q_A(x) < 0$ *for all non-zero* $x \in \mathbb{R}^n$.

5. **indefinite** *if there exist* $x_1, x_2 \in \mathbb{R}^n$ *such that* $Q_A(x_1) > 0$ *and* $Q_A(x_2) < 0$.

Example 7.3 *Here are two examples illustrating these definition.*

1. *Consider the quadratic form already introduced, namely* $Q_A(x_1, x_2) = x_1^2 - 4x_1x_2 + 5x_2^2 = (x_1 - 2x_2)^2 + x_2^2$. *If* $x_1 \neq 2x_2$, *then* $Q_A(x_1, x_2) > 0$, *and if* $x_1 = 2x_2$ *and* $(x_1, x_2) \neq (0, 0)$, *then* $Q_A(x_1, x_2) = x_2^2 > 0$. *Hence,* Q_A *is positive definite.*

2. *If* $A = \begin{bmatrix} -1 & 0 & 0 \\ 0 & 1 & 0 \\ 0 & 0 & 3 \end{bmatrix}$, *then* $Q_A(x_1, x_2, x_3) = -x_1^2 + x_2^2 + 3x_3^2$. *Notice that* $Q_A(1, 0, 0) = -1 < 0$ *while* $Q_A(0, 1, 0) = 1 > 0$. *Thus,* Q_A *is indefinite.*

EXERCISES

1. Compute the associated quadratic form for each of the following symmetric matrices:

   a. $A = \begin{bmatrix} 8 & 1 \\ 1 & 2 \end{bmatrix}$

   b. $B = \begin{bmatrix} -6 & 1 \\ 1 & 0 \end{bmatrix}$

c. $C = \begin{bmatrix} 2 & -1 & 2 \\ -1 & 0 & 0 \\ 2 & 0 & -3 \end{bmatrix}$

2. Recover the symmetric matrix for each of the following symmetric forms:

   a. $Q_A(x_1, x_2) = x_1^2 - 6x_1x_2 - 2x_2^2$

   b. $Q_B(x_1, x_2) = 2x_1x_2 + 3x_2^2$

   c. $Q_C(x_1, x_2, x_3) = 2x_1^2 - 2x_1x_2 - x_2^2 + 4x_2x_3$

3. Decide whether each of the following quadratic forms are positive (or negative), semi-definite (or definite), or indefinite:

   a. $Q_A(x_1, x_2) = x_1^2 - 6x_1x_2 + 11x_2^2$

   b. $Q_B(x_1, x_2) = 2x_1^2 + 3x_2^2$

   c. $Q_C(x_1, x_2, x_3) = x_1^2 + 2x_2^2 + 4x_3^2$

   d. $Q_D(x_1, x_2, x_3) = x_1^2 + 2x_3^2$

4. State and prove a result regarding diagonal matrices as to whether they are positive (or negative), semi-definite (or definite), or indefinite.

5. Prove that the diagonal entries of a positive definite symmetric matrix must be positive.

6. Let $A \in M_{mn}$.

   a. Prove that $A^T A$ is symmetric.

   b. Prove that $A^T A$ is positive semi-definite.

   c. Prove that if the columns of A are linearly independent, then A is positive definite.

7.2   PRINCIPAL MINOR CRITERION

In this section, we develop a test for determining whether a symmetric matrix and its associated quadratic form is positive or negative, definite or semi-definite or indefinite. This test involves computing certain minors in the matrix A called the principal minors. We start off with a lemma which will be useful as an inductive step in the verification of the test.

Lemma 7.1 *Consider the 2×2 symmetric matrix $A = \begin{bmatrix} a & b \\ b & c \end{bmatrix}$*

   a. A is positive definite iff $a > 0$ and $|A| > 0$.

   b. A is negative definite iff $a < 0$ and $|A| > 0$.

Proof 7.1 *Consider the quadratic form associated with A, namely $Q_A(x_1, x_2) = ax_1^2 + 2bx_1x_2 + cx_2^2$. We prove only part a, since the proof of part b has an very similar argument.*

We first assume that $a > 0$ and $|A| > 0$. We wish to show for any $(x_1, x_2) \neq (0, 0)$ we have $Q_A(x_1, x_2) > 0$. We do this in two cases.
Case 1: $x_2 = 0$.
In this case $x_1 \neq 0$ and so $Q_A(x_1, x_2) = ax_1^2 > 0$.
Case 2: $x_2 \neq 0$.
In this case we can write $(x_1, x_2) = (tx_2, x_2)$, where $t = x_1/x_2$. Then

$$Q_A(x_1, x_2) = a(tx_2)^2 + 2b(tx_2)x_2 + cx_2^2 = (at^2 + 2bt + c)x_2^2$$

Set $\phi(t) = at^2 + 2bt + c$. Notice that $\phi'(t) = 2at + 2b$ and so ϕ has a critical value at $t^* = -b/a$. Since $\phi''(t) = 2a > 0$ for all t, this implies that t^* is a global minimum. In other words, for all $t \in \mathbb{R}$ we have

$$\phi(t) \geq \phi(t^*) = a(-b/a)^2 + 2b(-b/a) + c = c - b^2/a = (1/a)|A| > 0$$

Therefore, $Q_A(x_1, x_2) = \phi(t)x_2^2 > 0$.
For the reverse direction, assume now that Q_A is positive definite. So in particular, $Q_A(1, 0) > 0$, i.e. $a > 0$. Now we show $|A| > 0$. Note that $Q_A(x_1, x_2) > 0$ when $x_2 \neq 0$, so borrowing work from the reverse direction we know that $\phi(t) > 0$ for all $t \in \mathbb{R}$. This implies that the quadratic polynomial $at^2 + 2bt + c$ has no real roots. This in turn implies that the discriminant in the quadratic formula, $(2b)^2 - 4ac < 0$. But this inequality is equivalent to $-4|A| < 0$ and so $|A| > 0$. □

Example 7.4 *Now it's easy to check the earlier example $A = \begin{bmatrix} 1 & -2 \\ -2 & 5 \end{bmatrix}$ is positive definite, since $a = 1 > 0$ and $|A| = 1 > 0$.*

Our goal is to generalize this lemma to arbitrary $n \times n$ symmetric matrices. First, we define some terminology and notation.

Definition 7.3 *Let $A = [a_{ij}]$ be an $n \times n$ matrix. The kth **principal minor** of A, written*

$$\Delta_k = \begin{vmatrix} a_{11} & a_{12} & \cdots & a_{1k} \\ a_{21} & a_{22} & \cdots & a_{2k} \\ \vdots & \vdots & \ddots & \vdots \\ a_{k1} & a_{k2} & \cdots & a_{kk} \end{vmatrix}$$

Theorem 7.1 *Let A be an $n \times n$ symmetric matrix.*

1. *A is positive definite iff $\Delta_k > 0$ for all $k = 1, 2, \ldots, n$.*

2. *A is negative definite iff $(-1)^k \Delta_k > 0$ for all $k = 1, 2, \ldots, n$.*

3. If $\Delta_k > 0$ for $k = 1, 2, \ldots, n-1$ and $\Delta_n = 0$ then A is positive semi-definite.

4. If $(-1)^k \Delta_k > 0$ for $k = 1, 2, \ldots, n-1$ and $\Delta_n = 0$ then A is negative semi-definite.

5. If for some $i \neq j$ we have $a_{ii} a_{jj} - a_{ij}^2 < 0$, then A is indefinite.

Proof 7.2 *Part 5, we leave as an exercise. For the rest we will only show the proof of part 1, since the other parts are done in a similar manner. The proof of part 1 is a proof by induction on n, but in order to simplify our notation without losing the depth of the argument, we will simply show that $n = 2$ implies $n = 3$. Therefore, we wish to prove the result for*

$$A = \begin{bmatrix} a_{11} & a_{12} & a_{13} \\ a_{12} & a_{22} & a_{23} \\ a_{13} & a_{23} & a_{33} \end{bmatrix} \quad \text{with associated quadratic form}$$

$$Q_A(x_1, x_2, x_3) = a_{11} x_1^2 + a_{22} x_2^2 + a_{33} x_3^2 + 2a_{12} x_1 x_2 + 2a_{13} x_1 x_3 + 2a_{23} x_2 x_3.$$

For one direction, we assume Δ_1, Δ_2, $\Delta_3 > 0$ and assume part 1. holds for symmetric 2×2 matrices. We need to show that $Q_A(x_1, x_2, x_3) > 0$ for any $(x_1, x_2, x_3) \neq (0, 0, 0)$. We consider two cases.
Case 1: $x_3 = 0$.

Set $\hat{A} = \begin{bmatrix} a_{11} & a_{12} \\ a_{12} & a_{22} \end{bmatrix}$. *Notice that in this case, $Q_{\hat{A}} = Q_A$ and since Δ_1, $\Delta_2 > 0$, by assumption $Q_{\hat{A}}$ is positive definite. Therefore, so is Q_A.*
Case 2: $x_3 \neq 0$.

In this case we can write $x_1 = sx_3$ and $x_2 = tx_3$ for some $s, t \in \mathbb{R}$. But then

$$Q_A(x_1, x_2, x_3) = a_{11}(sx_3)^2 + a_{22}(tx_3)^2 + a_{33} x_3^2 + 2a_{12}(sx_3)(tx_3) + 2a_{13}(sx_3)x_3$$
$$+ 2a_{23}(tx_3)x_3 =$$

$$(a_{11} s^2 + a_{22} t^2 + 2a_{12} st + 2a_{13} s + 2a_{23} t + a_{33}) x_3^2.$$

Set $\phi(s, t) = a_{11} s^2 + a_{22} t^2 + 2a_{12} st + 2a_{13} s + 2a_{23} t + a_{33}$. We show now that $\phi(s, t) > 0$ for all ordered pairs (s, t). This will imply $Q_A(x_1, x_2, x_3) = \phi(s, t) x_3^2 > 0$ which proves the result in Case 2. The argument follows from content covered in multivariate calculus (something which will be generalized later on in this chapter). First, we look for critical points of ϕ by solving $\nabla \phi = 0$. This leads to the following linear equations:

$$\begin{aligned} \phi_s &= 2a_{11} s + 2a_{12} t + 2a_{13} &= 0 \\ \phi_t &= 2a_{12} s + 2a_{22} t + 2a_{23} &= 0 \end{aligned}$$

Rewriting this as a matrix equation yields

$$\begin{bmatrix} a_{11} & a_{12} \\ a_{12} & a_{22} \end{bmatrix} \begin{bmatrix} s \\ t \end{bmatrix} = \begin{bmatrix} -a_{13} \\ -a_{23} \end{bmatrix}$$

Notice that the coefficient matrix of this linear system is $\hat{A}$. Since $|\hat{A}| = \Delta_2 > 0$ this implies the system has a unique solution, say (s^*, t^*). To classify this critical point we compute the second order partials $\phi_{ss} = 2a_{11}$, $\phi_{st} = 2a_{12}$ and $\phi_{tt} = 2a_{22}$. Since $\phi_{ss} = 2\Delta_1 > 0$ and $\phi_{ss}\phi_{tt} - \phi_{st}^2 = 4\Delta_2 > 0$ this implies that (s^*, t^*) is a global minimum for ϕ (you cannot have a unique critical point which is local but not global). In other words, for all ordered pairs (s, t) we have $\phi(s, t) \geq \phi(s^*, t^*)$. We will now show that $\phi(s^*, t^*) > 0$ from which we get our result, since

$$Q_A(x_1, x_2, x_3) = \phi(s, t)x_3^2 \geq \phi(s^*, t^*)x_3^2 > 0$$

Claim: $\phi(s^*, t^*) > 0$

To see this we return to the system for which (s^*, t^*) was the unique solution. In other words

$$a_{11}s^* + a_{12}t^* + a_{13} = 0$$
$$a_{12}s^* + a_{22}t^* + a_{23} = 0$$

Multiplying the top equation by s^* and the bottom equation by t^* yields

$$a_{11}(s^*)^2 + a_{12}s^*t^* + a_{13}s^* = 0$$
$$a_{12}s^*t^* + a_{22}(t^*)^2 + a_{23}t^* = 0$$

Utilizing these last equations we compute and simplify

$$\phi(s^*, t^*) = a_{11}(s^*)^2 + a_{22}(t^*)^2 + 2a_{12}s^*t^* + 2a_{13}s^* + 2a_{23}t^* + a_{33} =$$

$$[(s^*)^2 + a_{12}s^*t^* + a_{13}s^*] + [a_{12}s^*t^* + a_{22}(t^*)^2 + a_{23}t^*] + [a_{13}s^* + a_{23}t^* + a_{33}] =$$

$$0 + 0 + a_{13}s^* + a_{23}t^* + a_{33} = a_{13}s^* + a_{23}t^* + a_{33}$$

Using Cramer's Rule we can compute directly the values of s^* and t^* as

$$s^* = \frac{\begin{vmatrix} -a_{13} & a_{12} \\ -a_{23} & a_{22} \end{vmatrix}}{\Delta_2} \quad and \quad t^* = \frac{\begin{vmatrix} a_{11} & -a_{13} \\ a_{21} & -a_{23} \end{vmatrix}}{\Delta_2}$$

Putting all that we derived so far together we have

$$\Delta_2 \phi(s^*, t^*) = a_{13}s^*\Delta_2 + a_{23}t^*\Delta_2 + a_{33}\Delta_2 =$$

$$a_{13} \begin{vmatrix} -a_{13} & a_{12} \\ -a_{23} & a_{22} \end{vmatrix} + a_{23} \begin{vmatrix} a_{11} & a_{13} \\ a_{21} & a_{23} \end{vmatrix} + a_{33} \begin{vmatrix} a_{11} & a_{12} \\ a_{12} & a_{22} \end{vmatrix} =$$

$$a_{13} \begin{vmatrix} a_{12} & a_{13} \\ a_{22} & a_{23} \end{vmatrix} - a_{23} \begin{vmatrix} a_{11} & -a_{13} \\ a_{21} & -a_{23} \end{vmatrix} + a_{33} \begin{vmatrix} a_{11} & a_{12} \\ a_{12} & a_{22} \end{vmatrix} = \Delta_3$$

Therefore, $\phi(s^*, t^*) = \frac{\Delta_3}{\Delta_2} > 0$, by assumption. Thus, the claim is proved and one direction of the argument is done.

We now assume that Q_A is positive definite and follow the same two cases. First, consider (x_1, x_2, x_3) with $x_3 = 0$. As above $Q_{\hat{A}} = Q_A$ and is positive definite. Therefore, by induction $\Delta_1, \Delta_2 > 0$. Now consider (x_1, x_2, x_3) with $x_3 \neq 0$. Since $\Delta_2 \neq 0$, using some of the work from the reverse direction, the critical point (s^*, t^*) once again exists for ϕ and

$$\frac{\Delta_3}{\Delta_2} = \phi(s^*, t^*) = \phi(s^*, t^*)(1)^2 = Q_A(s^*, t^*, 1) > 0$$

Therefore, since $\Delta_2 > 0$, it follows that $\Delta_3 > 0$. □

Example 7.5 *Here are several examples for which we apply the Principal Minor criterion.*

1. Let $A = \begin{bmatrix} 2 & 1 & 0 \\ 1 & 2 & 0 \\ 0 & 0 & 2 \end{bmatrix}$.

$$\Delta_1 = 2 > 0, \qquad \Delta_2 = \begin{vmatrix} 2 & 1 \\ 1 & 2 \end{vmatrix} = 3 > 0 \ \text{and} \ \Delta_3 = |A| = 6 > 0.$$

Therefore, by Theorem 7.1.a, A is positive definite.

2. Let $A = \begin{bmatrix} -1 & 2 & 0 \\ 2 & -5 & 1 \\ 0 & 1 & -2 \end{bmatrix}$.

$$\Delta_1 = -1 < 0, \qquad \Delta_2 = \begin{vmatrix} -1 & 2 \\ 2 & -5 \end{vmatrix} = 1 > 0 \ \text{and} \ \Delta_3 = |A| = -1 < 0.$$

Therefore, by Theorem 7.1.b, A is negative definite.

3. Let $A = \begin{bmatrix} 2 & 1 & 0 \\ 1 & 2 & 0 \\ 0 & 0 & 0 \end{bmatrix}$.

$$\Delta_1 = 2 > 0, \qquad \Delta_2 = \begin{vmatrix} 2 & 1 \\ 1 & 2 \end{vmatrix} = 3 > 0 \ \ and \ \ \Delta_3 = |A| = 0.$$

Therefore, by Theorem 7.1.c, A is positive semi-definite.

4. Let $A = \begin{bmatrix} -1 & 1 & 1 \\ 1 & -2 & 0 \\ 1 & 0 & -2 \end{bmatrix}$.

$$\Delta_1 = -1 < 0, \qquad \Delta_2 = \begin{vmatrix} -1 & 1 \\ 1 & -2 \end{vmatrix} = 1 > 0 \ \ and \ \ \Delta_3 = |A| = 0.$$

Therefore, by Theorem 7.1.d, A is negative semi-definite.

5. Let $A = \begin{bmatrix} 1 & -1 & 3 & -2 \\ -1 & 3 & 2 & 1 \\ 3 & 2 & 5 & 0 \\ -2 & 1 & 0 & 1 \end{bmatrix}$. *Notice that*

$$\begin{vmatrix} a_{11} & a_{13} \\ a_{31} & a_{33} \end{vmatrix} = \begin{vmatrix} 1 & 3 \\ 3 & 5 \end{vmatrix} = -4 < 0,$$

Therefore, by Theorem 7.1.e, A is indefinite.

EXERCISES

1. Use Theorem 7.1 to classify the following matrices as positive or negative, definite or semi-definite, or indefinite:

a. $A = \begin{bmatrix} 2 & 4 & 0 \\ 4 & 1 & 0 \\ 0 & 0 & -6 \end{bmatrix}$.

b. $B = \begin{bmatrix} -1 & 1 & 1 \\ 1 & -2 & 0 \\ 1 & 0 & -3 \end{bmatrix}$.

c. $C = \begin{bmatrix} 2 & 0 & 2 \\ 0 & 2 & -1 \\ 2 & -1 & 4 \end{bmatrix}$.

d. $D = \begin{bmatrix} 1 & 1 & 1 \\ 1 & 2 & 1 \\ 1 & 1 & 1 \end{bmatrix}$.

e. $E = \begin{bmatrix} -2 & 1 & 1 \\ 1 & -1 & 1 \\ 1 & 1 & -5 \end{bmatrix}$.

2. Explain why the following example does not contradict Theorem 7.1:

Let $A = \begin{bmatrix} 1 & -4 \\ 1 & 1 \end{bmatrix}$. Notice that $\Delta_1 = 1 > 0$ and $\Delta_2 = 5 > 0$, however

$$Q_A(1,1) = \begin{bmatrix} 1 & 1 \end{bmatrix} \begin{bmatrix} 1 & -4 \\ 1 & 1 \end{bmatrix} \begin{bmatrix} 1 \\ 1 \end{bmatrix} = -1 < 0.$$

3. Prove Lemma 7.1.b

4. Prove Theorem 7.1.b

5. Prove Theorem 7.1.c

6. Prove Theorem 7.1.d

7. Prove Theorem 7.1.e

7.3 EIGENVALUE CRITERION

In this section we derive the second test for determining if a symmetric matrix is positive or negative, definite or semi-definite, or indefinite called the Eigenvalue Criterion. This result relies on the fact that we can perform this determination for diagonal matrices.

Theorem 7.2 *Let A be a symmetric real-valued matrix. Then*

1. *A is positive definite iff all the eigenvalues of A are positive.*

2. *A is positive semi-definite iff all the eigenvalues of A are non-negative.*

3. *A is negative definite iff all the eigenvalues of A are negative.*

4. *A is negative semi-definite iff all the eigenvalues of A are non-positive.*

5. *A is indefinite iff A has both positive and negative eigenvalues.*

Proof 7.3 *By Corollary 5.3 in Section 5.7, $P^T A P = D$ where $P^T P = I$ and D is a diagonal matrix consisting of real eigenvalues, say $\lambda_1, \lambda_2, \ldots, \lambda_n$ of A. For any $x = (x_1, x_2, \ldots, x_n) \in \mathbb{R}^n$ there exists $y = (y_1, y_2, \ldots, y_n) \in \mathbb{R}^n$ such that $x = Py$ (indeed, $y = P^T x$). Therefore,*

$$Q_A(x) = x^T A x = (Py)^T A(Py) = y^T P^T A P y = y^T D y = \lambda_1 y_1^2 + \lambda_2 y_2^2 + \cdots + \lambda_n y_n^2.$$

To prove part 1 (parts 2,3,4,5 are done similarly), first assume that A is positive definite. Set $x = Pe_i$. Then $0 < Q_A(x) = \lambda_i$ for $i = 1, 2, \ldots, n$. Now assume $\lambda_1, \lambda_2, \ldots, \lambda_n > 0$. Then certainly, $Q_A(x) = \lambda_1 y_1^2 + \lambda_2 y_2^2 + \cdots + \lambda_n y_n^2 > 0$ when $x \neq 0$.

□

Example 7.6 *(revisited from Section 7.1)*

1. Let $A = \begin{bmatrix} 2 & 1 & 0 \\ 1 & 2 & 0 \\ 0 & 0 & 2 \end{bmatrix}$. *We compute*

$$p_A(t) = \begin{vmatrix} 2-t & 1 & 0 \\ 1 & 2-t & 0 \\ 0 & 0 & 2-t \end{vmatrix} = (2-t)\begin{vmatrix} 2-t & 1 \\ 1 & 2-t \end{vmatrix} = (2-t)[(2-t)^2-1] =$$

$$(2-t)(t^2-4t+3) = -(t-2)(t-1)(t-3).$$

Thus, the roots of $p_A(t)$ are $2, 1, 3$ (all positive). Therefore, by Theorem 7.2.a, A is positive definite.

2. Let $A = \begin{bmatrix} -1 & 2 & 0 \\ 2 & -5 & 1 \\ 0 & 1 & -2 \end{bmatrix}$.

One can show that the eigenvalues of A are approximately -6.04, -1.87 and -0.09 (all negative). Therefore, by Theorem 7.2.c, A is negative definite.

3. Let $A = \begin{bmatrix} 2 & 1 & 0 \\ 1 & 2 & 0 \\ 0 & 0 & 0 \end{bmatrix}$.

One can show that the eigenvalues of A are $1, 3, 0$ (all non-negative). Therefore, by Theorem 7.2.b, A is positive semi-definite.

4. Let $A = \begin{bmatrix} -1 & 1 & 1 \\ 1 & -2 & 0 \\ 1 & 0 & -2 \end{bmatrix}$.

One can show that the eigenvalues of A are $-3, -2, 0$ (all non-positive). Therefore, by Theorem 7.2.d, A is negative semi-definite.

5. Let $A = \begin{bmatrix} 1 & -1 & 3 & -2 \\ -1 & 3 & 2 & 1 \\ 3 & 2 & 5 & 0 \\ -2 & 1 & 0 & 1 \end{bmatrix}$.

One can show that the eigenvalues of A are approximately -1.98, 0.47, 4.41 and 7.10 (both positive and negative values). Therefore, by Theorem 7.2.e, A is indefinite.

We finish this section with a very nice classification of positive semi-definite symmetric matrices.

Proposition 7.1 *A symmetric matrix $A \in M_{nn}$ is positive semi-definite iff there exists a matrix $B \in M_{mn}$ such that $A = B^T B$.*

Proof 7.4 *One direction is found in Exercise 6.b of Section 7.1. For the other direction, assume A is symmetric and positive definite. Since A is symmetric, by Corollary 5.3, there exists an orthogonal matrix P and diagonal matrix D such that $P^T A P = D$, where the diagonal entries in D are (real) eigenvalues of A. Since A is positive semi-definite, by Theorem 7.2.b, A has only non-negative eigenvalues. Therefore, the matrix $D^{1/2}$ is a real matrix. Set $B = D^{1/2} P^T$. Then*

$$B^T B = P D^{1/2} D^{1/2} P^T = P D P^T = A.$$

□

Example 7.7 *We point out that the proof of Proposition 7.1 is constructive and yields an algorithm for expressing A as $B^T B$ which we illustrate now. Consider the positive semi-definite symmetric matrix from Example 7.5.3.*

$$A = \begin{bmatrix} 2 & 1 & 0 \\ 1 & 2 & 0 \\ 0 & 0 & 0 \end{bmatrix}.$$

We have seen that the eigenvalues of this matrix are $0, 1, 3$. The corresponding normalized eigenvectors are $[0, 0, 1]$, $[-1/\sqrt{2}, 1/\sqrt{2}, 0]$ and $[1/\sqrt{2}, 1/\sqrt{2}, 0]$. Therefore, the orthogonal matrix which diagonalizes A is

$$P = \begin{bmatrix} 0 & -1/\sqrt{2} & 1/\sqrt{2} \\ 0 & 1/\sqrt{2} & 1/\sqrt{2} \\ 1 & 0 & 0 \end{bmatrix}.$$

According to Proposition 7.1, set

$$B = D^{1/2} P^T = \begin{bmatrix} 0 & 0 & 0 \\ 0 & 1 & 0 \\ 0 & 0 & \sqrt{3} \end{bmatrix} \begin{bmatrix} 0 & 0 & 0 \\ -1/\sqrt{2} & 1/\sqrt{2} & 0 \\ 1/\sqrt{2} & 1/\sqrt{2} & 0 \end{bmatrix} = \begin{bmatrix} 0 & 0 & 0 \\ -1/\sqrt{2} & 1/\sqrt{2} & 0 \\ \sqrt{3/2} & \sqrt{3/2} & 0 \end{bmatrix}.$$

One can check that $B^T B$ does indeed equal A.

One final note on Proposition 7.1. The matrix B is not uniquely determined and furthermore there is a choice of B which is upper triangular. Such a decomposition always exists and is called the Cholesky decomposition.

EXERCISES

1. Redo Exercise 1 in Section 7.1 using the Eigenvalue Criterion, namely

   a. $A = \begin{bmatrix} 2 & 4 & 0 \\ 4 & 1 & 0 \\ 0 & 0 & -6 \end{bmatrix}$.

   b. $B = \begin{bmatrix} -1 & 1 & 1 \\ 1 & -2 & 0 \\ 1 & 0 & -3 \end{bmatrix}$.

   c. $C = \begin{bmatrix} 2 & 0 & 2 \\ 0 & 2 & -1 \\ 2 & -1 & 4 \end{bmatrix}$.

   d. $D = \begin{bmatrix} 1 & 1 & 1 \\ 1 & 2 & 1 \\ 1 & 1 & 1 \end{bmatrix}$.

   e. $E = \begin{bmatrix} -2 & 1 & 1 \\ 1 & -1 & 1 \\ 1 & 1 & -5 \end{bmatrix}$.

2. Prove for any real number a the following matrix is positive semi-definite:

$$A = \begin{bmatrix} a^4 & a^3 & a^2 \\ a^3 & a^2 & a \\ a^2 & a & 1 \end{bmatrix}$$

   Note: Principal Minor Criterion does not work here.

3. Prove Theorem 7.2.e.

7.4 APPLICATION: UNCONSTRAINED NON-LINEAR OPTIMIZATION

In this section, we apply what we have gone over thus far in the chapter in order to analyze critical points of a multivariate function and decide whether or not they are extrema. This material falls under the topic of unconstrained nonlinear optimization. Some of this material would be covered in a multivariable calculus course, however not to the extent to which we do now. In reality multivariable calculus only covers functions of two variables when it comes to unconstrained nonlinear optimization. This section will generalize the method shown in multivariable calculus to any number of variables. First, we need to introduce some definitions and terminology.

Definition 7.4 *A multi-variable real-valued function $f : \mathbb{R}^n \to \mathbb{R}$ has as input an n-tuple $x = (x_1, x_2, \ldots, x_n) \in \mathbb{R}^n$ and as output a real number $f(x) = f(x_1, x_2, \ldots, x_n)$.*

Example 7.8 *Define* $f : \mathbb{R}^4 \to \mathbb{R}$ *by* $f(x_1, x_2, x_3, x_4) = 2x_1^2 x_3 - \cos(x_1 x_4) - \frac{5x_4}{x_3}$.

Definition 7.5 *For an n-tuple x and positive real number r, an* **open ball centered at x of radius** r, *written*

$$B(x, r) = \{y \in \mathbb{R}^n \ : \ |x - y| < r\}.$$

Example: *In $\mathbb{R}^3$ an open ball is a solid sphere without its outer surface. In $\mathbb{R}^2$ an open ball is a solid disk without its circular edge, and in $\mathbb{R}$ an open ball is an open interval.*

Definition 7.6 *Let $D \subseteq \mathbb{R}^n$.*

- $x \in D$ *is an* **interior point** *of D if $\exists r > 0$ such that $B(x, r) \subseteq D$.*

- *The* **interior** *of D, written D° is the collection of all interior points of D.*

- *The* **boundary** *of D, written $\partial D = D - D^\circ$.*

- *D is* **open** *if $D = D^\circ$.*

- *D is* **closed** *if its complement D^c is open.*

- *D is* **bounded** *if $\exists r > 0$ such that $D \subseteq B(0, r)$.*

For the remainder of this section, we shall assume that our functions have continuous first and second order partial derivatives on its domain, which we shall denote by D.

Definition 7.7 *Let $D \subseteq \mathbb{R}^n$ and $f : D \to \mathbb{R}$ with $x^* \in D^\circ$.*

- *x^* is a* **global minimizer** *with* **global minimum** *$f(x^*)$ if $\forall x \in D$ we have $f(x^*) \leq f(x)$.*

- *x^* is a* **strict global minimizer** *with* **strict global minimum** *$f(x^*)$ if $\forall x \in D$ and $x \neq x^*$ we have $f(x^*) < f(x)$.*

- *x^* is a* **local minimizer** *with* **local minimum** *$f(x^*)$ if $\exists \delta > 0$ such that $\forall x \in D \cap B(x^*, \delta)$ we have $f(x^*) \leq f(x)$.*

- *x^* is a* **strict local minimizer** *with* **strict local minimum** *$f(x^*)$ if $\exists \delta > 0$ such that $\forall x \in D \cap B(x^*, \delta)$ and $x \neq x^*$ we have $f(x^*) < f(x)$.*

- *For the definitions of maximizer/maximum simply reverse all the inequalities in the definitions above.*

- *All the above definitions are referred to collectively as* **extrema** *of f.*

- *x^* is a* **saddlepoint** *if there exist $y, z \in \mathbb{R}^n$ such that $f(x + ty)$, $t \in \mathbb{R}$, has a strict local minimum and $f(x + tz)$, $t \in \mathbb{R}$, has a strict local maximum.*

- x^* is a **critical point** if $\frac{\partial f}{\partial x_i}$ exists and equals 0 for $i = 1, 2, \ldots n$. Recall the gradient of f is defined as

$$\nabla f = \left[\frac{\partial f}{\partial x_1}, \frac{\partial f}{\partial x_2}, \ldots, \frac{\partial f}{\partial x_n} \right]$$

Thus, a critical point satisfies $\nabla f(x^*) = 0$.

We shall omit some of the proofs of the following facts, since their proofs are more appropriate in a real analysis course.

Facts and Definitions:

- Let $D \subseteq \mathbb{R}^n$ and $f : D \to \mathbb{R}$ with $x^* \in D^\circ$. Assuming $\frac{\partial f}{\partial x_i}$ exist for $i = 1, 2, \ldots n$, if x^* corresponds to a local extremum of f, then x^* is a critical point (proof omitted).

- For $x, x^* \in \mathbb{R}^n$ we define the interval $[x^*, x] = \{x^* + t(x - x^*) : 0 \le t \le 1\}$. This is basically a line segment (with orientation) connecting x^* to x.

- If $f(x)$ has continuous first and second order partials, then $\frac{\partial^2 f}{\partial x_i x_j} = \frac{\partial^2 f}{\partial x_j x_i}$ for all $i, j \in \{1, 2, \ldots, n\}$.

- **Generalized Taylor's Theorem:** Let $f(x)$ have continuous first and second order partials. There exists $z \in [x^*, x]$ such that

$$f(x) = f(x^*) + \nabla f(x^*)(x - x^*) + \frac{1}{2}(x - x^*)^T H f(z)(x - x^*),$$

where

$$H f = \begin{bmatrix} \frac{\partial^2 f}{\partial x_1^2} & \frac{\partial^2 f}{\partial x_1 x_2} & \cdots & \frac{\partial^2 f}{\partial x_1 x_n} \\ \frac{\partial^2 f}{\partial x_1 x_2} & \frac{\partial^2 f}{\partial x_2^2} & \cdots & \frac{\partial^2 f}{\partial x_2 x_n} \\ \vdots & \vdots & \ddots & \vdots \\ \frac{\partial^2 f}{\partial x_1 x_n} & \frac{\partial^2 f}{\partial x_2 x_n} & \cdots & \frac{\partial^2 f}{\partial x_n^2} \end{bmatrix}$$

The symmetric matrix $H f$ is called the **Hessian** of f. In multivariable calculus, its determinant is typically called the **discriminant** and is denoted $D(x, y)$, i.e.

$$D(x, y) = |H f(x, y)| = \begin{vmatrix} \frac{\partial^2 f}{\partial x^2} & \frac{\partial^2 f}{\partial x y} \\ \frac{\partial^2 f}{\partial x y} & \frac{\partial^2 f}{\partial y^2} \end{vmatrix}$$

The proof of this result (omitted) relies heavily on the single variable Taylor's Theorem.

From all these facts listed one can obtain the following results which follow almost immediately from the Generalized Taylor's Theorem (proofs omitted).

Theorem 7.3 *Let $x^* \in D^\circ$ be a critical point of f where $f : D \subseteq \mathbb{R}^n \to \mathbb{R}$ has continuous first and second order partials.*

- *Global Information*

  1. *If $Hf(x)$ is positive definite for all $x \neq x^*$, then x^* is a strict global minimizer.*
  2. *If $Hf(x)$ is negative definite for all $x \neq x^*$, then x^* is a strict global maximizer.*
  3. *If $Hf(x)$ is positive semi-definite, then x^* is a global minimizer.*
  4. *If $Hf(x)$ is negative semi-definite, then x^* is a global maximizer.*

- *Local Information*

  1. *If $Hf(x^*)$ is positive definite, then x^* is a strict local minimizer.*
  2. *If $Hf(x^*)$ is negative definite, then x^* is a strict local maximizer.*
  3. *If $Hf(x^*)$ is indefinite, then x^* is a saddlepoint.*

Before we dive into some detailed examples, it might be good to see the connection between Theorem 7.3 and Theorem 7.1 and multivariate calculus. In that class when critical points were analyzed students were basically applying the Principal Minor Criterion in the case of local extrema in Theorem 7.3. Indeed, given $f : \mathbb{R}^2 \to \mathbb{R}$ and a critical point x^* students evaluated $f_{xx}(x^*)$ and $f_{xx}(x^*)f_{yy}(x^*) - f_{xy}(x^*)^2$ to determine if they had a relative maximum, relative minimum or a saddlepoint. But these two quantities are precisely $\Delta_1(x^*)$ and $\Delta_2(x^*)$ of the Principal Minor Criterion!

Example 7.9 *In these examples, we will apply either the Principal Minor Criterion or Eigenvalue Criterion in order to analyze critical points.*

1. *Let $f(x, y) = e^{x^2+y^2}$. First, we compute $\nabla f = [f_x, f_y] = [2xe^{x^2+y^2}, 2ye^{x^2+y^2}]$ and find critical points by solving $\nabla f = 0$, i.e.*

$$\begin{cases} 2xe^{x^2+y^2} & = & 0 \\ 2ye^{x^2+y^2} & = & 0 \end{cases}$$

Therefore, the only critical point is $x^* = (0, 0)$. To analyze the critical points we compute the Hessian. Note that

$$f_{xx} = 2e^{x^2+y^2} + 2x(2xe^{x^2+y^2}) = 2(1+2x^2)e^{x^2+y^2}, \quad f_{xy} = 2x(2ye^{x^2+y^2}) = 4xye^{x^2+y^2}$$

and by symmetry $f_{yy} = 2(1 + 2y^2)e^{x^2+y^2}$. Therefore,

$$Hf(x,y) = \begin{bmatrix} f_{xx} & f_{xy} \\ f_{xy} & f_{yy} \end{bmatrix} = \begin{bmatrix} 2(1+2x^2)e^{x^2+y^2} & 4xye^{x^2+y^2} \\ 4xye^{x^2+y^2} & 2(1+2y^2)e^{x^2+y^2} \end{bmatrix}.$$

Notice that $\Delta_1 = 2(1+2x^2)e^{x^2+y^2} > 0$ for all (x,y) and

$$\Delta_2 = 2(1+2x^2)e^{x^2+y^2}2(1+2y^2)e^{x^2+y^2} - \left(4xye^{x^2+y^2}\right)^2$$
$$= \left(e^{x^2+y^2}\right)^2 (4(1+2x^2)(1+2y^2) - (4xy)^2) =$$

$$4e^{2x^2+2y^2}(1+2x^2+2y^2) > 0 \quad \text{for all } (x,y).$$

Therefore, $Hf(x,y)$ is positive definite which implies $(0,0)$ is the strict global minimizer. Figure 7.1 shows the graph of the surface with its strict global minimizer.

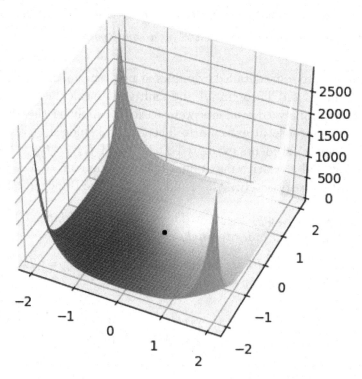

Figure 7.1 The graph of $z = e^{x^2+y^2}$ with strict global minimizer $(0,0,0)$.

2. Let $f(x,y) = -\frac{1}{3}x^3 + xy + \frac{1}{2}y^2 - 12y$. Then $\nabla f = [f_x, f_y] = [-x^2+y, x+y-12]$. To find critical points we solve $\nabla f = 0$, i.e.

$$\begin{cases} -x^2 + y & = 0 \\ x + y - 12 & = 0 \end{cases}$$

One can show that the solutions to this non-linear system are $x^* = (-4, 16)$ and $(3, 9)$. Next we compute the Hessian,

$$Hf(x, y) = \begin{bmatrix} f_{xx} & f_{xy} \\ f_{xy} & f_{yy} \end{bmatrix} = \begin{bmatrix} -2x & 1 \\ 1 & 1 \end{bmatrix}.$$

It's clear that Hf can not be used to determine global information, since $\Delta_1 = -2x$ and its sign can be both positive and negative for different values of x. Therefore, we will plug in each critical point and determine local information.

We show that $Hf(-4, 16) = \begin{bmatrix} 8 & 1 \\ 1 & 1 \end{bmatrix}$ is positive definite. For comparison, we will verify this using each of the two criteria. First, since $\Delta_1 = 8 > 0$ and $\Delta_2 = 7 > 0$. Second, the eigenvalues of $Hf(-4, 16)$ turn out to be both positive. Therefore, $(-4, 16)$ is a strict local minimizer.

We show that $Hf(3, 9) = \begin{bmatrix} -6 & 1 \\ 1 & 1 \end{bmatrix}$ is indefinite. One way to see this is since the determinant of $Hf(3, 9) = -7 < 0$. Another way to see this is to compute the eigenvalues of $Hf(3, 9)$ which has one positive and one negative eigenvalue. Therefore, $(3, 9)$ is a saddlepoint. Figure 7.2 shows the graph of the surface with its strict local minimizer and saddlepoint.

3. *Let $f(x, y, z) = x^2 + y^2 + z^2 + xy$. Then $\nabla f = [f_x, f_y, f_z] = [2x + y, x + 2y, 2z]$. To find critical points we solve $\nabla f = 0$, i.e.*

$$\begin{cases} 2x + y &= 0 \\ x + 2y &= 0 \\ 2z &= 0 \end{cases}$$

Since the determinant of the coefficient matrix of this linear system equals zero, the system has only the trivial solution, $(0, 0, 0)$. Next we compute the Hessian,

$$Hf(x, y, z) = \begin{bmatrix} f_{xx} & f_{xy} & f_{xy} \\ f_{xy} & f_{yy} & f_{yz} \\ f_{xz} & f_{yz} & f_{zz} \end{bmatrix} = \begin{bmatrix} 2 & 1 & 0 \\ 1 & 2 & 0 \\ 0 & 0 & 2 \end{bmatrix}.$$

Now $Hf(x, y, z)$ is positive definite, since $\Delta_1 = 2 > 0$, $\Delta_2 = 3 > 0$ and $\Delta_3 = 6 > 0$. Therefore, $(0, 0, 0)$ is a strict global minimizer (or since the eigenvalues of $Hf(x, y, z)$ are 1, 2 and 3, which are all positive).

4. *Let $f(x, y, z) = -x^4 - y^4 - z^4$. One can compute that the only critical point for f is $(0, 0, 0)$ and the Hessian is*

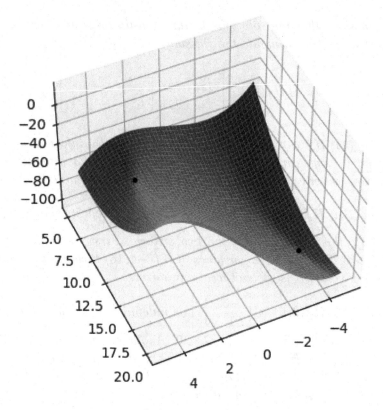

Figure 7.2 The graph of $z = -\frac{1}{3}x^3 + xy + \frac{1}{2}y^2 - 12y$ with its local minimizer $(-4, 16, -320/3)$ and saddlepoint $(3, 9, -99/2)$.

$$Hf(x, y, z) = \begin{bmatrix} -12x^2 & 0 & 0 \\ 0 & -12y^2 & 0 \\ 0 & 0 & -12z^2 \end{bmatrix}.$$

Since Hf is a diagonal matrix with diagonal entries ≤ 0 we have seen this implies Hf is negative semi-definite (or since the eigenvalues of $Hf(x, y, z)$ are $-12x^2$, $-12y^2$ and $-12z^2$, which are all non-negative). Therefore, $(0, 0, 0)$ is a global maximizer. Now we cannot conclude Hf is negative definite, but there are two ways we can argue that $(0, 0, 0)$ is a strict global maximizer. One argument is that since $(0, 0, 0)$ is the only critical point it must be the strict global maximizer. Another argument is "by inspection". Notice that for $(x, y, z) \neq (0, 0, 0)$ we have $f(x, y, z) < 0 = f(0, 0, 0)$ and this shows that $(0, 0, 0)$ is the strict global maximizer.

EXERCISES

1. Use the Principal Minor Criterion to classify the critical points of f.

   a. $f(x, y, z) = x^2 + \frac{1}{2}y^2 + z^2 + \frac{1}{12}y^4 + xz - yz$.

b. $f(x, y) = x^3 + y^2 - 2xy + 2$

c. $f(x, y, z) = x^2 + 2y^2 - 3z^2 + 4xy - 5x - 6y - 2z$

d. $f(x, y, z) = -\frac{1}{2}x^2 - y^2 - \frac{3}{2}z^2 + xy + xz$

e. $f(x, y) = x^4 + y^4 - 4xy + 1$

f. $f(x, y) = x^2 y - 2y^2 - (1/24)x^4 - 2y^4$

g. $f(x, y) = x^2 e^y + x^4 + e^{2y}$

h. $f(x, y, z) = e^{x^2 + y + z}$

i. $f(x, y) = 3x^4 + 2y^3 - 6x^2 y$

j. $f(x, y) = 6xy^2 - 2x^3 - 3y^4$

k. $f(x, y, z) = e^x + e^y + e^z + 2e^{-x-y-z}$

2. Use the Eigenvalue Criterion to classify the critical points of f (you can at times use your work in Problem 1).

a. $f(x, y, z) = x^2 + \frac{1}{2}y^2 + z^2 + \frac{1}{12}y^4 + xz - yz.$

b. $f(x, y) = x^3 + y^2 - 2xy + 2$

c. $f(x, y, z) = x^2 + 2y^2 - 3z^2 + 4xy - 5x - 6y - 2z$

d. $f(x, y, z) = -\frac{1}{2}x^2 - y^2 - \frac{3}{2}z^2 + xy + xz$

e. $f(x, y, z) = x^3 y + y^3 z + z^2 + xy.$

f. $f(x, y, z) = x^2 + y^2 + 2yz + (1/2)z^2 + z$

g. $f(x, y, z) = x^2 + 2xy + (1/2)y^2 + y + z^2$

h. $f(x, y, z) = x + xz - 2x^2 - xy - y^2 - z^2$

3. Use either criterion to classify the critical points of f.

a. $f(x, y, z) = x^2 + y^2 + xyz + 4z^2$

b. $f(x, y, z) = 6xy^2 - 2x^3 - 3y^4 - z^4 + 4z.$

4. Apply both criteria to classify the critical points of

$$f(x, y, z) = e^{x^2 + y^2 + z^2}$$

7.5   GENERAL QUADRATIC FORMS

In this section, we generalize the notion of a quadratic form with the ultimate goal of proving an elegant result of J. J. Silvester.

Definition 7.8 *Let V be a vector space with a symmetric bilinear map $(*|*)$. The* **quadratic form associated with** $(*|*)$ *is a function $q : V \to \mathbb{R}$ defined by $q(v) = (v|v)$.*

The way in which quadratic form has been generalized is the fact that our symmetric bilinear map is no longer an inner product, since it does not necessarily have the positive definite property of an inner product.

Example 7.10 *The first three examples represent quardatic forms derived from an inner product, however the fourth example defines a symmetric bilinear map which is not an inner product.*

1. *Let $V = \mathbb{R}^n$ with the usual dot product, then*

$$q(a_1, a_2, \ldots, a_n) = a_1^2 + a_2^2 + \cdots + a_n^2.$$

2. *Let $V = C[a, b]$ with inner product $(f|g) = \int_a^b f(x)g(x) \, dx$, then*

$$q(f) = \int_a^b f(x)^2 \, dx.$$

3. *Let $V = M_n$ with inner product $(A|B) = tr(B^T A)$, then for $A = [a_{ij}]$,*

$$q(A) = a_{11}^2 + a_{12}^2 + \cdots + a_{nn}^2,$$

   i.e. the sum of the squares of all the entries in A.

4. *Let $V = \mathbb{R}^2$ with inner product $([a_1, a_2] \mid [b_1, b_2]) = a_1 b_1 - a_1 b_2 - a_2 b_1 + 4 a_2 b_2$, then*

$$q(a_1, a_2) = a_1^2 - 2 a_1 a_2 + 4 a_2^2.$$

We have already seen in Section 7.1 how any quadratic form in $\mathbb{R}^n$ can be represented by a symmetric matrix in the sense that $q(v) = v^T Q v$ for Q a symmetric matrix. The same still holds in this more general context. Instead of getting bogged down in these details, let's prove a more general result regarding symmetric bilinear maps on $\mathbb{R}^n$.

Theorem 7.4 *Consider the vector space $\mathbb{R}^n$.*

1. *If $\mathbb{R}^n$ equipped with a symmetric bilinear map $(*|*)$, then there exists a unique symmetric matrix C such that $(u|v) = u^T C v$ for all $u, v \in \mathbb{R}^n$.*

2. *If C is a symmetric matrix, then $(u|v) = u^T C v$ defines a symmetric, bilinear map.*

Proof 7.5 *First we show the existence of C. For $u, v \in \mathbb{R}^n$ set $u = [a_1, \ldots, a_n]$ and $v = [b_1, \ldots, b_n]$. Then*

$$(u|v) = (a_1e_1 + \cdots + a_ne_n|b_1e_1 + \cdots + b_ne_n) = \sum_{i,j=1}^{n} a_i(e_i|e_j)b_j.$$

Set $c_{ij} = (e_i|e_j)$ and $C = [c_{ij}]$. Then $u^T C v = (u|v)$. Furthermore, C is symmetric since the bilinear map is symmetric. Indeed, $c_{ij} = (e_i|e_j) = (e_j|e_i) = c_{ji}$. To prove uniqueness, suppose C' was another symmetric matrix satisfying $(u|v) = u^T C' v$ for all $u, v \in \mathbb{R}^n$. Set $D = [d_{ij}] = C - C'$. Then $u^T D v = 0$ for all $u, v \in \mathbb{R}^n$ and in particular, for all $1 \leq i, j \leq n$ we have $d_{ij} = e_i^T D e_j = 0$. Hence, $D = 0_{nn}$ and $C = C'$.

Statement 2 is left as an exercise. □

Example 7.11 *Refer back to Example 7.10.4 where $V = \mathbb{R}^2$ with symmetric bilinear map $([a_1, a_2]|[b_1, b_2]) = a_1b_1 - a_1b_2 - a_2b_1 + 4a_2b_2$. Following the proof and looking at the coefficients of q we see that*

$$C = \begin{bmatrix} 1 & -1 \\ -1 & 4 \end{bmatrix}.$$

Example 7.12 *Suppose $C = \begin{bmatrix} 1 & 2 & -1 \\ 2 & 0 & 3 \\ -1 & 3 & -2 \end{bmatrix}$ for a symmetric bilinear map on $\mathbb{R}^3$.*

Then

$$([a_1, a_2, a_3] \mid [b_1, b_2, b_3]) = a_1b_1 + 2a_1b_2 - a_1b_3 + 2a_2b_1 + 3a_2b_3 - a_3b_1 + 3a_3b_2 - 2a_3b_3.$$

The reader should take note of the relationship between entries in C and the coefficient of each term a_ib_j, namely c_{ij}.

A more general result of the above theorem (left as an exercise), is that for any finite dimensional vector space V with a symmetric bilinear map $(*|*)$ and any basis B for V there exists a unique symmetric matrix C such that $(u|v) = [u]_B^T C [v]_B$.

Another important observation is that given a quadratic form q we can recover the symmetric bilinear map. This follows from the fact that $(u|v) = \frac{1}{4}(q(u+v) - q(u-v))$ (left as an exercise, but see Lemma 5.1.c). We also point out that an easier way to find the symmetric bilinear map is to note that the matrix C for a symmetric bilinear map is one and the same as the matrix Q for the associated quadratic form (exercise).

Example 7.13 *Let $V = \mathbb{R}^2$ with quadratic form $q(a_1, a_2) = 3a_1^2 + 4a_1a_2 - a_2^2$. We could find the symmetric bilinear map by computing*

$$([a_1, a_2] \mid [b_1, b_2]) = \frac{1}{4}(q(a_1 + b_1, a_2 + b_2) - q(a_1 - b_1, a_2 - b_2)) \qquad etc.,$$

*but the easier way to find the symmetric bilinear map is to follow the second obser-
vation made above, namely that*

$$C = Q = \begin{bmatrix} 3 & 2 \\ 2 & -1 \end{bmatrix}, \quad and \ so \quad ([a_1, a_2] \mid [b_1, b_2]) = 3a_1b_1 + 2a_1b_2 + 2a_2b_1 - a_2b_2.$$

Our goal now is to prove Sylvester's Theorem which uncovers invariants regarding symmetric bilinear maps. Let V be a finite dimensional vector space with symmetric bilinear map $(*|*)$ and $v_1, v_2, \ldots, v_n$ be an orthogonal basis (which exists via Gram-Schmidt). Set $c_i = (v_i|v_i)$ for $i = 1, 2, \ldots, n$ and reorder the basis so that $c_1, \ldots, c_r > 0$, $c_{r+1}, \ldots, c_s < 0$ and $c_{s+1} = \cdots = c_n = 0$. Sylvester's Theorem states that the value of r and s are invariant, i.e. they are the same values regardless of our choice of orthogonal basis.

A couple of remarks are in order. First, if $B = (v_1, \ldots, v_n)$ is an ordered orthogonal basis, then the matrix C representing the symmetric bilinear map is a diagonal matrix, where $(u|v) = [u]_B^T C[v]_B$. If in addition B is orthonormal, then $c_1 = \cdots = c_r = 1$ and $c_{r+1} = \cdots = c_s = -1$. We prove Sylvester's Theorem in two separate theorems

Theorem 7.5 *Let V be a finite dimensional with symmetric bilinear map $(*|*)$. Then the value of s as defined above is invariant.*

Proof 7.6 *Set $n = \dim(V)$ and consider the set of vectors*

$$V_0 = \{v \in V \mid (v|w) = 0, \ \forall w \in V\}.$$

We need to show that V_0 is a subspace and that $\dim(V_0) = n-s$. This will show that s is invariant, since we make no mention of an orthogonal basis in V_0's definition. We leave verifying V_0 is a subspace of V as an exercise. We will show $\dim(V_0) = n - s$ as follows: For any orthogonal basis $v_1, \ldots, v_n$ we have $V_0 = \text{span}(v_{s+1}, \ldots, v_n)$. To show this we prove containment both ways. For one inclusion, take any v_i with $s < i \leq n$ and $w \in V$ and write $w = a_1v_1 + \cdots + a_nv_n$. Because of orthogonality we have $(v_i|w) = a_i(v_i|v_i) = a_i \cdot 0 = 0$ and so $v_{s+1}, \ldots, v_n \in V_0$. But then by Lemma 3.3.iii, $\text{span}(v_{s+1}, \ldots, v_n) \subseteq V_0$. For the reverse inclusion, Take any $v \in V_0$ and write $v = a_1v_1 + \cdots + a_nv_n$. Note that for $1 \leq i \leq s$ we have $0 = (v|v_i) = a_i(v_i|v_i)$ and since $(v_i|v_i) \neq 0$ it must be that $a_i = 0$ which means $v = a_{s+1}v_{s+1} + \cdots + a_nv_n \in \text{span}(v_{s+1}, \ldots, v_n)$. □

Definition 7.9 *In the proof of Theorem 7.5, the quantity $n - s = \dim(V_0)$ is called the* **index of nullity**.

Theorem 7.6 *Let V be a finite dimensional vector space with symmetric bilinear map $(*|*)$. Then the value of r as defined above is invariant.*

Proof 7.7 *Let $v_1, v_2, \ldots, v_n$ be an orthogonal basis with $c_1, \ldots, c_r > 0$, $c_{r+1}, \ldots, c_s < 0$ and $c_{s+1} = \cdots = c_n = 0$, where each $c_i = (v_i|v_i)$. Let $w_1, w_2, \ldots, w_n$ be an orthogonal basis with $d_1, \ldots, d_{r'} > 0$, $d_{r'+1}, \ldots, d_{s'} < 0$ and $d_{s'+1} = \cdots = d_n = 0$, where* each $d_i = (v_i|v_i)$. We show that $r \le r'$ which is enough to prove the theorem, since by symmetry $r' \le r$ and thus $r = r'$. If we show that $v_1, \ldots, v_r, w_{r'+1}, \ldots, w_n$ are linearly independent, then $r + (n - r') \le n$ which gets us our result that $r \le r'$ and the theorem is complete. To this end, suppose that $a_1 v_1 + \cdots + a_r v_r + a_{r'+1} w_{r'+1} + \cdots + a_n w_n = 0$. Then

$$a_1 v_1 + \cdots + a_r v_r = -a_{r'+1} w_{r'+1} + \cdots - a_n w_n.$$

Set $v = a_1 v_1 + \cdots + a_r v_r$. On the one hand,

$$(v|v) = (a_1 v_1 + \cdots + a_r v_r | a_1 v_1 + \cdots + a_r v_r) = c_1 a_1^2 + \cdots + c_r a_r^2.$$

On the other hand,

$$(v|v) = (-a_{r'+1} w_{r'+1} + \cdots - a_n w_n | -a_{r'+1} w_{r'+1} + \cdots - a_n w_n) = d_{r'+1} a_{r'+1}^2 + \cdots + d_{s'} a_{s'}^2.$$

Now, $c_1 a_1^2 + \cdots + c_r a_r^2 \ge 0$ while $d_{r'+1} a_{r'+1}^2 + \cdots + d_{s'} a_{s'}^2 \le 0$, therefore $c_1 a_1^2 + \cdots + c_r a_r^2 = 0 = d_{r'+1} a_{r'+1}^2 + \cdots + d_{s'} a_{s'}^2 \le 0$ and the only way that is possible (since $c_1, \ldots, c_r > 0$ and $d_{r'+1}, \ldots, d_{s'} < 0$) is if $a_1 = \cdots = a_r = 0 = a_{r'+1} = \cdots = a_{s'}$. This implies that $a_{s'+1} w_{s'+1} + \cdots + a_n w_n = 0$, but $w_{s'+1}, \ldots, w_n$ are linearly independent and so $a_{r'+1} = \cdots = a_n = 0$. □

Definition 7.10 In the proof of Theorem 7.6, the quantity r is called the **index of positivity**.

Example 7.14 Suppose $V = \mathbb{R}^3$ with symmetric bilinear map $(u|v) = u^T C v$, where

$$C = \begin{bmatrix} -1 & 0 & 0 \\ 0 & 2 & 3 \\ 0 & 2 & 1 \end{bmatrix}.$$

We will now find the values of r and s. We will do this in two very different ways. For the first method, we will create an orthogonal basis using Gram-Schmidt. We will start with the standard basis $\hat{i}, \hat{j}, \hat{k}$. First $w_1 = \hat{i} = [1, 0, 0]$. Then

$$w_2 = [0, 1, 0] - \left(\frac{[0, 1, 0]^T C [1, 0, 0]}{[1, 0, 0]^T C [1, 0, 0]} \right) [1, 0, 0] = [0, 1, 0] - \left(\frac{0}{-1} \right) [1, 0, 0] = [0, 1, 0].$$

$$w_3 = [0, 0, 1] - \left(\frac{[0, 0, 1]^T C [1, 0, 0])}{[1, 0, 0]^T C [1, 0, 0]} \right) [1, 0, 0] - \left(\frac{[0, 0, 1]^T C [0, 1, 0])}{[0, 1, 0]^T C [0, 1, 0]} \right) [0, 1, 0] =$$

$$[0, 0, 1] - \left(\frac{0}{-1} \right) [1, 0, 0] - \left(\frac{2}{2} \right) [0, 1, 0] = [0, -1, 1].$$

Now that we have an orthogonal basis (the order may need to change) we can now compute

$$(w_1|w_1) = [1,0,0]^T C[1,0,0] = -1.$$

$$(w_2|w_2) = [0,1,0]^T C[0,1,0] = 2.$$

$$(w_3|w_3) = [0,-1,1]^T C[0,-1,1] = -2.$$

Hence, $c_1 = 2$, $c_2 = -1$ and $c_3 = -2$ with $r = 1$ and $s = 3$.

Another way we can compute r and s is to first diagonalize C by finding a basis of eigenvectors for C (which we know exists, since C is real symmetric).

$$p_C(t) = \begin{vmatrix} -1-t & 0 & 0 \\ 0 & 2-t & 3 \\ 0 & 2 & 1-t \end{vmatrix} = -(t-4)(t+1)^2.$$

Therefore, C has eigenvalues $4, -1, -1$ and one can compute the corresponding eigenvector basis to be $B = ([0,3,2], [1,0,0], [0,-1,1])$. By dropping these vectors in the columns of a matrix and calling that matrix P, we know that

$$(u|v) = u^T C v = u^T P^T \begin{bmatrix} 4 & 0 & 0 \\ 0 & -1 & 0 \\ 0 & 0 & -1 \end{bmatrix} Pv =$$

$$(Pu)^T \begin{bmatrix} 4 & 0 & 0 \\ 0 & -1 & 0 \\ 0 & 0 & -1 \end{bmatrix} (Pv) = [u]_B^T \begin{bmatrix} 4 & 0 & 0 \\ 0 & -1 & 0 \\ 0 & 0 & -1 \end{bmatrix} [v]_B.$$

Thus, we see that $d_1 = 4$, $d_2 = -1$ and $d_3 = -1$ with $r = 1$ and $s = 3$.

Therefore, in hindsight, it's enough to simply compute the eigenvalues of C to determine the signs of the c_i's and consequently the values of r and s.

Let's now find explicitly the connection between the index of nullity and positivity and our earlier discussion about positive and negative definite, semi-definite and indefinite. This should be easy to understand from our work in the above example. Remember that C for our symmetric bilinear map is identical to our Q for the associated quadratic form. Therefore, when we find the eigenvalues of Q we are simultaneously determining whether C is positive or negative, definite or semi-definite, or indefinite as well as determining the values of r and s. Hence, we can deduce the following result:

Proposition 7.2 Let V be a finite dimensional non-trivial vector space with non-degenerate symmetric bilinear map $(u|v) = u^T C v$.

1. C *is positive definite* iff $r = n$.

2. C *is positive semi-definite* iff $0 < r = s < n$.

3. C *is negative definite* iff $r = 0$ *and* $s = n$.

4. C *is negative semi-definite* iff $r = 0$ *and* $s < n$.

5. C *is indefinite* iff $0 < r < s \leq n$.

This proposition can in fact serve as an alternative proof of Sylvester's Theorem (assuming, of course, we have proved the Eigenvalue Criterion).

Example 7.15 *Referring to Example 7.14, since* $0 < r = 1 < s = 3 = n$ *we see that* C *must be indefinite.*

EXERCISES

1. Compute the quadratic form on $\mathbb{R}^n$ for each of the following symmetric bilinear maps:

   a. $([a_1, a_2] \mid [b_1, b_2]) = 2a_1b_1 + a_1b_2 + a_2b_1 - 3a_2b_2$

   b. $([a_1, a_2, a_3] \mid [b_1, b_2, b_3]) = 2a_1b_1 + a_1b_2 - a_1b_3 + a_2b_1 - a_2b_2 + 4a_2b_3 - a_3b_1 + 4a_3b_2 - 3a_3b_3$

   c. $([a_1, a_2, a_3] \mid [b_1, b_2, b_3]) = 2a_1b_1 + a_1b_3 + a_3b_1 - 3a_2b_3 - 3a_3b_2 - a_3b_3$

2. Compute the C for each of the following symmetric bilinear maps on $\mathbb{R}^n$:

   a. $([a_1, a_2] \mid [b_1, b_2]) = 2a_1b_1 + a_1b_2 + a_2b_1 - 3a_2b_2$

   b. $([a_1, a_2, a_3] \mid [b_1, b_2, b_3]) = 2a_1b_1 + a_1b_2 - a_1b_3 + a_2b_1 - a_2b_2 + 4a_2b_3 - a_3b_1 + 4a_3b_2 - 3a_3b_3$

   c. $([a_1, a_2, a_3] \mid [b_1, b_2, b_3]) = 2a_1b_1 + a_1b_3 + a_3b_1 - 3a_2b_3 - 3a_3b_2 - a_3b_3$

3. Recover the symmetric bilinear map on $\mathbb{R}^n$ for each symmetric matrix C:

   a. $C = \begin{bmatrix} 2 & -1 \\ -1 & 1 \end{bmatrix}$

   b. $C = \begin{bmatrix} 4 & 1 & -2 \\ 1 & 1 & 0 \\ -2 & 0 & 1 \end{bmatrix}$

   c. $C = \begin{bmatrix} 1 & 1 & -2 \\ 1 & 2 & -1 \\ -2 & -1 & 1 \end{bmatrix}$

4. Recover the symmetric bilinear map for each of the following quadratic forms on $\mathbb{R}^n$:

   a. $V = \mathbb{R}^2$ and $q(a_1, a_2) = 2a_1^2 - 3a_1a_2 - a_2^2$.

   b. $V = \mathbb{R}^3$ and $q(a_1, a_2, a_3) = 3a_1^2 + 2a_1a_2 - 4a_1a_3 + a_3^2$.

5. Consider the following matrix:

$$C = \begin{bmatrix} 4 & 1 & -2 \\ 1 & 1 & 0 \\ -2 & 0 & 1 \end{bmatrix}.$$

   a. Find the values of r and s by repeating the computations done in the last example of the section.

   b. Determine if C is positive or negative, definite or semi-definite, or indefinite.

6. Repeat the previous exercise for the following matrix:

$$C = \begin{bmatrix} 1/2 & -\sqrt{12}/12 & -\sqrt{6}/6 \\ -\sqrt{12}/12 & -1/2 & \sqrt{18}/18 + \sqrt{2}/3 \\ -\sqrt{6}/6 & \sqrt{18}/18 + \sqrt{2}/3 & 0 \end{bmatrix}.$$

7. Prove that for any finite dimensional vector space V with the symmetric bilinear map $(*|*)$ and any basis B for V there exists a unique symmetric matrix C such that $(u|v) = [u]_B^T C[v]_B$.

8. In the previous exercise, show that if B is an orthonormal basis, then $C = I$.

9. Prove that if V is a vector space with the symmetric bilinear map $(*|*)$ and associated quadratic form q, then $(u|v) = \frac{1}{4}(q(u+v) - q(u-v))$.

10. Prove Theorem 7.4.b

11. Prove that V_0 in the proof of Theorem 7.5 is a subspace of V.

Regular Matrices

We now formally prove the result that a Markov chain with a regular transition matrix converges to a steady state. This is a proof adapted from one presented in [6]. The technical work which is required for this result is done in the following lemma:

Lemma A.1 *Let $P = [\pi_{ij}] \in M_{mm}$ be a transition matrix with no zero entries and define the following constants:*

$\alpha =$ *the value of the smallest entry in the matrix P.*

For any row vector $v \in \mathcal{R}^m$,

$\alpha_0 =$ *the minimum component of v.*

$\omega_0 =$ *the maximum component of v.*

$\alpha_1 =$ *the minimum component of vP.*

$\omega_1 =$ *the maximum component of vP.*

The following are then true:

i. $\omega_1 \leq \omega_0$ and $\alpha_0 \leq \alpha_1$.

ii. $\omega_1 - \alpha_1 \leq (1 - 2\alpha)(\omega_0 - \alpha_0)$.

Proof A.1 *Let v' be the vector obtained by replacing every component of v by ω_0 except α_0. Then certainly for $1 \leq j \leq m$, the jth component of v' is greater than or equal to the jth component of v and thus the jth component of $v'P$ is greater than or equal to the jth component of vP. Notice also that the jth component of $v'P$ has the form*

$$\pi_{1j}\omega_0 + \cdots + \pi_{i-1j}\omega_0 + \pi_{ij}\alpha_0 + \pi_{i+1j}\omega_0 + \cdots + \pi_{mj}\omega_0 = (1 - \pi_{ij})\omega_0 + \pi_{ij}\alpha_0.$$

Set $\beta = \pi_{ij} \geq \alpha$. Hence, the jth component of $v'P$ equals

$$(1 - \beta)\omega_0 + \beta\alpha_0 = \omega_0 - \beta(\omega_0 - \alpha_0) \leq \omega_0 - \alpha(\omega_0 - \alpha_0).$$

DOI: 10.1201/9781003217794-A

By our earlier comment we have

$$w_1 \leq w_0 - \alpha(w_0 - \alpha_0). \tag{A.1}$$

Since $\alpha > 0$ and $w_0 - \alpha_0 \geq 0$, it follows from this last equation that $w_1 \leq w_0$. Now apply the same reasoning to the vector $-v$. The jth component of $-v$ is greater than or equal to the jth component of $-v'$ and note that the maximal component of $-v$ is $-\alpha_0$, etc. In the end we obtain the inequality

$$-\alpha_1 \leq -\alpha_0 - \alpha(w_0 - \alpha_0). \tag{A.2}$$

Again, since $\alpha > 0$ and $w_0 - \alpha_0 \geq 0$, it follows that $\alpha_0 \leq \alpha_1$. Furthermore, adding equations C.1 and C.2 yields

$$w_1 - \alpha_1 \leq w_0 - \alpha_0 - 2\alpha(w_0 - \alpha_0) = (1 - 2\alpha)(w_0 - \alpha_0).$$

$\square$

With Lemma A.1 we can now prove the main result.

Theorem A.1 *If P is a regular transition matrix, then P^n converges to a matrix Q as $n \longrightarrow \infty$ and has the following properties:*

i. The columns of Q are identical probability vectors.

ii. Each entry of Q is positive.

Proof A.2 *Set α equal to the minimal entry in the matrix P and for $n = 0, 1, 2, \ldots$ set α_n to be the minimal component of the vector $e_i P^n$, w_n to be the maximal component of the vector $e_i P^n$ and $\delta_n = w_n - \alpha_n$. Keep in mind that $e_i P^n$ is the ith row of P^n.*

Claim: *$\delta_n \to 0$ as $n \to \infty$*

To prove the claim, let's first consider the case when P has no zero entries. By Lemma A.1,

$$w_0 \geq w_1 \geq w_2 \geq \cdots \quad and \quad \alpha_0 \leq \alpha_1 \leq \alpha_2 \leq \cdots.$$

Therefore,

$$\delta_0 \geq \delta_1 \geq \delta_2 \geq \cdots.$$

In addition, by Lemma A.1,

$$w_n - \alpha_n \leq (1 - 2\alpha)(w_{n-1} - \alpha_{n-1}), \quad (n = 1, 2, 3, \ldots),$$

or equivalently,

$$\delta_n \le (1 - 2\alpha)\delta_{n-1} \le (1 - 2\alpha)^2 \delta_{n-2} \cdots \le (1 - 2\alpha)^n \delta_0 = (1 - 2\alpha)^n,$$

since the maximal component of e_i is 1 and the minimal component is 0. Taking limits, since $|1 - 2\alpha| < 1$, we have

$$0 \le \lim_{n \to \infty} \delta_n \le \lim_{n \to \infty} (1 - 2\alpha)^n = 0.$$

Hence, $\lim_{n \to \infty} \delta_n = 0$.

   Now. we prove the claim in the general case of P being any regular transition matrix. Let k be such that P^k has no zero entries and set α' to be the smallest entry of P^k. By our work above we have

$$0 \le \delta_{nk} \le (1 - 2\alpha')^n \quad and \quad \lim_{n \to \infty} \delta_{nk} = 0.$$

   Hence, $\{\delta_n\}$ is a monotone sequence (non-increasing) which has a subsequence $\{\delta_{nk}\}$ which tends towards 0. It follows that $\delta_n \to 0$ as $n \to \infty$. This finishes the proof of the claim.

   The claim implies that $\lim_{n \to \infty} w_n = \lim_{n \to \infty} \alpha_n$, but this can only mean that in the limit the entries of the ith row of the limiting matrix exist and are identical. Let's call the limiting matrix Q and set

$$Q = \begin{bmatrix} \pi_1 & \pi_1 & \cdots & \pi_1 \\ \pi_2 & \pi_2 & \cdots & \pi_2 \\ \vdots & \vdots & \ddots & \vdots \\ \pi_m & \pi_m & \cdots & \pi_m \end{bmatrix}.$$

   Note that, for $1 \le i \le m$, we have $0 < \alpha_n \le \pi_i \le w_n < 1$, so that $0 < \pi_i < 1$. Since the columns of P^n sum to 1 for all n, then in the limit this must be true as well for Q. □

Rotations and Reflections in Two Dimensions

We present the derivations of the formulas for the linear operators on $\mathbb{R}^2$ which rotate and reflect the plane while preserving length.

Theorem B.1 *The map* $T[a,b] = \begin{bmatrix} \cos\theta & -\sin\theta \\ \sin\theta & \cos\theta \end{bmatrix} \begin{bmatrix} a \\ b \end{bmatrix}$ *is the linear operator which rotates vectors in* $\mathbb{R}^2$ *through an angle* θ *while preserving their lengths.*

Proof B.1 *Consider Figure B.1 depicting the vector* $[a,b]$ *and the resulting rotation through an angle* θ.

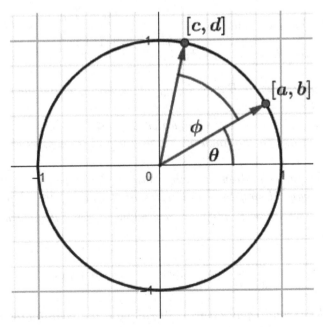

Figure B.1 two points on a circle. The $[a,b]$ is a rotation through an angle θ and $[c,d]$ is an additional rotation through an angle ϕ.

DOI: 10.1201/9781003217794-B

Notice by the definition of circular functions that the coordinates of the rotated vector

$$[c, d] = [\sqrt{a^2 + b^2} \cos(\theta + \phi), \sqrt{a^2 + b^2} \sin(\theta + \phi)].$$

Now, using trigonometric identities, we get

$$\sqrt{a^2 + b^2} \cos(\theta + \phi) = \sqrt{a^2 + b^2}(\cos\theta \cos\phi - \sin\theta \sin\phi)$$

$$= \sqrt{a^2 + b^2}\left(\cos\theta\left(\frac{a}{\sqrt{a^2 + b^2}}\right) - \sin\theta\left(\frac{b}{\sqrt{a^2 + b^2}}\right)\right) = a\cos\theta - b\sin\theta.$$

Similarly, one can show that

$$\sqrt{a^2 + b^2} \sin(\theta + \phi) = a\sin\theta + b\cos\theta.$$

Hence,

$$T[a, b] = [a\cos\theta - b\sin\theta, a\sin\theta + b\cos\theta] = \begin{bmatrix} \cos\theta & -\sin\theta \\ \sin\theta & \cos\theta \end{bmatrix}\begin{bmatrix} a \\ b \end{bmatrix}.$$

□

Theorem B.2 *The map* $T[a, b] = \begin{bmatrix} \frac{1-m^2}{m^2+1} & \frac{2m}{m^2+1} \\ \frac{2m}{m^2+1} & \frac{m^2-1}{m^2+1} \end{bmatrix}\begin{bmatrix} a \\ b \end{bmatrix}$ *is the linear operator which reflects vectors in* $\mathbb{R}^2$ *across the line* $y = mx$.

Proof B.2 *Consider Figure B.2 depicting the vector* $[a, b]$ *and the resulting reflection* $[x, y]$ *across the line* $y = mx$.

Think of x *and* y *as unknowns for which we wish to uncover the values. We will generate two equations in the unknowns* x *and* y *and then proceed to solve for these variables.*

To obtain the first equation we will compute the slope of the line segment $\overline{AB}$ *in two different ways. Using the coordinates of* A *and* B, *the slope is* $\frac{y-b}{x-a}$. *On the other hand, since* $\overline{AB}$ *is perpendicular to the line* $y = mx$ *it must have slope* $-1/m$. *Equating yields*

$$\frac{y - b}{x - a} = \frac{-1}{m} \quad \text{or} \quad x + my = a + mb.$$

To obtain the second equation we will compute the coordinates of the midpoint, M, *of the line segment* $\overline{AB}$. *Using the formula for midpoint, we obtain the coordinates* $M = ((1/2)(x+a), (1/2)(y+b))$. *On the other hand, since* M *lies on the line* $y = mx$, *the y-coordinate of* M *is also* $(m/2)(x + a)$. *Equating yields*

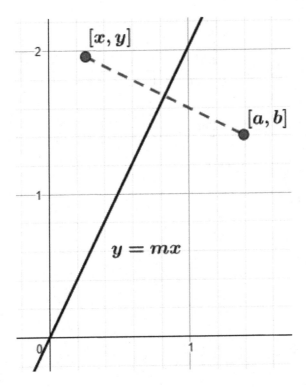

Figure B.2   Vector $[x, y]$ is the result of reflecting the vector $[a, b]$ across the line $y = mx$.

$$\frac{1}{2}(y + b) = \frac{m}{2}(x + a) \quad or \quad mx - y = b - ma.$$

Hence, the problem is reduced to solving the following system of equation:

$$\begin{cases} x + my & = & a + mb \\ mx - y & = & b - ma \end{cases}.$$

The corresponding augmented matrix is

$$\left[\begin{array}{cc|c} 1 & m & a + mb \\ m & -1 & b - ma \end{array}\right].$$

We then row-reduce the augmented matrix.

$$\xrightarrow{-mR_1 + R_2} \left[\begin{array}{cc|c} 1 & m & a + mb \\ 0 & -1 - m^2 & b - 2ma - m^2b \end{array}\right]$$

$$\xrightarrow{\left(\frac{-1}{1+m^2}\right)R_2} \left[\begin{array}{cc|c} 1 & m & a + mb \\ 0 & 1 & \frac{2ma + m^2 b - b}{m^2 + 1} \end{array}\right] \xrightarrow{-mR_2 + R_1} \left[\begin{array}{cc|c} 1 & 0 & \frac{2mb - m^2 a + a}{m^2 + 1} \\ 0 & 1 & \frac{2ma + m^2 b - b}{m^2 + 1} \end{array}\right].$$

Hence,

$$T[a, b] = \left[\frac{2mb - m^2a + a}{m^2 + 1}, \frac{2ma + m^2b - b}{m^2 + 1} \right]$$

$$= \left[\frac{(1 - m^2)a + (2m)b}{m^2 + 1}, \frac{(2m)a + (m^2 - 1)b}{m^2 + 1} \right] = \left[\begin{array}{cc} \frac{1-m^2}{m^2+1} & \frac{2m}{m^2+1} \\ \frac{2m}{m^2+1} & \frac{m^2-1}{m^2+1} \end{array} \right] \left[\begin{array}{c} a \\ b \end{array} \right].$$

□

Answers to Selected Exercises

CHAPTER 1

Section 1.1

2. $u + v = [4, -3, 1]$, $2v = [6, -2, 2]$, $\frac{1}{2}u = [1/2, -1, 0]$ and $2u - 3v = [-7, -1, -3]$

Section 1.2

3. (a) $|u| = 2\sqrt{2}$, $u \cdot v = \sqrt{3} + 3$, $2(v \cdot v)v = [12, -12, 24]$, $(u \cdot v)|3u - 2v| = 48\sqrt{3} + 216$ (b) $u/|u| = \left[\frac{\sqrt{3}}{2\sqrt{2}}, \frac{1}{2\sqrt{2}}, \frac{2}{2\sqrt{2}}\right]$ and $v/|v| = \left[\frac{1}{\sqrt{6}}, \frac{-1}{\sqrt{6}}, \frac{2}{\sqrt{6}}\right]$. (c) $\theta \approx 0.8189$

Section 1.4

1. $A + B = \begin{bmatrix} 3 & -3 & 4 \\ 1 & 0 & 3 \\ 0 & 0 & 3 \end{bmatrix}$, $-3C = \begin{bmatrix} -3 & 3 \\ -6 & 0 \\ 0 & 9 \end{bmatrix}$, $2A - 3B = \begin{bmatrix} -4 & -1 & 3 \\ -3 & 5 & -9 \\ 5 & -5 & 6 \end{bmatrix}$,

$(A - B)^T = \begin{bmatrix} -1 & -1 & 2 \\ -1 & 2 & -2 \\ 2 & -3 & 3 \end{bmatrix}$, $E^T + 2F = \begin{bmatrix} 3 & -2 & 1 \end{bmatrix}$.

2. B through K are square. G through K are diagonal. D, F and G through K are upper triangular. E and G through K are lower triangular. B and G through K are lower triangular. B and G through K are symmetric. C and H are skew-symmetric.

Section 1.5

1. (a) $BA = \begin{bmatrix} 7 & 9 & 2 \\ -4 & -6 & -2 \end{bmatrix}$ (b) $B^3 = \begin{bmatrix} 1 & 9 \\ 0 & -8 \end{bmatrix}$

6. $A = \begin{bmatrix} 2 & 3 \\ 0 & 0 \end{bmatrix}$, $B = \begin{bmatrix} 2 & 4 \\ 0 & 0 \end{bmatrix}$ and $C = \begin{bmatrix} 1 & 0 \\ 0 & 0 \end{bmatrix}$. $AC = BC$ while $A \neq B$

DOI: 10.1201/9781003217794-C

CHAPTER 2

Section 2.1

1. a, d and e.

Section 2.2

1. a,b,c and f

2. **(a)** $\{\,[-9/7,\ 3/7]\,\}$ **(b)** $\{\,[-4,\ -3/2,\ 2]\,\}$ **(c)** No solution **(e)** $\{\,[2x_2 - x_4 - 4, x_2, 2x_4 + 5, x_4, 2]\mid x_2, x_4 \in \mathbb{R}\,\}$

Section 2.9

1. $\begin{bmatrix} 0 & 0 & 1 \\ 0 & 1 & 0 \\ 1 & 0 & 0 \end{bmatrix}$, $\begin{bmatrix} 1 & 0 & 0 \\ 0 & -3 & 0 \\ 0 & 0 & 1 \end{bmatrix}$, $\begin{bmatrix} 1 & 0 & 0 \\ 0 & 1 & 0 \\ 0 & -2 & 1 \end{bmatrix}$

3. **(c)** $\text{rref}(A) = \text{rref}(B) = \begin{bmatrix} 1 & 0 & 0 \\ 0 & 1 & 0 \\ 0 & 0 & 0 \end{bmatrix}$

Section 2.6

1. **(a)** $\begin{bmatrix} 3/7 & 1/7 \\ -1/7 & 2/7 \end{bmatrix}$ **(b)** $\begin{bmatrix} -2 & 1 & 1 \\ -1 & 1/4 & 1/2 \\ 1 & -1/2 & 0 \end{bmatrix}$ **(c)** no inverse

2. **(a)** $\{\,[-9/7,\ 3/7]\,\}$ **(b)** $\{\,[-4,\ -3/2,\ 2]\,\}$ **(c)** no unique solution

3. (solutions can vary) **(a)** $E_1 = \begin{bmatrix} 1 & 3 \\ 0 & 1 \end{bmatrix}$, $E_2 = \begin{bmatrix} 1 & 0 \\ 0 & -7 \end{bmatrix}$, $E_3 = \begin{bmatrix} 1 & 0 \\ 2 & 1 \end{bmatrix}$, $E_4 = \begin{bmatrix} 0 & 1 \\ 1 & 0 \end{bmatrix}$ and $A = E_4 E_3 E_2 E_1$.

Section 2.8

1. **(e)** $[2x_2 - x_4 - 4, x_2, 2x_4 + 5, x_4, 2] = [2x_2 - x_4, x_2, 2x_4, x_4, 0] + [-4, 0, 5, 0, 2] = X_h + X_p$

2. **(a)** $rk(A) = 3 < 4$ **(b)** not invertible **(c)** infinite solutions

3. **(a)** $rk(A) = 3$ **(b)** only trivial solution

6. **(a)** $rk(B) \le 3 < 4$ **(b)** no solution or infinite solutions

Section 2.9

1. **(a)** $42 \ne 0$ **(b)** only trivial solution

2. (a) $rk(B) = 2 < 3$ **(b)** $|B| = 0$

3. (a) $4 \neq 0$ **(b)** C^{-1} exists

5. (a) $rk(A) = 3 < 4$ **(b)** $|A| = 0$ **(c)** no solution or infinite solutions **(d)** not invertible

6. (a) $6 \neq 0$ **(b)** $6 \neq 0$ **(c)** only trivial solution

8. (a) $24 \neq 0$ **(b)** $24 \neq 0$ **(c)** not invertible

Section 2.10

1. (a) $\{ [-1, 2] \}$

4. (a) $-1 \neq 0$ **(b)** unique solution **(c)** $\{[2, -1, 0]\}$ **(d)** A^{-1} exists **(e)**

$$A' = \begin{bmatrix} 7 & -3 & -3 \\ -5 & 2 & 2 \\ -3 & 2 & 1 \end{bmatrix} \text{ and } A^{-1} = \begin{bmatrix} -7 & 5 & 3 \\ 3 & -2 & -2 \\ 3 & -2 & -1 \end{bmatrix}$$

11. (a) $2 \neq 0$ **(b)** invertible **(c)** $a'_{31} = -4$, $a'_{23} = -1$ **(d)** $A^{-1} =$

$$\begin{bmatrix} -3 & -3 & -2 \\ -2 & -1 & -1 \\ 1 & -1/2 & 0 \end{bmatrix}$$

CHAPTER 3

Section 3.1

1. Property 3,4,5,6,8 Fail

7. Property 0,1,4,5,6 Fail

11. Property 4,5,7,8 Fail

Section 3.2

1. (c) is not **(d)** is

3. (c) is not **(h)** is

4. (a) is **(h)** is not

5. (a) is **(d)** is not

Section 3.3

1. (a) $\begin{cases} 2a + 3b &= -5 \\ -3a - 2b &= 0 \end{cases}$ **(b)** $w = 2v_1 - 3v_2$

4. (a) linearly independent **(b)** linearly dependent **(c)** linearly dependent **(d)** linearly independent

5. (a) linearly dependent **(c)** linearly dependent

6. (a) linearly independent **(e)** linearly dependent **(h)** linearly dependent

10. one possible answer is $[1,0,0]$, $[0,1,0]$, $[1,1,0]$

Section 3.4

1. (a) i. $[1,0,1/2]$, $[0,1,1/2]$ **ii.** $\{\,[a,b,(1/2)a+(1/2)b]\;:\;a,b \in \mathbb{R}\}$ **iii.** in span **1. (f) i.** $\begin{bmatrix} 1 & 0 \\ -1 & 0 \end{bmatrix}$, $\begin{bmatrix} 0 & 1 \\ 1 & -1 \end{bmatrix}$ **ii.** $\left\{\begin{bmatrix} a & b \\ -a+b & -b \end{bmatrix} : a,b \in \mathbb{R}\right\}$ **iii.** not in span **(i) i.** $1 + 2x^2$, $x - x^2$ **ii.** $\{\,a + bx + (2a-b)x^2 : a,b \in \mathbb{R}\}$ **iii.** in span

Section 3.5

1. (b) i. Y **ii.** Y **iii.** Y **(d) i.** Y **ii.** N **iii.** N **(e) i.** Y **ii.** N **iii.** N

3. (a) $\begin{bmatrix} 1 & 0 \\ 1 & 0 \end{bmatrix}$, $\begin{bmatrix} 0 & 1 \\ -1 & 0 \end{bmatrix}$, $\begin{bmatrix} 0 & 0 \\ 2 & 1 \end{bmatrix}$ **(c)** 3

4. (a) $1 - 2x^3$, x, $x^2 + 3x^3$ **(c)** 3

5. (a) $\{[-2a, a-b, a, b] : a,b \in \mathbb{R}\}$ **(b)** $[-2,1,1,0]$, $[0,-1,0,1]$ **(c)** 2

8. (c) $\begin{bmatrix} 1 & -2 \\ -1 & 0 \end{bmatrix}$, $\begin{bmatrix} 2 & 0 \\ 1 & -1 \end{bmatrix}$, $\begin{bmatrix} 1 & 0 \\ 0 & 0 \end{bmatrix}$, $\begin{bmatrix} 0 & 1 \\ 0 & 0 \end{bmatrix}$

9. (a) $[-16/15, 4/3, 13/15]$ **(b)** $[1,1,0]$ **(c)** $[1/2,2,1,-1]$

10. (a) $P = \begin{bmatrix} 6/5 & -2/5 \\ 7/5 & -4/5 \end{bmatrix}$, $[2,-3]_{B'} = [2,7/2]$, $[2,-3]_B = [1,0]$ **(b)** $P = \begin{bmatrix} 2/3 & -1/3 \\ 1/3 & 4/3 \end{bmatrix}$, $[1-x]_{B'} = [1,0]$, $[1-x]_B = [2/3,1/3]$ **(c)** $P = \begin{bmatrix} 0 & 1 \\ 2 & 2 \end{bmatrix}$, $\begin{bmatrix} 2 & 0 \\ 0 & -1 \end{bmatrix}_{B'} = [-1/2,2]$, $\begin{bmatrix} 2 & 0 \\ 0 & -1 \end{bmatrix}_B = [2,3]$

Section 3.7

2. (a) A basis for V is $E_{11}, E_{12}, E_{21}, E_{22}$ so $\dim V = 4$. A basis for U is E_{11}, E_{12}, E_{22} so $\dim U = 3$. A basis for W is $\begin{bmatrix} 0 & 1 \\ -1 & 0 \end{bmatrix}$ so $\dim W = 1$. **(b)** $U \cap W = \{0_{22}\}$ so it has no basis and its dimension is 0. **(c)** 4, Y

4. (a) A basis for V is $1, x, x^2$ so $\dim V = 3$. A basis for U is x, x^2 and $\dim U = 2$. A basis for W is $-1+x, -x+x^2$ and $\dim U = 2$. **(b)** $U \cap W = \{-ax + ax^2 \mid a \in \mathbb{R}\}$, basis is $-x + x^2$ and $\dim(U \cap W) = 1$. **(c)** 3, Y

12. (a) A basis for V is $\hat{i}, \hat{j}, \hat{k}$ so $\dim V = 3$. A basis for U is $[1, 0, -3], [0, 1, 2]$ and $\dim U = 2$. A basis for W is $[1, 0, 1], [0, 1, 1]$ and $\dim W = 2$. **(b)** $U \cap W = \{[a, 4a, 5a] \mid a \in \mathbb{R}\}$. A basis for $U \cap W$ is $[1, 4, 5]$ and $\dim(U \cap W) = 1$. **(c)** 3, Y

CHAPTER 4

Section 4.1

1. (a) is **(b)** is not

2. (e) is **(l)** is not **(o)** is

Section 4.2

1. (a) $\{(2c) + (-3c)x + cx^2 : c \in \mathbb{R}\}$ **(b)** is not **(c)** $2 - 3x + x^2$ **(d)** 1 **(e)** 2 **(f)** it does

4. (a) $\{0\}$ **(b)** is **(c)** no basis for $\ker T$ and $\dim(\ker T) = 0$ **(d)** 2, does not

Section 4.3

8. (a) $\begin{bmatrix} 7 & -4 \\ 6 & -4 \\ -10 & 6 \end{bmatrix}$ **(b)** $S(a + bx) = (a + b)[0, 2, 1] + (a + 2b)[1, 0, 0] = [a + 2b, 2a + 2b, a + b]$

11. (a) $\begin{bmatrix} 2 & 1 & 4 \\ -1/2 & 3/2 & -4 \end{bmatrix}$ **(b)** $\begin{bmatrix} -4 & 0 \\ 0 & 4 \end{bmatrix}$ **(c)** $S[a, b, c] = \begin{bmatrix} a - c & 0 \\ 0 & 4a + 5b - 7c \end{bmatrix}$

Section 4.4

1. (a) $\begin{bmatrix} 2/3 & 1/3 \\ -1/3 & 1/3 \end{bmatrix}$ **(c)** $T^{-1}[a, b] = \begin{bmatrix} \frac{2}{3}a + \frac{1}{3}b & 0 \\ 0 & \frac{1}{3}a + \frac{1}{3}b \end{bmatrix}$

12. (a) $[3, 10, 15]$ **(b)** $|A| = 6 \neq 0$, $A^{-1} = \begin{bmatrix} 0 & 1/2 & -1/2 \\ -1/3 & -1/3 & 5/3 \\ 1/3 & -1/6 & -1/6 \end{bmatrix}$ **(c)** $T[a, b, c] = [a, 2a - b/2, c/3 - b/2]$

Section 4.5

3. (a) $P = \begin{bmatrix} 1 & 0 \\ 1 & -1 \end{bmatrix}$, $[T]_{B'} = \begin{bmatrix} 0 & 1 \\ -3 & 3 \end{bmatrix}$ **(b)** $T(a + bx) = (3a + b) - 3ax$

5. (a) $\begin{bmatrix} 1 & -1 & 2 \\ 0 & 1 & -2 \\ 1/2 & 1/2 & 1 \end{bmatrix}$ **(b)** $\begin{bmatrix} 1 & -2 & 0 \\ 1/2 & 1 & 1/2 \\ -1 & 2 & 1 \end{bmatrix}$ **6. (e)** $\begin{bmatrix} -1 & -3/2 \\ 2 & 3 \end{bmatrix}$ **(f)**

$\begin{bmatrix} 2 & 0 \\ 1 & 0 \end{bmatrix}$

Section 4.6

1. (a) $\begin{bmatrix} -1 & 0 & 0 \\ 0 & 2 & 3 \\ 0 & 2 & 1 \end{bmatrix}$ **(b)** $|[T]_{ST}| = 4 \neq 0$ **(c)** $T^{-1}(a + bx + cx^2) = -a +$

$\left(-\frac{1}{4}b + \frac{3}{4}c\right)x + \left(\frac{1}{2}b - \frac{1}{2}c\right)x^2$ **(d)** T has eigenvalues -1 and 4, $E_{-1} = \{a - cx + cx^2 :$
$a, c \in \mathbb{R}\}$, $E_4 = \{(3/2)cx + cx^2 : a, c \in \mathbb{R}\}$ **(e)** A basis for E_{-1} is $1, -x + x^2$
and $\dim(E_{-1}) = 2$ and a basis for E_4 is $(3/2)x + x^2$ and $\dim(E_4) = 1$. **(f)** Y,

$B = (1, -x + x^2, (3/2)x + x^2)$, $[T]_B = \begin{bmatrix} -1 & 0 & 0 \\ 0 & -1 & 0 \\ 0 & 0 & 4 \end{bmatrix}$

3. (a) T has eigenvalues 1 and 4 **(b)** N **(c)** $E_1 = \left\{ \begin{bmatrix} 0 & b \\ 0 & c \end{bmatrix} : b, c \in \mathbb{R} \right\}$, $E_4 = $
$\left\{ \begin{bmatrix} -(3/2)c & -(1/2)c \\ 0 & c \end{bmatrix} : c \in \mathbb{R} \right\}$ **(d)** A basis for E_1 is $\begin{bmatrix} 0 & 1 \\ 0 & 0 \end{bmatrix}, \begin{bmatrix} 0 & 0 \\ 0 & 1 \end{bmatrix}$
and so $\dim(E_1) = 2$. A basis for E_4 is $\begin{bmatrix} -(3/2) & -(1/2) \\ 0 & 1 \end{bmatrix}$ and so $\dim(E_4) = 1$.
(e) Since $\dim(E_1) + \dim(E_4) = \dim(U_{22})$, T is diagonalizable with basis $B = $
$\left(\begin{bmatrix} 0 & 1 \\ 0 & 0 \end{bmatrix}, \begin{bmatrix} 0 & 0 \\ 0 & 1 \end{bmatrix}, \begin{bmatrix} -(3/2) & -(1/2) \\ 0 & 1 \end{bmatrix} \right)$, $[T]_B = \begin{bmatrix} 1 & 0 & 0 \\ 0 & 1 & 0 \\ 0 & 0 & 4 \end{bmatrix}$

4. (a) T are -2 and 3 **(b)** N **(c)** $E_{-2} = \{-b + bx + cx^2 : b \in \mathbb{R}\}$, $E_3 = \{-2b + bx :$
$b \in \mathbb{R}\}$ **(d)** A basis for E_{-2} is $-1 + x, x^2$, so $\dim(E_{-2}) = 2$ while a basis for E_3 is
$-2 + x$, so $\dim(E_3) = 1$ **(e)** Since $\dim(E_{-2}) + \dim(E_3) = 2 + 1 = 3 = \dim(P_2)$,
T is diagonalizable. $B = (-1 + x, x^2, -2 + x)$ and $[T]_B = \begin{bmatrix} -2 & 0 & 0 \\ 0 & -2 & 0 \\ 0 & 0 & 3 \end{bmatrix}$.

7. (a) $\begin{bmatrix} -2 & -1 \\ 1 & 2 \end{bmatrix}$

8. (a) $\begin{bmatrix} 1 & -1 \\ 1 & 1 \end{bmatrix}$ **(b)** $\begin{bmatrix} 122 & 121 \\ 121 & 122 \end{bmatrix}$

Section 4.8

1. (b) $1 + U$, $x + U$

Section 4.9

3. **(b)** $\phi_1 \begin{bmatrix} a & b \\ c & d \end{bmatrix} = a - b$, $\quad \phi_2 \begin{bmatrix} a & b \\ c & d \end{bmatrix} = b - c$, $\quad \phi_3 \begin{bmatrix} a & b \\ c & d \end{bmatrix} = c - d$,

$\phi_4 \begin{bmatrix} a & b \\ c & d \end{bmatrix} = d$

CHAPTER 5

Section 5.1

1. **(a)** 22 **(b)** 15/4 **(c)** 8

2. **(a)** $\sqrt{5}$ **(b)** 1 **(c)** $\sqrt{6}$

4. **(a)** is **(b)** is not

Section 5.2

2. **(a)** $[1/\sqrt{2}, 1/\sqrt{2}, 0]$, $[1/\sqrt{6}, -1/\sqrt{6}, \sqrt{2/3}]$, $[-1/\sqrt{3}, 1/\sqrt{3}, 1/\sqrt{3}]$ **(b)** 5

4. **(a)** $\begin{bmatrix} 1 & 0 \\ 0 & 0 \end{bmatrix}$, $\begin{bmatrix} 0 & 1/\sqrt{2} \\ 0 & 1/\sqrt{2} \end{bmatrix}$, $\begin{bmatrix} 0 & -1/\sqrt{2} \\ 0 & 1/\sqrt{2} \end{bmatrix}$ **(b)** 1

5. $\begin{bmatrix} 1/\sqrt{2} & 1/\sqrt{2} \\ 0 & 0 \end{bmatrix}$, $\begin{bmatrix} -\sqrt{2}/2\sqrt{3} & \sqrt{2}/2\sqrt{3} \\ 0 & \sqrt{2}/\sqrt{3} \end{bmatrix}$, $\begin{bmatrix} \sqrt{3}/3 & -\sqrt{3}/3 \\ 0 & \sqrt{3}/3 \end{bmatrix}$

7. **(a)** $\{[-(5/4)c, (3/4)c, c] \; : \; c \in \mathbb{R}\}$ **(b)** $\{(1/6)c - cx + cx^2 \; : \; c \in \mathbb{R}\}$ **(c)** $\left\{ \begin{bmatrix} -(1/3)c + (2/3)d & (2/3)c + (2/3)d \\ c & d \end{bmatrix} \; : \; c, d \in \mathbb{R} \right\}$

Section 5.6

1. **(a)** $\frac{8}{\sqrt{10}}$, $[4/5, 0, -8/5, -4/5, 8/5]$ **(b)** $\frac{1}{\sqrt{13}}$, $\begin{bmatrix} 1/13 & 0 & -1/13 \\ 2/13 & 1/13 & -1/13 \\ 0 & -1/13 & 2/13 \end{bmatrix}$ **(c)**

$\frac{7\sqrt{645}}{860}$, $-\frac{21}{172}x^2 + \frac{21}{86}x + \frac{21}{172}$

2. **(a)** $[1, 1, 1]$ **(b)** $[1, 1, 1, 0]$ **(d)** $\frac{94349097310435973}{198158383604301824} - \frac{31378641891135303}{198158383604301824}x - \frac{2857448618217529}{4503599627370496}x^2$ **(e)** E_{11}

Section 5.7

1. **(a)** $P = \begin{bmatrix} -1/\sqrt{2} & 1/\sqrt{2} \\ 1/\sqrt{2} & 1/\sqrt{2} \end{bmatrix}$ and $P^T A P$ is a diagonal matrix with diagonal

entries 2 and 4 **(b)** $P = \begin{bmatrix} -\sqrt{6}/6 & -\sqrt{2}/2 & \sqrt{3}/3 \\ -\sqrt{6}/6 & \sqrt{2}/2 & \sqrt{3}/3 \\ \sqrt{6}/3 & 0 & \sqrt{3}/3 \end{bmatrix}$ and $P^T B P$ is a diagonal

matrix with diagonal entries 2, 3 and 6

Section 5.8

1. (d) $U \approx \begin{bmatrix} 0.2850 & -0.8315 & 0.3895 & 0.2750 \\ -0.2413 & -0.5292 & -0.5000 & -0.6417 \\ -0.8758 & -0.1512 & -0.0055 & 0.4583 \\ -0.3058 & 0.0756 & 0.7735 & -0.5500 \end{bmatrix}$, $S \approx \begin{bmatrix} 3.3948 & 0 & 0 \\ 0 & 2.6458 & 0 \\ 0 & 0 & 1.2145 \\ 0 & 0 & 0 \end{bmatrix}$,

$V \approx \begin{bmatrix} -0.5932 & -0.6000 & 0.5368 \\ -0.6710 & -0.0000 & -0.7415 \\ 0.4449 & -0.8000 & -0.4026 \end{bmatrix}$ (e) $U = \begin{bmatrix} -\sqrt{6}/6 & \sqrt{3}/3 & -\sqrt{2}/2 \\ -\sqrt{6}/6 & \sqrt{3}/3 & \sqrt{2}/2 \\ \sqrt{6}/3 & \sqrt{3}/3 & 0 \end{bmatrix} =$

V and $S = \begin{bmatrix} 6 & 0 & 0 \\ 0 & 3 & 0 \\ 0 & 0 & 2 \end{bmatrix}$ (f) $U = \begin{bmatrix} 1/\sqrt{2} & 1/\sqrt{2} \\ 1/\sqrt{2} & -1/\sqrt{2} \end{bmatrix}$, $S = \begin{bmatrix} \sqrt{8} & 0 & 0 & 0 \\ 0 & 0 & 0 & 0 \end{bmatrix}$,

$V = \begin{bmatrix} 1/2 & -1/\sqrt{2} & -1/\sqrt{2} & 1/\sqrt{2} \\ -1/2 & 0 & 0 & 1/\sqrt{2} \\ 1/2 & 0 & 1/\sqrt{2} & 0 \\ 1/2 & 1/\sqrt{2} & 0 & 0 \end{bmatrix}$

Section 5.9

1. (b) $(10/7, 2/7)$

5. (a) $y = 1.1 - 0.2x$ (b) $y = 0.1 - 1.2x + x^2$ (c) $y = 1 - 2.5x - 0.5x^2 + x^3$

9. $(-1/7, 3/7, -2/7)$

CHAPTER 6

Section 6.2

1. (b) $\begin{bmatrix} 14/5 & 17/5 \\ 17/5 & 26/5 \end{bmatrix}$ (b) $[0.5776, 0.8163]$, $\theta \approx 0.9550$ (c) $\frac{14}{5} \cos^2 \theta +$ $\frac{34}{5} \cos \theta \sin \theta + \frac{26}{5} \cos^2 \theta$

Section 6.3

1. $P = \begin{bmatrix} -0.8163 & 0.5776 \\ 0.5776 & 0.8163 \end{bmatrix}$, $[0.5945, -1.3549]$, $[1.1720, -0.5386]$, $[0.3557, 0.0390]$, $[0.9333, 0.8553]$, $[0.6945, 2.2492]$

Section 6.6

1. $S \approx \begin{bmatrix} -0.2476 & -0.5820 \\ 0.6720 & -0.2859 \\ -0.6278 & -0.0765 \\ 0.2918 & 0.2195 \\ -0.0884 & 0.7249 \end{bmatrix}$

Section 6.7

1. (a) $[-11/4, 1/2]$ **(b)** $y = (11/2)x - (29/12)$

3. $y = -x + 1$

CHAPTER 7

Section 7.1

1. (a) $8x_1^2 + 2x_1x_2 + 2x_2^2$ **(b)** $-6x_1^2 + 2x_1x_2$ **(c)** $2x_1^2 - 2x_1x_2 + 4x_1x_3 - 3x_3^2$

2. (a) $\begin{bmatrix} 1 & -3 \\ -3 & -2 \end{bmatrix}$ **(b)** $\begin{bmatrix} 0 & 1 \\ 1 & 3 \end{bmatrix}$ **(c)** $\begin{bmatrix} 2 & -1 & 0 \\ -1 & -1 & 2 \\ 0 & 2 & 0 \end{bmatrix}$

Section 7.2

1. (a) indefinite **(b)** negative definite **(c)** positive definite **(d)** positive semi-definite **(e)** negative semi-definite

Section 7.3

1. (a) indefinite **(b)** negative definite **(c)** positive definite **(d)** positive semi-definite **(e)** negative semi-definite

2. eigenvalues are 0, 0 and $1 + a^2 + a^4$ all ≥ 0.

Section 7.4

1. (a) $(0,0,0)$ is a strict global minimizer **(g)** no critical points **(h)** no critical points **(k)** $((1/4)\ln 2, (1/4)\ln 2, (1/4)\ln 2)$ is a strict global maximizer

2. (a) $(0,0,0)$ is a strict global minimizer **(c)** $(1/2, 1, -1/3)$ is a saddlepoint **(e)** $(0,0,0)$ is a saddlepoint **(h)** $(1/3, -1/6, 1/6)$ is a strict global maximizer

3. (a) $(0,0,0)$ is a strict local (in fact, global) minimizer, $(-4, -4, -2)$, $(-4, 4, 2)$ and $(4, -4, 2)$ are saddlepoints **(b)** $(0,0,1)$ and $(1, -1, 1)$ are strict local maximizers

Section 7.5

1. (b) $2a_1^2 + 2a_1a_2 - 2a_1a_3 - a_2^2 + 8a_2a_3 - 3a_3^2]$

2. (b) $\begin{bmatrix} 2 & 1 & -1 \\ 1 & -1 & 4 \\ -1 & 4 & -3 \end{bmatrix}$

3. $4a_1b_1 + a_1b_2 - 2a_1b_3 + a_2b_1 + a_2b_2 - 2a_3b_1 + a_3b_3$

4. $3a_1b_1 + a_1b_2 - 2a_1b_3 + a_2b_1 - 2a_3b_1 + a_3b_3$

5. (a) $r = 2$ and $s = 3$ **(b)** indefinite

6. (a) $r = 1$ and $s = 2$ **(b)** indefinite

References

[1] S. Andrilli and D. Hecker. *Elementary Linear Algebra*. PWS Publishing Co., Boston, MA, USA, 1993.

[2] C. Blake, E. Keogh, and C. Merz. UCI Repository of Machine Learning Databases. `http://wwwicsuciedu/mlearn/MLRepository.html`, 1998.

[3] R.O. Duda, P.E. Hart, and D.G. Stork. *Pattern Classification*. Wiley-Interscience, NY, USA, 2nd edition, 2001.

[4] S.H. Friedberg, A.J. Insel, and L.E. Spence. *Linear Algebra*. Pearson Education, NJ, USA, 4th edition, 2003.

[5] K. Hoffman and R. Kunze. *Linear Algebra*. Prentice-Hall, Inc., Englewood Cliffs, NJ, USA, 2nd edition, 1971.

[6] J.G. Kemeny and J.L. Snell. *Finite Markov Chains*. D. van Nostrand Co. Ltd., London, UK, 1960.

[7] B. Kolman and Beck R.E. *Elementary Linear Programming with Applications*. Academic Press, Inc., San Diego, CA, USA, 2nd edition, 1995.

[8] Y. LeCun, C. Cortes, and C.J.C. Burges. THE MNIST DATABASE of handwritten digits. `http://yann.lecun.com/exdb/mnist/`.

[9] S.J. Leon. *Linear Algebra with Applications*. Pearson Education, NJ, USA, 9th edition, 2015.

[10] A. Peressini, F.E. Sullivan, and J.J. Uhl. *The Mathematics of Nonlinear Programming*. Springer-Verlag, NY, USA, 1988.

Index

Printed in the United States
by Baker & Taylor Publisher Services